One of the *Wall Street Journal*'s
"Best Biographies of 2015"

One of *Kirkus Reviews*'s
"Best Books of 2015"

D0954864

Pacific

PACIFIC

SILICON CHIPS AND SURFBOARDS, CORAL REEFS AND ATOM BOMBS, BRUTAL DICTATORS AND FADING EMPIRES

Simon Winchester

HARPER PERENNIAL

NEW YORK • LONDON • TORONTO • SYDNEY • NEW DELHI • AUCKLAND

HARPER ● PERENNIAL

FIRST HARPER PERENNIAL EDITION PUBLISHED 2016.

Designed by Leah Carlson-Stanisic

Engraving of the Pacific on title page and chapter openers by Marzolino/Shutterstock, Inc.

Maps on pages viii–ix, 256–57, and 390–91 by Nick Springer/Springer Cartographics LLC.

Grateful acknowledgment is made for permission to reproduce images:

Page 61: U.S. Department of Defense. Page 68: National Nuclear Security Administration. Pages 93, 168, 191: Associated Press. Page 122: Sony Pictures. Page 138: Los Angeles Times. Page 143: Hobie Archive. Pages 154, 416: U.S. Army. Pages 158, 387, 405: U.S. Navy. Page 179: Roman Harak. Page 191: Louis Gardella. Page 211: U.S. Marine Corps. Page 227: FormAsia. Pages 235, 286: Newspix/Getty Images. Page 240: Jacques Descloitres/NASA. Pages 258, 364: National Oceanic and Atmospheric Administration (NOAA). Page 275: National Library of Australia. Page 307: Woods Hole Oceanic Institute. Pages 313, 384: U.S. Geological Survey (USGS). Page 331: P. Rona/NOAA. Page 341: Acropora. Page 351: Wmpearl. Page 361: Hiroshi Hasegawa. Page 387: SteKrueBe. Page 394: UNCLOS and CIA. Page 399: Armed Forces of the Philippines. Page 411: Xinhua News Agency. Page 428: ® Image of Hōkūle'a used with permission by Polynesian Voyaging Society; © 2014 Polynesian Voyaging Society and 'Ōiwi TV; photographer: Justyn Ah Chong.

Library of Congress Cataloging-in-Publication Data is available upon request.

ISBN 978-0-06-231542-7 (pbk.)

16 17 18 19 20 OV/RRD 10 9 8 7 6 5 4 3 2 1

For Setsuko

Look East, where whole new thousands are!

—Robert Browning, *Waring*

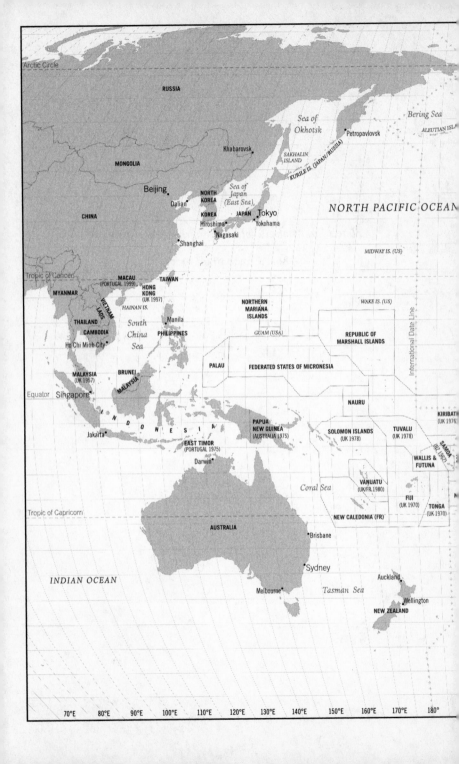

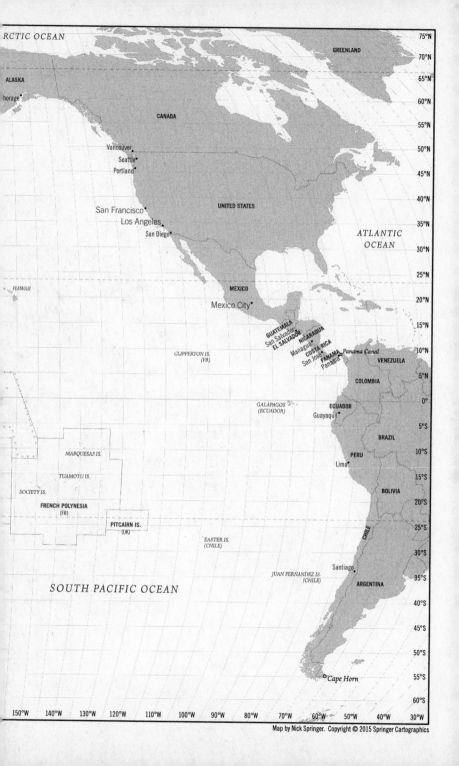

ARCTIC OCEAN

GREENLAND

ALASKA

Anchorage

CANADA

Vancouver
Seattle
Portland

UNITED STATES

San Francisco
Los Angeles
San Diego

ATLANTIC OCEAN

HAWAII

MEXICO

Mexico City

GUATEMALA
San Salvador
EL SALVADOR
NICARAGUA
Managua
COSTA RICA
San José
PANAMA
Panama

VENEZUELA

Panama Canal

COLOMBIA

CLIPPERTON IS.
(FR)

GALÁPAGOS
(ECUADOR)

ECUADOR
Guayaquil

BRAZIL

MARQUESAS IS.

TUAMOTU IS.

PERU
Lima

SOCIETY IS.

BOLIVIA

FRENCH POLYNESIA
(FR)

PITCAIRN IS.
(UK)

CHILE

EASTER IS.
(CHILE)

Santiago

JUAN FERNANDEZ IS.
(CHILE)

SOUTH PACIFIC OCEAN

ARGENTINA

Cape Horn

CONTENTS

MAPS AND ILLUSTRATIONS

Pacific

PROLOGUE:
THE LONELY SEA AND THE SKY

Here from this mountain shore, headland beyond stormy
   headland plunging like dolphins through the blue sea-
   smoke
Into pale sea—look west at the hill of water: it is half the
   planet . . .
arched over to Asia, Australia and white Antarctica: those are
   the eyelids that never close; this is the staring unsleeping
Eye of the earth; and what it watches is not our wars.
                    —Robinson Jeffers, from "The Eye," 1965

United Airlines Flight 154 leaves Honolulu International Airport just after dawn three times a week, bound ultimately for the city of Hagåtña, the capital of the island republic of Guam. If the northeast trades are blowing at their usual steady twelve knots, the jet will take off to the east, into the low morning sun over Waikiki, and those passengers on the aircraft's left side will see the wall of skyscraper hotels along the beachfront and be able to glimpse down at

Doris Duke's great seaside mansion, Shangri-La. Once the plane is two miles high above the crater of the dormant Diamond Head volcano, it will begin to make a long and lazy turn to the right.

If the morning haze is light, passengers on the right side now can sometimes glimpse the bombers and heavy transport planes lined up on the flight line of Hickam Field, and maybe a flotilla of sleek gray warships will be gliding slowly through the lochs of Pearl Harbor. There will be some suburbs clustered between the shore and the slopes; there will be a skein of rush-hour traffic crawling along on H-1, the main thruway into Honolulu; and behind these urban images will rise ranges of mountains, razor-sharp aiguilles dotted in places with white radar domes.

With every one of its seats invariably filled, the plane will then clear its throat and tilt its nose ever higher, and once at five miles high, it will set its autopilot to a southwesterly course, heading out initially over two thousand miles of clear blue, unpeopled ocean. As the climb flattens out, and the aircraft passes through a final stratum of small puffballs of cloud, in a blink the island behind fades, is suddenly gone from view, and all below is just sea, endless empty sea, with many hours of emptiness ahead.

The ocean beneath is almost unimaginably vast, and illimitably various. It is the oldest of the world's seas, the relic of the once all-encompassing Panthalassic Ocean that opened up seven hundred fifty million years ago. It is by far the world's biggest body of water—all the continents could be contained within its borders, and there would be ample room to spare. It is the most biologically diverse, the most seismically active; it sports the planet's greatest mountains and deepest trenches; its chemistry influences the world; and the planetary weather systems are born within its boundaries.

Most see this great body of sea only in parts—a beach here, an atoll there, a long expanse of deep water in between. Just a few,

mariners mostly, have the good fortune to confront the ocean
in its entirety—and by doing so, to win some understanding of
the immense spectrum of happenings and behaviors and people
and geographies and biologies that are to be found within and on
the fringes of its sixty-four million square miles. For those who
do, the experience can be profoundly humbling.

Captain Cook wrote that by exploring the Pacific he had gone
"as far as I think it is possible for man to go." To traverse it today,
two and a half centuries later—to set a course from Kamchatka
to Cape Horn, to pass between the Aleutians and Australia, to
make a ten-thousand-mile crossing from Panama to Palawan—is
to experience a sense of the frontier that is lacking almost everywhere else on the planet. And not simply for its immensity,
but also for the pervasive sense, even today, of confrontation
with the unknown and the unknowable. The British Admiralty's
revered chartroom bible, *Ocean Passages for the World*, still cau-
tions sailors embarking on a crossing, "Very large areas of the
Pacific Ocean are unsurveyed, or imperfectly so. In many areas
no sounding at all has been recorded . . . the only safeguards are
a good lookout, and careful sounding."

United 154, operated most days by one of the more battered old
planes from United's Hawaiian stable, is known locally as the
island hopper, makes its journey along almost six thousand
miles, and takes some fourteen shuddering hours to do so. It
skitters southwestward, then westward, then northward, stopping along the way at five places—all of them islands, scattered
among three different countries—that are even less familiar to
most than is Guam's one city of Hagåtña.

UA154's first stops, of half an hour or so, are on the flat atolls
of Majuro and Kwajalein in the Republic of the Marshall Islands;
it then does the same at runways that have been squeezed into
the more dramatically mountainous and jungle-draped topog-

raphies of the islands of Kosrae, Pohnpei, and Chuuk, in the Federated States of Micronesia.

A scant few passengers travel all the way to Guam. There is much getting off and getting on, and luggage of daunting sizes and bewildering shapes is brought on and taken off at each stop. The crew members, leather-skinned old-timers who have some passing acquaintance with the local island languages, are obliged by United to make the entire journey. They will have recited their seat-belt and tray-table hymns no fewer than twelve times before final touchdown, and seem almost comatose with relief on their arrival in Guam.

In the popular European imagination, the Pacific Ocean contains many of the elements that are to be found along that six-thousand-mile passage between Honolulu and Hagåtña. In every stopping place, it is invariably warm, tropical; both the sea and the sky are intensely blue, the air is sweetly breezy, and there are white sands and coral reefs and sparkling fish of vivid colors darting between the anemone fronds. The roads are decked with bougainvillea and flamboyants and orchids and parrots, with papaya trees and palms of incredible profligacy that drip with dates and bananas and coconuts. Palm trees are central to Pacific imagery: they are to be seen leaning slightly off the vertical, under the endless press of the trade winds, and thereby offering a picture-perfect and theatrically green backdrop for every beach scene; a frame for other equally familiar images of curling waves and spume; or as a border to an empty ocean panorama with its distant gatherings of surfers waiting patiently for the rollers to break and the seas to begin to run.

Such is in evidence everywhere, at every stop, on the United island hopper's run. Hawaii, the starting point, is of course the quintessential exemplar of the mixing of what outsiders see as Polynesian magic and transpacific migration. From Polynesia there is the plangent sound of the ukulele, the sight of the grass skirt, the blossom in thick black hair tipped behind the ear, the nut-brown skin, the dancing, the dancing, the dancing.

Yet when I lived for a while in Manoa, on the eastern side of Honolulu City, it was not so much Polynesian culture that lapped into every corner of my life as the cultural influences of the farther side of the Pacific Ocean.* There was a Japanese grocery store down the street, a Burmese restaurant next door, and every other person on the bus seemed to be from Manila. I kept hearing Korean spoken in the elevators, and met Chinese migrants in the most unexpected places: the elderly man who cut what remains of my hair was from Kowloon, and had been a waiter for decades on the *Coral Princess*, a Hong Kong–flagged liner that once shuttled between Sydney and Tokyo, by way of Jakarta, Port Moresby, Manila, and Shanghai.

Nonetheless, Polynesia rather than Pacific Asia is the current preferred affect of Hawaii. Polynesia is the impression one likes to carry away, as the islands fade astern or over the horizon. *Hula, luau, aloha, lei, ohana*—the best-known words from a lexicon constructed from the thirteen letters of the Hawaiian alphabet—hear them, and you know in an instant which ocean you're in.

This place—though an American state since 1959 (the other Pacific state, Alaska, was similarly admitted almost eight months before), and much changed as a consequence—is still in its perceived cultural essence the Pacific of its Western devotees. Hawaii manages still, deep down, to evoke some of what was felt by Gauguin during his time in the Marquesas, or by those who brought Omai of Ra'iatea to London, or by those gently compassionate scholar-administrators such as Arthur

---

* Statistics bear out the easily forgotten reality that whites—*haoli*, in the vernacular, a word uttered with some disdain—are a minority on the Hawaiian Islands. Their 336,000 (in 2010) are matched by almost half a million from countries around or within the Pacific Ocean, including 200,000 Filipinos and 185,000 Japanese. Eighty thousand only are native Hawaiians.

Grimble, famed for his once-favorite memoir of Pacific life, *A Pattern of Islands*.

Hawaii's shopping malls and warplanes and mountaintop telescopes and aircraft carriers and its legions of resident ocean-ographers and meteorologists may give the impression that the islands have fully entered and embraced the modern era. Yet, culturally, Hawaii is still Polynesia, linked firmly to Easter Island and the Cook Islands and Aotearoa ("the Land of the Long White Cloud"), New Zealand. Hawaii, for all its apparent inti-macy with the American mainland, still resonates with the old Pacific stories of Herman Melville, of Robert Louis Stevenson. It is still emotionally connected to the Pacific that so enchanted poets such as Rupert Brooke, who spent seven idyllic months two thousand miles farther south, in Tahiti, and memorialized it in lines that, once heard, are long remembered:

> Taü here, *Mamua*,
> *Crown the hair, and come away!*
> *Hear the calling of the moon,*
> *And the whispering scents that stray*
> *About the idle warm lagoon.*
> *Hasten, hand in human hand,*
> *Down the dark, the flowered way,*
> *Along the whiteness of the sand,*
> *And in the water's soft caress,*
> *Wash the mind of foolishness.*

They may well be. But the Pacific of Brooke's "soft caress" is soon swallowed up as United 154 soars ever westward each morning, into what swiftly becomes much darker territory— both metaphorically and actually.

Some fifteen hundred miles on from Honolulu, the flight crosses the International Date Line, and the day on board is instantly transmuted into tomorrow. Wristwatch date windows have to be changed, and diary pages flipped forward; the shadows at each airport stop lengthening, so the next day's afternoon steadily begins to take over from the past day's morning.

The date line, a necessary evil for maintaining a system of timekeeping in a spheroidal and interconnected world, was originally set down here, right in the middle of the Pacific. It could have arrowed straight from north to south, like an antimeridian to Greenwich, on the opposite side of the globe; but since it was going to make calendric mayhem in whatever populated place it passed through, the delegates to the International Meridian Conference, held in Washington, DC, in 1884, designed it to zigzag this way and that around the islands and minor republics and kingdoms that got in its way, or were near to it.

At the time this was done, the Pacific was a commercially comatose expanse of sea, with neither aircraft nor telephone cables in place, its international trade limited mainly to the peddling of copra, whale oil, and guano. The existence of an ephemeral line that magically decreed it to be noon on a nonworking Tokyo Saturday while it was still breakfast time on a potentially busy California Friday had scarcely any effect on such matters as the trade in guano futures. In the days long before arbitrage became all the rage, the date line was no more than an idle conceit, something for the stewards on long-haul ships to tell their passengers, who found "gaining a day" or "losing a day" to be greatly amusing.

Today, however, when snap decisions in Japan can urgently affect financial doings in San Francisco, the line has become something of an inconvenience. A Friday decision can't always be made, because where it needs to be made, it is Saturday. It need not have been so: back in 1884, because of arguments over the siting of the prime meridian, with the apoplectic French naturally opposed to its running through a suburb of London, a genial Canadian time zone expert named Sandford Fleming suggested that the prime meridian be placed in the middle of the Pacific instead, and the date line made to run from pole to pole, where the Greenwich line does today. His idea was ridiculed; but today some few remember him fondly, and would also like to see the date line moved into the Atlantic, to minimize the nuisance it causes in the commercially busiest ocean in the world.

Yet this is no more than technical stuff. There is quite another kind of darkness into which United 154 flies, a more metaphorical darkness, and one that helps paint a rather fuller picture of today's Pacific Ocean.

Once across the line, the plane leaves the ethnic limits of Polynesia. The Marshall Islands, of which Majuro is the capital, are part of the sprawling archipelagos of Micronesia. Together with the prettily named Carolines and the Marianas, these hundreds of small islands in the northwestern quadrant of the ocean— some of them tall and volcanic; most of the Marshalls low atolls of coral, and lethally vulnerable to the ever-rising seas—present en bloc a troubling example of the recent history of the ocean and its people. Of what can happen to a place and a population subjected to the casual exploitation of foreigners who, all too frequently in recent centuries, have seen the ocean's immense blue expanse as having been created for their convenience alone.

For the islanders, it has been a bewildering cavalcade of misfortune. The populated specks of Micronesia have existed under perhaps the most complex colonial history of anywhere in the ocean. This ancient and settled culture first felt the impress of outsiders in 1521, when Magellan landed in the Marianas and predestined Guam's seizure by Spain—leading to the establishment there of the very first formal European possession in the Pacific. The oldest native civilization in the region, the Chamorros—who had migrated from Southeast Asia as many as four millennia before—were all but destroyed, either by illness or at the hand of these European invaders. Their numbers dwindled to just a few thousand; their language briefly teetered on the edge of oblivion. And yet the sad saga of Micronesia had only just begun.

Germany was next in line to take an interest in the region, driven first by the needs of commerce and then by the pressing demands of pride, expansion, and empire. Spain initially balked at the impertinence of a Northern European rival trying to muscle in on its territory, but eventually acquiesced, and by the end of the nineteenth century, the Marshalls and the Carolines, settled by German traders who established copra and cotton plantations there, came to be administered by German governors. They were protected by a blue-water navy of sorts, the German East Asiatic Squadron, a cruiser force of six ships that eventually settled a fixed base in the docks of Tsingtao,* in China's Shandong province.

Meanwhile, in the aftermath of the Spanish-American War, the United States opted to take Guam from Spain, and placed a coaling station there for the convenience of its own fleet. To add to this confusing amalgam, Britain began to nibble away at the eastern parts of Micronesia and took the Gilbert Islands for its own, together with the Ellices, which were nearby (though ethnically Polynesian, like Hawaii and Tahiti and points south and west).

Thus might this cumbersome patchwork of island disarrangement have long remained, but for the outbreak of the Great War. Japan had joined forces with Britain from the very beginning of the conflict, and its navy moved quickly to police the sea-lanes

---

* Tsingtao beer, its brewery long overseen by a German brewmaster, remains the most visible reminder of the kaiser's historic influence in eastern China. While the beer retains the old Wade-Giles transliteration of its name, the city itself is now restyled Qingdao.

of the western Pacific, at the very least to stop all the German trade there. An immediate task was to chase away the German navy and to occupy what some in Tokyo saw as the commercially interesting and strategically important Micronesian islands. By October 1914, Japanese forces had almost all the islands firmly under their control—with long-term implications that would become much clearer only in the aftermath of the Second World War.

This was to be the third formal occupation of Micronesia. Tokyo newspapers of the time regularly showed photographs of bearded and be-medaled Japanese governors opening sugar-cane mills and schools and railway lines on island after island. Such images fed into a general Japanese affection for *Nan'yō*, the South Seas, and almost certainly helped foster the belief in Japan's inherent right to govern an even larger part of the region—governance that would eventually be employed, as it happened, for a wider and more sinister purpose. "It is our great task as a people to turn the Pacific into a Japanese lake," cried a noted historian and Diet member, Yosaburo Takekoshi, toward the end of the Great War. "Who controls the tropics controls the world."

Back then, few others knew why Tokyo so keenly wanted to occupy Micronesia, other than to sate the widespread nineteenth-century romantic yearning for the South Seas, fanned by the turn-of-the-century popularity of vernacular novels featuring *wahini* and tropical flowers and pink-hued coral beaches. It was a while before the sentiments of men such as Takekoshi took root, but when they did, such attitudes were swiftly hijacked by soldiers who were much more aggressively nationalistic.

Many in the West were uneasy with the knowledge that the Japanese had taken the Micronesian islands from the Germans by such stealth. It turns out there was ample reason for concern. After 1941 and the attacks on Pearl Harbor, on Malaysia, on Hong Kong, these now formidably well-equipped and well-defended islands (Pulau, Chuuk, Jaluit, and the two Marshall Island atolls to which United 154 was now heading, Majuro and Kwajalein) could be used as bases from which to attack Western forces. And it was clear the Japanese had been preparing them for such roles since back in the 1930s, confirming Western suspicions that Japan's intentions in taking the islands had been for military domination all along.

I once sailed close by the islands, aboard a Japanese ore-carrying ship. She was the *Africa Maru*, belonging to Sumitomo Metal, and she was carrying 135,000 tons of rich Australian iron ore from the Mount Newman mine in Western Australia up to the blast furnaces of Kashima, to the east of Tokyo. The steel that would be made soon would be pressed into the bodies and axles of Nissan Sentras and Toyota Corollas, no doubt.

The ship's captain, a friendly man named Shigetaka Takanaka, asked me up to the bridge one day as we threaded our way through the islands, with Yap and Palau far away to port, Chuuk similarly distant to starboard. He was poring over his nautical chart, a large-scale sea map that indicated the position of all these islands, and of his homeland to the north, and which was titled—whether out of optimism or nostalgia, I couldn't tell— "The Whole Nippon." He pointed out those smaller islands that were close enough to see on the radar. One we spotted through binoculars, if hazily: a lone mass of green, possibly the islet of Ifalek or the three-island atoll of Lamotrek. Captain Takanaka gestured around the horizon. "Ours, once upon a time," he said softly. "All of it. We were given them by the League of Nations. But then they were taken away." He seemed genuinely regretful.

They had been taken away, indeed, and at a terrible price. American forces recaptured them in the spring of 1944, atoll by atoll, in a stuttering series of appallingly bloody set-piece battles that have since passed into military legend. Chuuk was one of the last outposts for the defeated Japanese: their surrender was not taken until September 1945, almost a month after the great formalized surrender ceremony staged in Tokyo Bay, aboard the USS *Missouri*. The islands of Micronesia—by some estimates nearly three thousand of them, a mere one thousand square miles of dry land peppered across fully three million square miles of sea—have been American, in essence, ever since. The fourth imperial occupation, some might argue.

And the native people have won precious few benefits from all the centuries of foreign attention. Critics claim, not unreasonably, that all that was brought by the years of foreign trespass in the Micronesian islands has been death, disease, and dependency; its residue remains, and it is not a pretty sight.

Particularly on the atoll known as Kwajalein.

United 154 lands there, its second stop out from Honolulu. Most people are forbidden to disembark, and must remain in the parked aircraft, trusting that its cooling system will survive the punishing afternoon heat. But I had left Honolulu that morning with a permit to land, issued by a forward outpost of the U.S. Army.

For Kwajalein is an army base. Since the 1960s it has operated as a center for mid-ocean rocketry, and it is currently the Ronald Reagan Ballistic Missile Defense Test Site. Few are allowed to get off the plane, fewer still to linger on the atoll, because the site is festooned with a costly array of ultrasecret high-technology apparatus, and is peopled with hundreds of high-technology staff, soldiers, and scientists, who are performing clandestine tasks with the equipment, all officially bent on helping keep America safe from whatever are deemed currently the world's most dangerous threats.

The atoll is one of the world's largest, with a lagoon of eight hundred square miles surrounded by almost one hundred islands, slivers of sand and coral that peek just a few feet above the surf. Inside the atoll, the sea is pale blue—the wreck of the German heavy cruiser *Prinz Eugen*, which turned turtle as it was being towed to haven there, a damaged war prize,* is plainly visible. Outside the atoll, the sea is night dark, as the waters off the reef edge plunge thousands of feet down. There are the leaning palm trees, the endless rollers, the sound of seabirds, the roasting sun, the white-hot sand.

The actual Kwajalein Island is at the atoll's southern end, and it serves as the headquarters for this cruelly isolated base. The island extends three miles or so from tip to tip, is a quarter of a mile wide, and sports the base's aerodrome, water tower, and softball field. In other ways, it is as cheerless and institutional a place as any army base. The majority of those stationed there are civilian contractors, mostly from Alabama, many of them employed by an Alaska-based company that won the management contract from the Pentagon.

Some half dozen times a year, clients, customers, users of the Kwajalein facilities—they are called by many names—fire missiles from pads at Vandenberg in California and Kodiak in Alaska toward the atoll, to see how well they work. The rockets, of many different kinds and weights and speeds and newness, with many different kinds of warheads, all dummies of course, roar in and are tracked, measured, noted, and scored by the long-range telescopes and radars with which Kwajalein is equipped. It's all a multimillion-dollar game of darts, the accuracy of the weapons' splashdowns measured in inches, after a four-thousand-mile flight.

Once in a while specialist soldiers on Kwaj, as most call their

---

* There can be few more impressive examples of German engineering than this eighteen-thousand-ton, thirty-two-knot (and exceptionally beautiful) warship, since she survived not only innumerable strikes by RAF bombers during the war, but also two nuclear tests in the Bikini atoll lagoon, where she was placed as a target for one air-dropped bomb and for a massive underwater weapon called *Helen of Bikini*. Still floating after the second test, but fiercely radioactive—all her crew would have died—she was towed to Kwajalein, developed a leak, and capsized, her enormous guns falling out of their barbettes and onto the seafloor. One of her screws has now been placed in a museum; the others remain visible at low tide. *Prinz Eugen* will never be salvaged, however, since her steel is still lethally contaminated.

unlovely home, will fire their own missiles up toward the incoming warheads, and score how well they knock them down, creating their own multimillion-dollar skeet shoot—that has implications, all are assured, for the preservation of world peace.

The rocketry on Kwajalein is impressive, beautiful even—a nighttime test in particular can be quite memorable, with streaks of what looks like orange tracer fire lighting up the sky, and enormous plumes of phosphorescent water where the missiles hit the ocean. But what is seldom seen by visitors is the other side of Kwajalein—where the islanders themselves are obliged to live.

For almost no Marshall Islanders are permitted to stay overnight on Kwajalein Island. They have to leave each evening on a U.S. Army ferryboat, and are shuttled three miles northward up the lagoon, to the island of Ebeye—twelve thousand men, women, and children compelled to live on a squalid eighty acres of slum houses, in what is one of the most densely populated places on earth.

It is a pullulating, smelly, fetid, and degrading place, in appearance more like a slum in Bombay or Calcutta than a community in a country that enjoys "free association" with the United States. There is little by way of a proper sewage system. The schools are ill-equipped; the children—and half the population is under eighteen—ill-educated. The most common disease is diabetes, type 2; the one supermarket, run by a genial Irishman on contract to a company in Guam, sells improbably vast tonnages of sodas and Spam.

Melancholy sights are everywhere. Since there is no Laundromat on Ebeye, those wishing to wash their clothes must come on the ferryboat to Kwajalein and use machines set up in a wire-enclosed compound outside the security fence, while the workers who have permission, and the uniformed Americans who live on the base, cycle past just feet away. There is no mortuary on Ebeye, either; the dead must be brought to Kwajalein, placed in a freezer outside the chain-link fence, and stored there beyond America proper, awaiting burial.

Seldom are the realities of the first and third worlds placed so tantalizingly close to one another—even the realities of Mexican poverty are largely out of sight to most passersby beyond the iron fences in Arizona. But on Ebeye, the separation of the two cultures is cruelly and harshly visible, just inches apart—and with the shame of it made all the more damning because the islands— the entire atoll, the neighboring islands, the entire republic— make up the country that is the birthright of the very people who are now being denied access to it.

This is the Marshall Islands, and these are the Marshall Islanders. Yet international agreements signed in U.S. government offices thousands of miles away have decreed that these men and women and their thousands of children are now forbidden to inhabit large parts of their own island homeland. They must perforce wash their clothes and attend to their dead and be otherwise separated behind chain-link fences, while Alabamans and other strangers on the far side of the same fences come and go on the Marshall Islands quite as they please.

Moreover, there is also the matter of the Money. The U.S. government has a long-term agreement with the government of the Marshall Islands to lease eleven of the islets that rise around the Kwajalein lagoon. Kwajalein Island itself is the biggest and most important of these; but at the north end of the atoll, the outcrop of Roi-Namur has a large airfield also; and many gigantic radar installations and telescopes; and impossibly large long-range-missile-spotting cameras, too—and a small, dignified cemetery for the Japanese war dead (with bones still being found, often frustratingly indistinguishable from the bones of dead American marines). Nine other islands also have launch pads and sensor arrays and target crosses painted on thick concrete pads, which some of the Vandenberg birds, as they are called, are supposed to hit.

For all this, the Marshall Islands government receives many millions of U.S. taxpayer dollars each year—$18 million under the current agreement, on a lease that expires in 2066, with a twenty-year extension. There is a meta-agreement, too: a much more substantial sum is paid by Washington based on what is known as the Compact of Free Association, which essentially gives all the component island groups of Micronesia (of which the Marshalls are one) a guarantee of additional financial aid in exchange for the United States' being able to make use of the islands more or less as Washington wishes (the "more or less" being somewhat open to negotiation).

But the $18 million annual payment for the use of the Kwajalein islands is earmarked specifically for Kwajalein, which leads any visitor to the pressing question: where does all the money go and why are places such as Ebeye Island such sinks of squalor and poverty? It is a mystery, and it hints at much that is wrong, and has been wrong for generations. Every islander to whom I spoke about it tended to look at the ground and try to change the subject. I got only vague answers that usually mentioned something about the peculiar tribal arrangements of the Marshall Islanders, the traditional structures of power and authority, and the behaviors of the local kings and tribal chiefs, men who are widely respected and deferred to and who often hold elected positions of power.

There are mutterings, too, about the cost of maintaining large houses in Hawaii, which many of the senior figures in the Marshall Islands are said to own. But little is said or written directly about the dispiriting situation of so much poverty in a place where such an abundance of American government money has been so liberally sprinkled about. One University of Hawaii academic who wrote a book purporting to expose this very evident corruption was delayed for years, awash in threatened lawsuits. And though the book was eventually offered for sale, the publisher insisted it be heavily bowdlerized so as not to offend any of the local chieftains.

To add to the general wretchedness of the Marshalls, there is also the nagging matter of Bikini, a lonely but infamous atoll two hundred fifty miles farther north, but also part of the republic. In its pre-atomic existence Bikini was so typically Pacific that it could well have been on a *National Geographic* cover or a cafeteria poster or the jacket of an old South Seas novel—but it has since become notorious around the planet, ever since the first nuclear weapons were exploded there, back in the late 1940s. The evident misappropriation of funds on Kwajalein, and the fate of the Kwajalein islanders who are forbidden to live on most of their own atoll, constitutes a sorry enough story. But it pales in comparison with the fate of the Bikinians, on whose homeland most of America's atomic weapons were to be exploded. It is a continuing saga of dispossession also, since these islanders, too, are exiled from home, compelled to live hundreds of miles away, and (as we shall see in the pages that follow) in all too many cases suffering from a florid array of ailments of body and mind. Theirs is a tale—dispiritingly familiar in this corner of the ocean—from which no one in authority emerges with any measureable degree of credit.

The Pacific is an oceanic behemoth of eye-watering complexity. A near-limitless range of human and natural conditions exists within its borders, as one would expect of something so unimaginably enormous. Arthur C. Clarke once remarked, with droll prescience, that a space traveler, upon seeing our planet, would say that calling it Earth was a grave misnomer, since most of it is so obviously Sea. He must surely have been thinking of the Pacific, since its blue expanse entirely dominates the planet.

Its dimensions are staggering. It occupies almost one entire hemisphere. Looking westward, from Panama, and from where Balboa stood on his high peak in the Isthmus of Darién, across to the first encountered landmass of the eastern Malaysian coast, there are more than 10,600 miles of uninterrupted sea. From north to south, from the fogs and shivering waters of the Bering Strait down to the ice cliffs of Marie Byrd Land in Antarctica, is nearly nine thousand miles. The sixty-four million square miles in between fill almost one-third of the planet's surface. Forty-five percent of the planet's total surface waters are found in the Pacific Ocean, and seven miles down, it has the earth's deepest trenches. In short: everything about the Pacific, the last ocean to be found by Western man, presents an unchallengeable superlative.

It is also not easy to get into, or out of, and this difficulty insulates it from the rest of the world's oceans. Except for vessels bold enough to try the Bering Strait, between Russia and Alaska, or the gale-whipped seas that fringe the Antarctic, the Pacific enjoys no entranceway that is more than three hundred miles wide. Ships trying to enter from the Indian Ocean must make their way through a litter of islands scattered between Malaysia and Australia, the so-called Maritime Continent. Except for the Strait of Magellan, far down south, there is no natural entrance whatsoever on the American side. Only the Panama Canal, that narrow funnel gouged artificially through the isthmus in the early twentieth century, permits carefully sized ships from the Atlantic the luxury of a quick and easy transit.

The vast distances inherent in the Pacific's geography have consequences seldom known elsewhere. Consider the Republic of Kiribati, for instance, once the British-run Gilbert Islands. Its one hundred thousand inhabitants are spread over fully 1.35 million square miles of ocean. Two thousand miles from Tarawa, its administrative capital, lies Kiritimati Island—the Christmas Island where the British tested their atom bombs back in the 1960s, without evacuating any of the locals. Not only are these five thousand Kiribatians inconveniently distant from their country's capital, but they also are on the other side of the equator from Tarawa, and live on the day-before side of the International Date Line. So a summer Sunday on Tarawa is a winter Saturday on Christmas. Small wonder Kiribati struggles, having to cope with such logistical madness. It is one of the world's poorest countries: the seaweed and copra and fish it harvests are too expensive for most locals, so many of its menfolk are obliged to work abroad or to crew on long-distance cargo ships, remitting their paychecks home, both to keep their families in victuals and in the hope of keeping their country's pitiful economy alive. Size can be impressive, or it can be an impressive nuisance.

The Pacific is an ocean of secrets. Castaways, runaways, fugitives of all kinds people its recent history: the first islands that a sailor heading north from the Strait of Magellan encounters are those volcanic relics of the Juan Fernández group—where the Fifeshire buccaneer Alexander Selkirk was marooned for four years, his adventures later memorialized by Daniel Defoe in *Robinson Crusoe*.

Much of the dirty business of the modern world has been conducted within the confines of the Pacific. The nuclear testing—by the United States in the Marshall Islands, by Britain on the Gilbert Islands, by France in its remaining Polynesian possessions—is well enough known. In 2008, when a secret American reconnaissance satellite got into trouble in orbit and needed to be shot down, the military ordered a navy ship specializing in missile attacks to bring it down over the Pacific, supposing the ocean to be so vast that it would be harmless there. The Marshall Islanders complained about what they saw as the Pentagon's cavalier attitude, insisting that they lived in the ocean, as did millions of other islanders, and it was not some empty space on which any dangerous test might be performed at will. As it happened, the satellite was brought down locally, and the hydrazine rocket fuel hurt no one.

Still other experiments of equivalent unpleasantness have taken place on the U.S. Navy's tiny islets of Johnston Atoll, some seven hundred miles southwest of Hawaii. United 154 flies above them, though they are seldom pointed out, and there is even less notice of what has gone on there. For years, any passing yachts-man, in mid-ocean, was confronted by huge signs warning him to move on—nothing to see here, "Deadly Force Authorized"—and patrol boats stiff with heavily armed naval guards would cruise offshore, keeping all the curious at bay.

Strange things went on there. Rockets carrying atomic weapons accidentally exploded, contaminating the island with plutonium and americium. Almost two million gallons of Agent Orange from Vietnam were stored there, and then their storage carboys split open, adding to the toxic mix. The atoll was next used for testing biological weapons, and after another accident a quantity of the bacilli that cause tularemia and anthrax were released upwind, and the island was contaminated once again. Then, in 1990, a huge incinerator was built that would destroy such chemical weapons as the United States still admitted to possessing. According to Pentagon statements, the list included: "412,000 bombs, mines, rockets and projectiles . . . four million pounds of nerve and blister agents . . . only one recorded incident for every 200,000 man-hours worked." Then, in 2000, all work was stopped, the plant was broken up and carted away, the remaining pollution was said to have been cleaned up, and Johnston Island, ten times as big (thanks to landfilling) as it was when it was first found, was abandoned and offered up for sale. That's when it was invaded by a vicious type of ant. Nowadays, passing yachtsmen—initially lured to the island out of curiosity, since they would no longer be confronted by armed police—like to stop there, if briefly. The island has become a National Wildlife Refuge, a memorial to the utter despoliation of the sea.

All this, of course, relates just to the Pacific's islands, to some of the speckles and shards of territory that lie within the confines of the ocean. The stories to be found here and in a thousand other islands not so far mentioned are complicated enough, as the strange island-hopping odyssey of United 154 suggests. But to get a sense of the ocean as a whole, into this patchwork need to be stitched the fantastic cultural diversity and enormous power and scale of the countries that lie around the ocean, along its famous—and famously volcanic, and so infamously unstable— Pacific Rim.

Thanks to the dominant cultural bias of modern history, there is a topsy-turvydom inherent in any description of the rim. On its western side are the Eastern peoples: Chinese, Koreans, Japanese, Indonesians, Filipinos, and countless more if one chooses to push the ocean's frontiers westward toward Indochina and India. On the eastern side are the national admixtures of the various migrated and now notionally Western peoples: Canadians, Americans, Central Americans, Colombians, Ecuadoreans, Peruvians, Chileans. Around the south and beyond to Oceania are the more newly settled outsiders of modern New Zealand and Australia. The aboriginal peoples—Native Americans, Aleuts, Inuit, Maori, Australian indigenes, the Canadian First Nation, and a host of others, all genetically Pacific peoples as we now know—remain dotted around or within the rim, where their recent experiences have become conjoined with those of the Polynesian islanders, the inhabitants of Melanesia and Micronesia, and so have been protected or decimated, exploited or revered (but never left alone), as the various histories of the newcomers have unfolded.

And to be stirred into this mix of peoples and cultures and politics and ambitions is a formidable array of other phenomena besides. There is the constant, complex, and often spectacularly violent interplay of tectonics, with volcanoes, earthquakes, and tsunamis in deadly abundance. The lands, seas, and seafloors of the Pacific are home to all manner of exotic and unfamiliar and yet-to-be-discovered wildlife. Mineral wealth exists in and around the Pacific in quantities that are beyond the most extravagant of human dreams—wealth that, if exploited, would bring a raft of unanticipated consequences. The environment, fragile everywhere, seems somehow to display a more obvious fragility in the Pacific, with its thousands of square miles of delicate corals, with its all-too-vulnerably low-lying atolls, with its cyclones and typhoons that are so much more vicious and destructive than even the worst to be found elsewhere.

For all its apparent placidity, the Pacific seems today to be positioned at the leading edge of any number of potential challenges and crises—whether they relate to politics or economics, to geology, to weather, to the supply of food, or to the most basic questions about the number of people that this planet can support.

The future, in short, is what the Pacific Ocean is now coming to symbolize. For if one accepts that the Mediterranean was once the inland sea of the Ancient World; and further, that the Atlantic Ocean was, and to some people still remains, the inland sea of the Modern World; then surely it can be argued that the Pacific Ocean is the inland sea of Tomorrow's World. What transpires across these sixty-four million square miles of ultramarine ocean matters, and to all of us. Hence the need to write about it.

But *how*? How best to corral all that is important about the new Pacific—and by this I mean the Pacific in toto, the ocean considered shore to shore, pole to pole, and not just the modern and very limited convention of "Asia-Pacific," which I do *not* want this book solely to be about—into a single comprehensible and digestible volume? What structure might suit this ocean's story best?

For months it all seemed to me to be too overwhelmingly confusing—the body of water too monstrous, its narrative too insuperably challenging in its variety and vastness. Until one day, and quite by chance, when I came across a slim volume which was published in Germany almost a century ago, and which seemed to offer me a way, a structural vade mecum.

I came across it while I was reading as much as I could about those two very early Europeans who had reached out first to see, and then to cross, the ocean that had hitherto existed only in their respective imaginations. The men have become legends: there was Vasco Núñez de Balboa, who first sighted its blue gleam below him from Darién in September 1513; and there was Ferdinand Magellan, who seven years later made transit across Patagonia through the strait that now bears his name, entered the Pacific, named it *mare Pacifico*, and then sailed clear across it into history (and, as it happens, to his own unanticipated and violent death).

Not surprisingly, the stories of both men have been told by many. As I was doing my own reading I came quite by chance and in quick succession across two accounts, both by the same author, that seemed possessed of a certain luminous quality, and were markedly different from all the others in scope, in scale, and in style. Both were by the Austrian author Stefan Zweig—a 1920s writer mostly forgotten now, though his appearance as "The Author" in Wes Anderson's 2014 movie *The Grand Budapest Hotel* has lately revived his memory, somewhat.

It was Zweig's succinct and poetic account of Balboa that I found particularly alluring. It was contained in a book, published in 1927, that has perhaps suffered under more titles than any book deserves. In its original German it was *Sternstunden des Menschheit*, or *Great Moments of Humanity*. The first English translation was *The Tide of Fortune*, and then came *Decisive Moments in History*, and most recently, *Shooting Stars*. But whatever the book was called, its essence remains the same: it was a slender collection of ten ruminative essays, each one turned to perfection, about what Zweig considered to be seminal moments in the tide of human experience.

The writer chose his subjects eccentrically, but always interestingly. Balboa's expedition across the isthmus was first; the defeat of Napoleon at Waterloo another; Scott's failed expedition to the South Pole, a third. He then wrote of Handel's composing *The Messiah*, of the fall of Byzantium, the death of Cicero and the writing of "La Marseillaise." He told also an extraordinary version of the saga of Lenin and the sealed train that had taken him through war-torn Europe to Finland Station, to stage his Russian Revolution. The result is a tumbling mélange of a book, quite charming and, even if perhaps lacking in academic rigor, one that quite transfixed me.

So after many attempts to corral the wealth of material about the Pacific Ocean into one manageable whole, I chose for this account to attempt to ape the master, to create a modified version of Stefan Zweig's approach of almost a century ago. I decided I would sift through the events of the modern Pacific to try to find my own galaxy of shooting stars—of truly pivotal *moments* in the story of this vast acreage of ocean. I would choose a scattering of happenings each of which, to me at least, seemed to betoken some greater trend, and which might tell in microcosm a larger truth about the Pacific than the moments themselves suggest.

From a mass of such occurrences I could distill and weed and cull, and then decide which of the remainder truly said something significant about how the ocean, and the perception of the ocean, was developing and perhaps evolving: which events, in other words, indicated the direction in which the Pacific was shifting, as it moved to play its defining role in the future of the world.

So I made a list. I scoured newspapers and history books and databases and academic papers, and came up with some hundreds of more or less notable occurrences between January 1, 1950—a date chosen for a reason I will explain later—and the time I began to write this book, in the summer of 2014.

The cutting room floor was eventually to be buried under a blizzard of possibilities. I was fascinated, for instance, by the postwar reincorporation of the Japanese into the mainstream of Pacific life, and thought that stories from the aftermath of the notorious internment of Japanese American citizens might illustrate their welcome return to the world stage. I thought, more trivially, of the opening of a slew of Pacific Disneylands, first in Anaheim in 1955, then in Tokyo in 1983, and in Hong Kong in 2005, and wondered about the spread of America's cultural impress on the people around and within the ocean. I thought about the lasting social consequences of the staging of Olympic Games in such Pacific cities as Melbourne, Tokyo, and Seoul. I wondered similarly about the social effects of invention of the Boeing 747-400, a plane made on the Pacific coast, an aircraft built specially to cross the entire ocean without refueling, in one bound.

The list went on and on. What of the significance of pollution—so poignantly symbolized by the Minamata disease that was identified in May 1956? What of the impact of the first television service that began in Australia, in September, four months later? Of the Watts Riots in Los Angeles that erupted in 1965? The My Lai Massacre that took place in Vietnam in March 1968, and the shooting of Bobby Kennedy in Los Angeles on June 6 of that same year? The time when more than seven hundred members of the Unification Church were married in a ceremony in Seoul in 1970, and when twenty-six members of the Heaven's Gate cult neatly killed themselves in San Diego in 1997? What followed on from Richard Nixon's visit to the People's Republic of China in 1972—the first by an American president—and from Queen Elizabeth's visit later on, in 1986? What of the consequences of the Chilean Communist Party's formation in 1979? Or the disappearance of the "Dingo Baby" near Ayers Rock in Australia in 1980? The time when President Marcos was overthrown in the Philippines in 1985? The death of the last Hawaiian black-faced honeycreeper in 2004? News that a bridge had been built over the Eastern Bosphorus in Vladivostok in 2012? Or the curious time when Tonga, in 2014, signed a military agreement with the Nevada National Guard?

Rich though all this might be, there was still richer fare to consider. In the end, I chose just ten singular events, some of them portentous, some more trivial, but each appearing to me to herald some kind of trend. They showed, and in chronological order, a series of developments that, when accumulated as one, depicted an image, perhaps more pointillist than precise, of the ocean as it had arranged itself in the past sixty-five years; and that also hinted at the way the ocean might evolve through the near future. The wisdom or otherwise of these choices is what, of course, will determine whether this portrait of the ocean is judged to be fair and right. Naturally I hope it is considered so.

I decided to begin with the acceptance of a singular and distasteful reality: that the Pacific, despite what Ferdinand Magellan experienced when he first sailed into it, is anything but a pacific ocean; it is in fact, and more than any other, an *atomic ocean*. More dangerous than that, it is the ocean where most of the world's thermonuclear weapons have been tested, and has been ever since the beginning of the story, in January 1950. The story of Bikini and of the hydrogen bombs then tested there allows for neither praise nor admiration. At the same time, though, it serves as a reminder of the terrible toll atomic weaponry has already exacted on humankind—mostly to innocent Japanese civilians, of course—and so is part of the ocean's story, helping to define its past, present, and future.

Then, in a more cheerful vein, I chose the invention, four years later, of the transistor radio and the subsequent formation of the Sony Corporation. It seemed to me that these and other connected events of the early 1950s not only illustrated the energy, tenacity, and technical brilliance of the postwar Japanese, but also demonstrated the beginning of a pattern of eastbound transpacific trade that still dominates the ocean's story to this day. Maybe the early Japanese technicians were eventually to be overtaken in some fields by the Koreans, and the Koreans in turn by the Chinese, but the endless procession of heavily laden cargo ships passing eastbound, inbound, beneath the Golden Gate Bridge is testament to a trend that was started by Japan, and to the long-ago makings of a tiny wireless set that would fit inside a cunningly enlarged shirt pocket.

Surfing—not so trivial as it may initially seem, just like the charmingly slight American film that made it popular in what would shortly become the biggest of all surfing markets—is a gift from the Polynesian Pacific that is today worth billions. This ancient, graceful pastime of wave gliding, once so central to the aristocracies of Hawaii and Tahiti, deserves proper consideration—in the same way that football and cricket must not be overlooked elsewhere—to help fathom the ocean in which it was born, and the peoples who gave it birth.

I have traveled in North Korea several times, and in 1987 was sorely frustrated in an attempt to walk the entire length of the peninsula when, after three hundred miles, my way forward was firmly blocked by frontier fences and a pair of burly American sentries. Once, and very briefly, I harbored some small secret sympathy for a North Korean regime that in its earliest days was guided by an intense desire to follow its own path to economic and cultural independence; now the regime has become such a byword for domestic cruelty and international insanity that it can hardly be ignored. Its role as perpetual irritant is an unattractive aspect of today's Pacific, but a crucial one.

The reincorporation of Japan and the other native Pacific peoples—of American Indians, Australian aboriginals, Maori, and Pacific Islanders—into the running of their ocean's affairs has been matched, in more recent years, by the steady, though seldom peaceable, withdrawal of colonial powers from the region. To begin this chapter, I chose one symbolic event, the sabotage and fatal fire on a British ocean liner in the waters of a British Pacific colony, to allow me to tell the stories of a number of colonists' retreat: the French from Indochina; the Americans from Vietnam; the withdrawal elsewhere of the Dutch and the

Portuguese; and from everywhere remaining, the otherwise near-ubiquitous British. The Pacific peoples are more firmly on their own at last, as they should have been for centuries past.

Except, maybe, in Australia. The role that this one continental country will play in the future remains a mystery to most—in Australia as among its millions of neighbors. Australia is an overwhelmingly non-Pacific nation on the western edge of the ocean: Does it fit? Can it fit? Will it exert a regional force for good, in the short term or the long term, or ever?

Then there are the more technical considerations: of the Pacific as the *fons et origo* of the world's weather patterns; of the Pacific as the bellwether of the environmental dangers the planet as a whole must necessarily confront; of the Pacific as the epicenter of most of the world's lethally dangerous tectonic mayhem; and in part as a consequence of that, of the Pacific as the source of almost unimaginably vast undersea resources, which the world can plunder or preserve as it wills.

Then, most crucially of all, there is China—the world's most populous nation, fast ascending to the ladder tops, to the summits, of almost every measurable feature of modern humankind. This proud and ancient and imperturbable nation lies on the far side of the Pacific from America, the most powerful nation the world has ever known; it is easy to imagine that both are now glissading toward a rivalry and a possible confrontation that could easily end less than well for either party.

And then there is the sea itself, the fathomless expanse of vastness—of Robinson Jeffers's "staring unsleeping Eye of the earth"—which the islanders know and care for, even if so many outsiders manifestly do not. The Pacific Ocean—now almost freed from its former European control, yet brimming with new disputes, a region that is tectonically and meteorologically dangerous—is in serious environmental peril, is ringed with nations undergoing immense internal change, is unimaginably busy with commerce, has come to be at the forefront of science and self-discovery, but is at the same time also peopled by many clinging tenaciously to its old ways, as well as by civilizations, East and West, that seem steadily to be beginning to understand one another, and all this is occurring in a setting of philosophical and spiritual renewal and among fantastic yet threatened beauty.

It is the most turbulent ocean in the world, and an expanse of sea that should be central to all our thoughts. Is the ocean to be a place of coming war? Is it to be our eventual savior, a place so beautiful and fragile that its sheer vastness will one day demand that we pause in our careless and foolish behavior in the rest of the world? Or will it be something in between: a pillar of hope and example and good sense poised between East and West, on which, for good or ill, we construct humanity's future?

The book that follows is an account of this modern Pacific, the story of the development of the ocean in the sixty-five tumultuous years that began on January 1, 1950.

AUTHOR'S NOTE:
ON CARBON

New Year's Day 1950 was a Sunday, and by and large, as the clocks ticked and chimed and boomed their way into the first year of the century's fifth decade, the world seemed to have settled into a fairly stable place, with memories of the Second World War starting to fade, and scant suggestion of any of the turmoil soon to come.

The Japanese, still busily repairing their country and still occupied by American forces, had some small reason for good cheer that day, with the ending of their custom of declaring children to be one year old at birth and of everyone adding a year to his or her age each January 1. This change meant that all eighty million Japanese would not become numerically older on this day: a forty-year-old would wait until his next actual birthday before becoming forty-one. For a brief while that morning, all Japanese were said to have suddenly felt younger.

There was smaller cheer for New Yorkers. The canned music that had flooded the concourse of Grand Central Terminal for the previous three months, and that had whipped silence-loving commuters into a mutinous fury, was turned off, and forever. Riders on the New York Central regained their sanity; the calm of the everyday hubbub was resumed. For a while, some relieved New Yorkers were said to feel suddenly younger, too.

And in England, a teapot lid maker named Elizabeth Hulme and a man from Lancashire named James Jackson, whose job was listed as "mule spinner," said to be a textile-related occupation, were each given awards for the contributions they had made to these crafts, in a ceremony at Buckingham Palace. In Britain nothing else of great significance occurred, according to the day's newspaper accounts.

Beyond, in the outside world, most of the men of note then charged with running the world could count on remaining some time yet in power. Truman, Atlee, Stalin, Adenauer, Franco, Tito, Perón—even comparative newcomers such as Kim Il Sung and Mao Zedong—were all, for now, lying easy in their beds. There were similarly complacent kings and queens and princes in abundance, from Egypt to Tonga and Kathmandu, together with an emperor in Japan, a shah in Persia, and a grand duchess in Luxembourg, mostly respected, occasionally revered, and all, for now, reigning in comparative comfort.

Yet there were shifts in the wind. The grandest of the world's monarchies was still England's, with the incumbent, King George VI, still technically able to say that he reigned over a quarter of the world's population, with his empire still in rude good health. Except that just three weeks into the New Year, his hold on one of those dominions would weaken when, as expected, Jawaharlal Nehru proclaimed India to be a republic. And farther still around the world, a little-known Vietnamese Communist named Ho Chi Minh would, at the year's beginning, commence a series of negotiations with China and the Soviet Union that would eventually ensure that the French would be turfed wholesale out of Indochina, and would leave Asia forever.

But these small hints aside, if cracks were beginning to appear in the settled order of the world, they were only hairline, visible to few, and troubling to almost no one.

Except, that is, for one development that is still marked indelibly on that New Year's Day of 1950, and that came to be regarded in a pair of ways: as being, first, of the gravest moment and, second, of lasting, perhaps even everlasting, scientific significance.

Three months previously, on September 3, 1949, a Geiger counter mounted in the nose of an American B-29 weather-monitoring plane that was flying reconnaissance missions in the western Pacific between Yokota in Japan and Eielson Air Force Base in Alaska, began to chatter furiously. Puzzled technicians swarmed to examine the records and soon determined that atomic radiation seemed to be pouring into the sky, from somewhere.

Two days later a second plane, based in Guam, flew over the same route and picked up signs of even more radioactivity: barium, cesium, and molybdenum fission isotopes were found in the upper atmosphere, signatures that suggested that either there had been a nuclear accident somewhere to the east of the plane's track or someone had exploded an atomic weapon.

It turned out to be the latter. An atomic bomb known in Russia as *First Lightning* and elsewhere, eventually, as *Joe 1*, had been exploded by Joseph Stalin's Soviet Union five days before, in an experiment conducted at a hitherto unknown and subsequently top-secret nuclear test site at Semipalatinsk, in Kazakhstan. The successful exploding of the bomb, which was modeled on, looked uncannily like, and was in fact slightly more powerful than the plutonium weapon dropped by the Americans four years previously over Nagasaki, stunned the outside world. Few Americans and few of their allies thought the Soviets would be able to catch up with the United States in terms of nuclear capability for many more years. But as was discovered a decade later, Moscow had a spy in Los Alamos, Klaus Fuchs; and though debate continues to this day about how valuable the information was that this brilliant young Briton passed to the Soviets, it is generally agreed that, perhaps more than any other spy before or since, Klaus Fuchs changed world history.

For by allowing the Soviet Union to construct nuclear weapons, and ultimately to make hydrogen bombs and all the other terrible paraphernalia of the nuclear age, his gift of secrets permitted the Cold War between East and West officially to commence—with the consequence that for the next half century, and perhaps for longer still, the planet lived in the shadow of the very real possibility of nuclear annihilation.

There was another consequence of this development, however—that to this day is of great significance to the scientific community and, as it happens, has some bearing on the structure of this book. It concerns radioactive pollution.

The explosion in the atmosphere over the coming Cold War years of hundreds upon hundreds of atomic bombs—big and small; fission and fusion; to be launched by missiles or dropped by aircraft or fired from guns; made by the United States, the Soviet Union, Britain, France, China, Israel, India, Pakistan, North Korea, or perhaps Iran—would contribute to the pollution of the earth's atmosphere by myriad poisonous and radioactive decay products. Until atmospheric testing was banned in 1963, the world was living under a blanket of increasingly radio-polluted air, with effects that would be likely to last for thousands of years.

Crucially, one of the many products created by atomic explosions is the unstable radioactive isotope of carbon known as carbon-14.

This isotope is already naturally present in the world (its

presence caused by cosmic ray bombardment), in extremely tiny but measurable amounts. Compared with the amount of normal, nonradioactive carbon-12 in the air, about one part in a trillion is carbon-14.

Plants absorb this carbon during photosynthesis, and animals that consume the plants absorb it, too. So while an animal or a plant is alive, its cells contain both carbon-12 and carbon-14, and in the same ratio as exists in the atmosphere.

However, once the plant or animal dies, its cells stop absorbing carbon—and at that precise moment, the ratio of the two isotopes begins to change, for the simple reason that carbon-14 is unstable, and begins to decay. The isotope has a half-life of 5,730 years, meaning that after that period, half of it will have vanished. After another 5,730 years, half of what remains will have disappeared, and so on and so on. And, it is important to note, the changing ratio of carbon-12 to carbon-14 in the dead animal or plant can be very accurately measured.

What followed this discovery—first made in 1946 by a University of Chicago chemist named Willard Libby, who would win the Nobel Prize for it—was the realization that by measuring the amount of carbon-14 remaining in a dead creature or plant, it should be possible to date, and with some precision, just when that plant or animal died. Thus was born the technique of *carbon dating*, and it has been in use ever since, a vital tool of archaeologists and geologists in determining the age of found organic materials.

The technique requires one constant, though: for any age calculation to be accurate, the baseline ratio of naturally occurring carbon-12 and carbon-14 has to be a real baseline—it must, in other words, stay the same as it always has been. The figure accepted by Libby and his colleagues and used as the base was the aforementioned one to a trillion: one atom of carbon-14 to one trillion atoms of carbon-12. With that figure firmly in place, the age calculations could be made, and reliably.

But then came the unexpected. As soon as the testing of atomic bombs began in earnest in the 1950s, that baseline figure suddenly began to change. The bombs created immense mushroom clouds of lethal chemistry. They thrust, among other things, a sizable amount of extra carbon-14 into the atmosphere, upsetting the baseline figure and causing the dating calculations suddenly to go awry.

Radiochemists around the world monitored the situation, and as the levels of new carbon kept increasing, test by test, year by year, they kept on writing algorithms to correct the distortions caused by the bombs. But as more and more bombs produced more and more unstable carbon, the situation was fast becoming complicated, irritating; and for a world that placed value on near-absolute precision, it threatened to render age determinations so inaccurate as to be useless.

To address this problem, a decision was reached that would unscramble matters. A date was chosen before which radiocarbon dating could be regarded as accurate, because the baseline was constant. Radiocarbon results that were achieved after that selected date would continue to be regarded with suspicion.

And the date selected—of what is now known as the start of the standard reference year, or the Index Year—was January 1, 1950. Before January 1950 the atmosphere was radiochemically pure. After January 1950 it was sullied, fouled by bomb-created isotopes. So this date, this otherwise unexceptional Sunday when Ho Chi Minh began his campaign in Vietnam, when the Japanese started recalculating how old they were, the day the music died in Grand Central Terminal, would become for scientists a new Year Zero.

The choice of the date was scientifically elegant, logical, and precise. And it would soon spread beyond the world of science alone. For it would have an impact on the entire question of what was meant by the use of the simple word *ago*.

Science had until this point never been involved in the creation of human calendars. The fact that these words are being written in the year 2015 has to do, not at all with science, but with the decidedly nonscientific and imprecise concepts of myth, faith, and belief. For, in refining the meaning of *ago*, most of the Western world would employ the initials BC and AD. It was said that something occurred a number of years "before Christ," or in the Year of the Lord, "Anno Domini," as in AD 2015.

But this was, of course, contentious to non-Westerners, to nonbelievers. It was a kind of notation that would fall foul of those for whom Jesus Christ meant little; and so in recent times other circumlocutions were offered to help soothe hurt feelings. There was BCE, most commonly, which referred to "before Christian Era" or, for the secular-minded, "before Common Era."

Yet even this was still a fudge, still woefully imprecise, still essentially based on myth. And BCE did not appeal to scientists, especially once carbon dating and other, more precise atomic dating techniques had been discovered. So they eventually came up with the idea of using the initials BP, "before present." The Wisconsin ice age, for instance, had its culmination fifty thousand years BP.

All that the acceptance of this new notation required was an agreement on *just when was present*? So, in the early 1960s, a pair of radiochemists came up with an answer. They suggested the use of the same standard reference year, the Year Zero moment of January 1, 1950.

Their suggestion seemed logical, neat, *appropriate*. Everyone, more or less, agreed. So that date is now accepted well-nigh universally among scientists for the ephemeral concept that is fleetingly known as the present. *And the present begins at the start of January 1950*.

And it seemed to me also the ideal date to use for beginning a description of the modern Pacific Ocean.

Other dates were briefly beguiling, to be sure. It could be argued that the new Pacific truly began its unfolding at the end of the Second World War—so I could have chosen the date of the Japanese surrender, September 2, 1945. Or else I could have selected Mao Zedong's declaration of the founding of the People's Republic of China, a momentous and solemn occasion that was staged on October 1, 1949, and that would eventually turn the Pacific into a cauldron of contention. I briefly also thought of using the date of the detonation of America's most powerful hydrogen bomb, the so-called Castle Bravo test of March 1, 1954, a moment of some symbolism.

Yet I kept coming back to the idea of the "beginning of present," which just seemed to have an elegant simplicity about it. The date has a strict scientific neutrality to it. It is an agnostic moment, agreed to and understood by all. And for this book, it turned out to have an added geographical bonus, a coincidence.

For nearly all the carbon-14 pollution that was sent up into the skies and that caused the scientific community to create the concept of "present" and "before present" in the first place came as the result of explosions that occurred in the Pacific. Bombs that went off in Bikini and Enewetak, Christmas Island and Woomera, Semipalatinsk and Lop Nor, Mururoa and Fangataufa, all in or around the ocean, were the prime pollutants, the original cause of the problem.

This made it all the more appropriate, it seemed to me, to choose that moment—the hinge, the dividing line, between purity and impurity—as the start line for this account. The story of the ocean of tomorrow, in other words, begins at the start of the present.

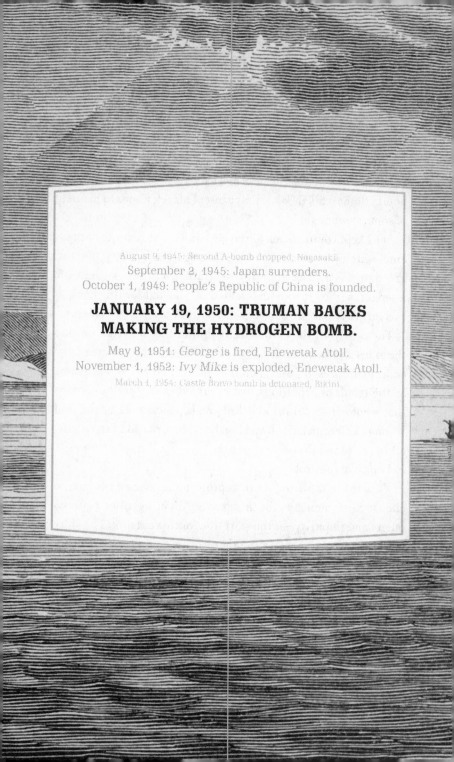

August 9, 1945: Second A-bomb dropped, Nagasaki.
September 2, 1945: Japan surrenders.
October 1, 1949: People's Republic of China is founded.

JANUARY 19, 1950: TRUMAN BACKS MAKING THE HYDROGEN BOMB.

May 8, 1951: *George* is fired, Enewetak Atoll.
November 1, 1952: *Ivy Mike* is exploded, Enewetak Atoll.
March 1, 1954: *Castle Bravo* bomb is detonated, Bikini.

Chapter 1

THE GREAT
THERMONUCLEAR SEA

—⚬—

I remembered the line from the Hindu scripture, the Bhagavad
Gita . . . *"I am become death, the destroyer of worlds."*
—J. ROBERT OPPENHEIMER, JULY 16, 1945, ON THE DETONATION OF
THE FIRST A-BOMB, NEW MEXICO

*The unleashed power of the atom has changed everything save
our modes of thinking, and thus we drift toward unparalleled
catastrophe.*
—ALBERT EINSTEIN, MAY 24, 1946, TELEGRAM SENT
TO PROMINENT AMERICANS

The first hint that the Pacific would be tragically transformed
into the world's first and only atomic ocean came at lunchtime
on January 4, when President Harry S. Truman uttered a single

cryptic sentence during his State of the Union address for 1950, to this effect: "Man has opened the secrets of nature and mastered new powers." He never mentioned the Pacific by name; nor did he mention it two weeks later, on January 19, when he finally made the fateful decision to which his congressional speech had alluded. Nor did he, two further weeks on, when he issued a formal directive and announced publicly what he had decided.

He didn't have to. So far as the United States was concerned, the sixty-four-million-square-mile expanse of the Pacific Ocean was the only place big enough and empty enough, and American enough, to allow the testing of the thermonuclear weapons the president had now finally committed his country to create.

The ocean already had had a taste of what was to come. Since 1946 the U.S. government had been secretly testing simple atomic fission bombs in the blue lagoons of its tropics. But these were quite modest weapons—deadly and terrible, to be sure, but nothing compared with what was to come next. The decision Truman made on that third Thursday of January, as well as his formal order to the Atomic Energy Commission that followed, was to start a program of work on a very different kind of device, and of a type both of unimaginable deadliness and theoretically limitless destructive power. It was a bomb that would forever change the nature of warfare, and would forever change the world. And its potential power was such that it could now be tried out, displayed, and demonstrated only in the empty middle of the Pacific.

Until the mid-1940s the ocean had been, in the popular imagination, just as Ferdinand Magellan described it four hundred years earlier. It had seemed a truly *pacific* sea, a place of maritime languor and quiet, of warm ultramarine waters and gentle trade winds. It suffered its ferocious storms, true, and its island peoples had not always lived lives of placid serenity, but it had not been a battle-scarred sea of churning and salt-stained gray, as the Atlantic was known. Just recently the war between the United States and Japan had seen violence on a gargantuan scale. But what was about to happen now was quite different, and by many orders of magnitude.

When President Truman authorized the 1950 budget of three hundred million dollars for the AEC to begin work on these quite different weapons (the "supers," as they were lightly called, the fusion bombs, the thermonuclear devices), they were little more than the blackboard musings of physicists' dreams—but musings well worth bringing to the attention of the Oval Office.

It had been several weeks earlier, on October 6, 1949, that the director of Central Intelligence, Admiral Sidney Souers, told Truman about some physicists' remarkable claims: that it might well be possible to employ the nuclear fusion of light gases to create explosions of tremendous force, unlike anything known before. Truman's interest was instantly piqued—driven in part by his knowledge that the Russians had exploded their first crude atomic fission bomb just a few weeks earlier. This had led to bitter and ferocious argument in the United States, principally between the military and the scientific communities, over the morality of making a new kind of weapon that could and probably would have the power to obliterate not merely thousands but millions. Many of the leading figures in the Pentagon, well aware that the now nuclear-capable Soviets would soon also be able to construct such bombs, insisted that the United States develop them, either to keep up or to keep ahead. But many scientists, more aware than most of the terrible powers of the proposed weapons, found the idea of their development utterly abhorrent. Many were gripped with a profound sense of guilt, even shame, for having ever provided the theoretical basis for their construction in the first place. Fission bombs were bad enough; fusion bombs were unimaginable in their potential for horror.

However, and so far as the U.S. government was concerned, this particular debate was officially ended on January 19, when Truman summoned Admiral Souers to the White House to tell him, in person, of what would come to be seen as one of the truly momentous decisions of his presidency. Developing the new superbomb, Truman told him, finally *"made a lot of sense . . . that was <u>what we should do</u>"* (my emphasis).

On January 31 the president made the necessary formal pronouncement that he had commanded the AEC to begin the necessary research. Enough money had been made available in the budget. America had to have the bomb, he said to his cabinet colleagues, because although no one ever wanted to use it, its possession would offer a bargaining chip during future negotiations with the Soviets. That alone was the pitiless rationale that finally squared the circle, at least for President Truman, in the moral debate.

The AEC duly began its work, in secret, and with great speed. Within a year the musings had become material. The technical challenges of fashioning a thermonuclear bomb were essentially solved. A first, small prototype device, known as *George*, was exploded three months later, on May 8, 1951. Then, on November 1, 1952, the first true thermonuclear test weapon, known as *Mike*, was detonated. Then the largest of them all—a weapon that was tested despite a memorable miscalculation that triggered results both unforgivable and unforgiven—was detonated sixteen months after that.

And owing to their daunting size, all these thermonuclear devices were exploded in the middle of the once pacific Pacific Ocean.

So far as the ocean was concerned, the journey to this point began in 1946, on the mid-sea atoll that shares its name with the much-reduced new style of bathing costume introduced that same year. A costume that a disconcerted *Le Monde* editorial archly described as displaying "the extreme minimization of modesty" and, rather presciently, as "quite as shocking as an explosion." The swimsuit's creator, Louis Réard, had said much the same thing, though intending his remark to be more PR than pejorative: "Like the bomb, the bikini is small and devastating."

As was the island story.

There was in the Pacific an Arcadian time, of course, when all its islands belonged, if *belonged* is the proper word, to those who had made their livings there for generations. But one by one, group by group, European discoverers happened upon these islands, and one by one, group by group, they lost their easy innocence. The islands that in due time would interest the American bomb testers were first spotted in the eighteenth century by an English seafarer named John Marshall: his fleet came across a vast scattering of atolls in an otherwise empty sea a thousand miles north of the great island of New Guinea. The island inhabitants—Micronesians, as they came to be called by anthropologists—were part Malay, part Polynesian. For thirty previous centuries, they had lived peaceably enough on the atolls that would soon be called the Marshall Islands. They had fished and gathered coconuts, and aside from occasional tussles and skirmishes among one another, they had seldom troubled anyone beyond.

But then came their "discovery," and in turn a bewildering succession of outsiders who claimed to own and then to rule them, and the Elysian order of old was rudely and permanently interrupted.

As mentioned in the prologue, the Spaniards were the first to arrive, and though they ruled large tracts of the western Pa-

cific from Manila in the Philippines from the sixteenth century onward, they considered the Marshalls too far away to be
of much interest. Moreover, the Spaniards' eventual loss of the
Philippines to the United States in the Spanish-American War
left their administration of these more distant islands well-nigh
impossible—there were an estimated six thousand of them, and
it was quite impractical to try to rule them from faraway Madrid.

A few American missionaries, who were busy converting the
Hawaiians to Christianity, had stopped by the Marshalls earlier in the nineteenth century, en route to Japan. They left the
islanders with a smattering of English, some vague awareness
of biblical teachings, and the occasional use of the all-covering
Christian version of the Muslim *niqab*, the Mother Hubbard
dress—all influences that remain today. (The Marshalls are
overwhelmingly Christian, and Protestant.) These missionaries were not acting as stalking horses for American colonists; that would come later. Instead, it was left to the then more
adventurous and imperially inclined Germans, who arrived in
the ocean in the later nineteenth century—stout Hamburg traders who discovered goods of one kind and another that could be
sent back home to Germany.

Unlike the Spaniards, the Germans believed it was practical
and commercially advantageous to try to rule here. They first set
up commercial trading stations on the atolls, then established
settlements, and finally, with the help of Lutheran missionaries,
so entirely convinced the Marshallese that their future would be
brighter under the kaiser's rule that the islands became German
colonies. A simple treaty, signed in 1899 with the Spanish and
accompanied by a payment of twenty-five million pesetas, transferred ownership of all Spain's Pacific islands from Madrid to
Berlin.* So, from 1906 onward, the islanders enjoyed an entirely

---

* Interestingly, the 1899 treaty never specifically mentioned the Marshall Islands,
leaving some to argue about their legal status still today—arguments that, considering

new status. They were no longer overlooked outposts of Spain, but subjects of the Imperial German Pacific Protectorate; were ruled from a Papuan city named Herbertshöhe, fifteen hundred miles away to the southwest; had governors who sported names such as Rudiger, Hahl, and Skopnik; and were persuaded that to get on in life, they had to forget any Spanish they might have known, and learn German instead.

It would have been a somewhat wasted effort. Just eight years later, in 1914, and though few locally were aware of the Great War raging on the far side of the world, its effects became immediately apparent. Japanese warships suddenly appeared on the horizon, Japanese troops—who at the time were allied to the faraway British—marched ashore, and all the Germans were commanded to leave. They were replaced this time by administrators plucked from the ministries in Tokyo. Once the Germans had been properly vanquished in Europe in 1918, an official League of Nations mandate allowed Japan to run the islands entirely, making the Marshall Islanders "subjects of the Empire of Japan resident in the South Seas Mandate." They were now to be ruled not from Papua, but from a new colonial headquarters in Saipan, fifteen hundred miles away to the northwest, and run by governors who sported names such as Tawara, Matsuda, and Hiyashi. The islanders were persuaded that to get on in life, they had to forget all their Spanish and German, and learn Japanese.

Then came the Second World War, and everything changed yet again. So far as the Marshall Islands were concerned, it did so most violently, during the last days of January 1944 and the Battle of Kwajalein, when a large force of American marines killed all but fifty-one of the thirty-five hundred Japanese in the garrison. That spring, governance of the Marshall Islands changed hands once more, with the puzzled locals accepting

---

the amount of money involved for aid and compensation, are of more than mere historic interest.

the rule of a third set of masters in forty years. They were now subjects of the faraway United States of America; were ruled in theory from Washington, DC; paid some kind of notional fealty to President Roosevelt (and soon to Truman); and were advised that to get on in the world, they had best forget all the Spanish and German and Japanese they might have remembered, and learn how to speak English.

They might have supposed that this was to be the final chapter. In fact, it was only the beginning. A new nightmare was about to unfold.

At the end of the war—though the Soviet Union was well on the way—the United States was the only nation to possess atomic weapons, and it had exploded three of them. All had exploited the physics of the fissioning of heavy metals. The first had been a test weapon in the New Mexico desert; the second and third were the live weapons dropped on Hiroshima and Nagasaki. Given the utter devastation of the two bombs that had been dropped in anger, and how quickly they helped end Japan's war-fighting abilities, President Truman had no doubts: these new devices, terrible though they might be, should now become a core element in America's arsenal. He instructed his Pentagon chiefs to make more of them, to test them, to perfect them, and to create ever better and more lethal versions—and so make quite certain that in matters atomic, the United States retained an absolute military lead over the rest of the world.

It was first decided that the U.S. Navy (and not the army) should be in charge of the tests. The rationale had all to do with the likely metrics of destruction. The success of the early atom bombs, even though their targets were cities, had quite spooked American admirals into suddenly believing that of all the main instruments of war, the surface ship at sea might be the most vulnerable to atomic destruction. Soldiers might perhaps hide in deep cement bunkers; aircraft might be swiftly flown out of harm's way; but a surface ship (especially an enormous lumbering vessel such as an aircraft carrier) was entirely vulnerable to nuclear attack, and could possibly be sunk by a single bomb, and within minutes. It could neither run nor hide from a bomber coming at it with a weapon of such power. Consequently, the future of the American navy—of navies in general, in fact—might be at stake: for if an atomic weapon could sink all ships with such ease, then the capital ship itself would soon be an obsolete entity, no better than a knight in armor on an iron-plated charger.

If, however, was key. No one knew if an atom bomb could actually sink an enormous naval vessel. It looked quite likely. But no one could be certain. So one of the guiding principles behind the early test program that President Truman now demanded was the need to elicit the truth. Could an atom bomb destroy a major capital ship: a battleship, an aircraft carrier, a heavy cruiser? The navy feared that its assets were the most vulnerable target; so the navy should conduct the tests.

The mechanics of the Manhattan Project, the secret wartime plan to make the first atom bombs, had left behind a well-oiled production line. As during the war, plutonium for the postwar tests would come from the giant plant at Hanford, in Washington State; the enriched uranium would come from the immense centrifuge farms in Oak Ridge, Tennessee; and the design and final assembly of the gadgets, as they were initially termed, would continue at the laboratory in Los Alamos, in New Mexico. But where best to test them? The White House charged a vice admiral named William Blandy with finding the best place "to permit the accomplishment of the tests with acceptable risk and minimum hazard."

Wherever the bombs were to be tested had, first, to be in territory that was firmly under American control. Since one of the main concerns at the Pentagon was the effect such weapons would have on large warships, it seemed prudent to carry out the test in a sheltered lagoon in which test vessels could be anchored as targets and blasted with bombs. The chosen place should also have a very limited local population—as Admiral Blandy remarked, "[I]t was important that the local population be small and cooperative so they could be moved to a new location with a minimum of trouble."

Weather had to be reliable—most especially the winds, which had to be predictable at a range of altitudes up to a dozen miles, the height of the mushroom cloud's pillar, since any sustained movements of air would determine where plumes of radiation from the pillars might end up. There was the question of remoteness: ideally it should be far away from shipping lanes and from the inquisitive, and yet not too remote, since it had to be within range of an airfield that could house the bombers that would carry any air-dropped weapons to be tested. The favored heavy bomber of the time was the B-29 Superfortress, with an average range of 3,700 miles. The perfect test site could thus be no more than half that number of miles from an airfield, to allow a journey out and back: 1,000 miles distant from the field seemed ideal.

The search for such a place began in October; the choice had become clear by January. After the Pentagon discounted a number of remote spots in the Atlantic Ocean and in the Caribbean Sea, the idea became more and more compelling to create a test site somewhere in the seemingly limitless expanses of the Pacific Ocean. There was brief but serious consideration of the biologically abundant Galápagos Islands—a move that, even in those relatively unsophisticated times, would have raised environmentalists' eyebrows. In the end it was the Marshall Islands (which had lately become de facto American territory) that seemed best suited.

The Marshalls were close to the ocean's midpoint, far removed from sightseers. There was a large airfield at Kwajalein, ideal for B-29 operations. And while almost any of the twenty-nine atolls and five islands that make up the Marshalls might fit the bill, one group of islands above all others looked ideal. Two hundred fifty-six miles north of Kwajalein, at the northern end of the so-called Ralik Chain—the "Sunset Chain," the western chain—of islands, there was Bikini.

It was the chosen site for the enactment of a sorry irony. For once the Pacific war was fully over—once the unbearable sounds of battle, and the landing craft and the tanks and the gun emplacements and trenches, had gone away; and once all these things had been replaced by a half-forgotten quietude called peace, and there were lapping blue waters once again, and multicolored fish and white sands and green parrots and thermal-dancing frigate birds and coral reefs and ranks of palm trees leaning into the endless trade winds; once all such things had reestablished themselves as the hallmarks of the South Seas; and once they had particularly done so on tiny, pretty, peaceful, caricaturedly Pacific Bikini—Admiral Blandy and his team devised a plan to end all this, and turn Bikini and all her islands and their lagoon once again into a hellish gyre of ruin and mayhem.

The ruin of this near-perfect paradise was quite deliberate, and it was achieved because the number of Marshallese was vanishingly small, while America, the victor in the recent conflict, was a huge and very visible nation of almost limitless power.

In 1946 only 167 Marshallese men, women, and children were living on the handful of habitable islands strung around Bikini's substantial shark-filled lagoon. Like all Marshallese communities, they had a local leader, a chief, an *iroij*, named Juda Kessibuki. But neither the islanders nor their paramount chief had much chance of avoiding the near-total destruction of their homeland, because they were pitted against the will and recommendation of Admiral Blandy, a New Yorker whose prominent beak had earned him the nickname Spike* and whose influence in the Pentagon and the White House was seemingly limitless. His motto was *Pax per Potestatem*, "Peace Through Power"—and this was essentially how he persuaded the Bikinians to leave their island and let the Americans ruin it forever.

Admiral Blandy had made his formal choice of Bikini in mid-January 1946. It was promptly approved; and on February 10, Ben Wyatt, the middle-aged U.S. Navy commodore who had been appointed military governor of the Marshalls, flew out to the atoll on a seaplane to deliver the news to the 167 islanders. They should meet him on a Sunday, he said. After church. He was going to use "gentle words" to tell them.

He would use a biblical story. Whether it was cleverly cynical manipulation or a sincere belief in the islanders' innocence may never be known, but it was decided that the U.S. Navy should appeal to the Bikinians' devotion to their Bibles, to the legacy of the Victorian missionaries who had passed by a century before.

Commodore Wyatt gathered the islanders around him in a semicircle, under the shadow of a grove of coconut trees. Movies of the event show the ocean surf beating steadily in the background, waves crashing on the outer reef, the sky filled with high cloudlets and with seabirds whirling lazily on the currents. A number of American soldiers stood around, idly half-listening, half on sentry-go for the visitors.

Wyatt took as his Sunday text the Book of Exodus, chapter 13, verse 14, which tells the story of God's leading the Israelites out of Egypt, during those tense moments shortly before the parting of the Red Sea. To get the refugees to the desert crossing point, Wyatt quoted, "The Lord went before them by day in a pillar of cloud, to lead them the way; and by night in a pillar of fire, to give them light; to go by day and night."

---

* Or, according to one Internet source, Waffle Nose. He had a remarkable similarity to the actor Karl Malden.

It was under the guidance of God that the United States of America, said the commodore, had constructed its own great pillars of fire and smoke, which could and would be used as a weapon "if in the future any nation attacked the peoples of God." His puzzled listeners smiled weakly, but were silent. Wyatt went on: To make sure that such pillars of fire and smoke worked properly in the service of the Lord, it was now necessary to test them. To test them on Bikini. You have been chosen, the officer went on, to help America develop something created under God's guidance, "for the good of mankind and to end all world wars."

It was thus necessary that for a short while everyone leave Bikini Atoll and go off with the navy to be housed elsewhere. "Would you be willing to sacrifice your island," Commodore Wyatt pleaded, "for the welfare of all men?"

One of the island chiefs said later that Wyatt's invocation of the Bible had been the clincher, the masterstroke: "We didn't feel we had any other choice but to obey the Americans." And Chief Juda, the local *iroij* who by custom led the community in all matters, reluctantly agreed.

It has never been fully explained just how or when the islanders acquiesced. The Pentagon later said Chief Juda had given his enthusiastic assent right away, and that he thought the bomb tests were a wonderful idea. What we know from a public relations film made some three weeks later, when the commodore tried to get the chief to repeat his enthusiasm in front of a camera, is that the story was somewhat different. The film's director had to suffer several takes before Juda performed with the degree of sincerity required. He agreed to make what today looks like a rehearsed and robotic utterance before a gathering of puzzled and miserable-looking islanders: "We will go. We will go believing that everything is in the hands of God."*

---

* This evacuation was to be echoed two decades later, in the Indian Ocean, when the Pentagon wanted to use the British colonial possession Diego Garcia as a military base. Denis (later Lord) Greenhill wrote in an infamous memo that there were just "a few

That was sufficient for the Americans. Later there was to be much keening and wailing. But initially the islanders did as they had been bidden. The islanders were duly out of Bikini within a month. They packed up their belongings, abandoned their modest houses and beloved outrigger canoes, and left the homes and gardens they had occupied and tended peaceably for scores of generations past, and they went off in a big and ungainly American naval vessel to an unknown island far away—and all at the behest of white men they'd never before seen, so these white men could perform tasks that they did not readily comprehend and that seemed to be of little value to them.

With enough food for an eight-week stay, they were herded into a single landing craft and bumped uncomfortably over the sea 125 miles east to a very much smaller atoll, Rongerik; it had just half a square mile of land compared with the three and a half of Bikini. Rongerik was already well known to the Bikinians—it took just a day and night's voyage on an outrigger to get there—and they didn't like the place. It had poor soil, precious little fresh water, and a wretched few coconut trees. More important, it was, according to local legend, home to a clutch of strange demonic spirits much feared in this corner of the Marshalls. Nonetheless, blithely trusting that the Americans were acting in good faith, they settled in on Rongerik as best they could. They tried to resume a semblance of their disrupted lives, while the testing program back on their home islands got fully under way.

The transformation of their former home was almost instantaneous. Just as soon as the islanders passed over the horizon, from demurrage stations far out at sea a vast armada of American ships started swiftly moving in to take their place.

Admiral Blandy had named his testing program Operation Crossroads—"it is apparent that warfare, perhaps civilization

---

Tarzans or Men Fridays" living there. In fact, a vibrant community of more than two thousand people was shipped off against its will to Mauritius. It has been fighting for compensation ever since.

itself, has been brought to a crossroads by this revolutionary weapon," he had said back in Washington—and it was to be run to a very tight schedule. Construction battalions, Seabees, moved onshore to erect blockhouses and barracks and steel towers for all the cameras and radiation sensors and telescopes and men that would be needed; flotillas of ships brought in heavy materials (cement, steel, bulldozers, backhoes, tons of protective lead shielding); and the navy started ferrying in scores of old and captured ships of every type imaginable, which were to be set down at fixed positions in the lagoon and used as targets.

More than forty thousand men were soon to be involved out in the western Pacific—eating, inter alia, twenty tons of meat and seventy thousand candy bars every single day—making sure the testing program went ahead as scheduled. For everyone knew that with the help of their spies, the Soviets were breathing hard down the Americans' necks. And those in the U.S. Navy knew, or suspected, that their very profession, their navy, could well be imperiled, because sinking their ships with atomic bombs was now, apparently, quite as easy as shooting fish in a barrel.

Operation Crossroads was to be the first of the 55 nuclear test programs (most of which involved several separate tests) that would be run by the United States over the next half century. The total of 1,032 atomic bombs that America has exploded since 1945 far exceeds the combined totals of all the other nuclear devices exploded by all other nuclear-capable countries in the world. In later years the United States would conduct tests of different kinds of delivery systems (of gravity drops, ballistic missiles, artillery shells, mines) and for different kinds of uses and customers (for the army; for use in outer space; even for peaceful uses, such as digging great trenches in the earth). Most of these later tests would be carried out in the deserts of Nevada, many of them underground. But the most impressive first 67 of these tests were carried out in the Pacific, and the biggest and most symbolic was on Bikini itself.*

Though only twenty-three tests were carried out there, the TNT-equivalent tonnage of each of the bombs was enormous; and because the Bikini weapons taken together were so huge (and because the tests of two of them, as we shall see, went so badly wrong) those twenty-three tests account for more than *15 percent* of the total power of all the atomic explosions triggered in the history of all American testing.

Crossroads was the very first of these tests, and it was specifically designed for the benefit of the navy: the ships being led into the anchorage in the weeks leading up to the first of the two main explosions were to be steel-clad guinea pigs, the first nonhuman victims of the Pacific's atomic age.

Navy crews first assembled a total of seventy-three ships toward the eastern end of the lagoon, some four miles southwest of Bikini Island. The vessels were clustered in concentric circles around a red-and-white-painted American superdreadnought battleship, the USS *Nevada*, the ship that had famously managed to get away during the Pearl Harbor attack, despite being hit by a torpedo and bombs during the raid. She was old, built in 1914, and the navy thought that choosing her to be the bull's-eye for the first A-bomb test would permit her to die with dignity, still in service to her country.

But she didn't die. In the end, the bombardier of the plane that carried the first bomb up from the airfield at Kwajalein—the Able shot, as it was termed—proved less than competent and missed her by seven hundred yards. She didn't sink, was nicely repaired, and limped back into service for two more years.

This bomb used for the Crossroads Able shot was almost identical in design and delivery to the weapon that had been

---

* A number of weapons were also exploded on the nearby atoll of Enewetak, an atoll that suffered similarly but that for many reasons has never attracted quite the same attention. "A Pacific Isle," a *New York Times* headline read in 2014, "Radioactive and Forgotten."

dropped on Nagasaki a year previously, was essentially the same as the first-ever test weapon exploded weeks beforehand in New Mexico: it was a *Fat Man*, with a plutonium core, and it was set to detonate in midair five hundred feet above the target. It did so, precisely on schedule if not precisely on target, at 9:00 a.m. on July 1, 1946.

Its explosion, and its effects, turned out to be only moderately spectacular. The press—more than a hundred reporters were gathered on ships moored outside the lagoon*—was seemingly compelled to display reverent ecstasies of purple prose. And who could blame them? After all, the first three atomic explosions had been witnessed only by American military personnel or by the victims. Almost no American civilians had ever seen such a thing—another reason that Bikini, as the mise-en-scène for the weapons' first public display, remains so symbolically important a place, and why the Pacific, as backdrop, remains the most nuclear of the world's oceans.

The *New York Times* reporter aboard the USS *Appalachian*, William Laurence, was dutifully awed, dictating over the ship's wireless:

> As I watched the pillar of cosmic fire from the sky-deck of this ship it was about eighteen miles to the northeast. It was an awesome, spine-chilling spectacle, a boiling, angry, super volcano struggling toward the sky, belching enormous masses of iridescent flames and smoke and giant rings of rainbow, at times giving the appearance of a monster tugging at the earth in an effort to lift it and hurl it into space.
>
> From this point I watched the atomic bomb as it burst. It was like watching the birth and the death of a star, born and

---

* The countdown and explosion were relayed by radio around the country and world. The BBC broadcast the test on the Light Programme, a station usually reserved for music and soap operas, but it was late at night in Britain, and static interference made the entire event well-nigh inaudible, with only "one word in ten" able to be understood.

disintegrated in the instant of its birth. The new-born star made its appearance in a flash so dazzling no human eye could look at it except through goggles that turned bright daylight over the Pacific into a pitch-dark night. When the flash came it lighted up the sky and ocean with the light of many suns, a light not of the earth.

The journalists' ardor cooled somewhat over the coming days. The *Times* triple-decker front-page headline on the morning after intimated, at least among the editors, an early degree of sobriety, verging on disappointment. "Blast Force Seems Less Than Expected," it read, and the lead paragraph's obligatory description of the bomb's initial dazzling flash as "ten times brighter than the sun" was followed by a cautionary "but," and by the news that of the seventy-three ships moored inside the atoll, only two had actually been sunk. The paper may have been a little hasty, because actually five went down: two old American destroyers, two transport ships, and, eventually, the graceful Japanese cruiser *Sakawa*, which had been seriously damaged and which foundered as she was being towed from her anchorage. She nearly dragged the towing tug down with her, though panicking crewmen cut the towline with acetylene torches in the nick of time.

Aside from being the first atomic bomb detonation ever seen publicly, the Able shot is now probably best remembered for what it failed to do—and because some of the failures were positive, and confirmed what Admiral Blandy had reassured everyone days beforehand: "The bomb will not start a chain-reaction in the water converting it all to gas and letting the ships on all the oceans drop down to the bottom. It will not blow out the bottom of the sea and let all the water run down the hole. It will not destroy gravity." It didn't set fire to any of Bikini's palm trees, either.

But it also didn't do what was hoped for. It didn't seem to stir much agitation among the immense fleet arrayed around the drop zone. It damaged generally only rather small ships that were very close to the explosion's center. It failed to sink the USS *Nevada*; it failed to sink the enormous Japanese battleship *Nagato*—a fate that many had hoped for, since *Nagato* had been Admiral Yamamoto's flagship during the raid on Pearl Harbor, and her destruction would have been rich in retributive symmetry. It also failed to sink the former German pocket battleship *Prinz Eugen*, which was at the time a commissioned ship of the U.S. Navy, having been claimed as a war prize and been brought all the way to Bikini from Wilhelmshaven by a German American crew.*

The bomb also didn't do as much damage to the animals that had been posted onto some of the ships as stand-ins for crewmen. There were goats in gun turrets, rats at the radar screens, pigs on the poop decks, mice by the mainmast, and rodents by the score just about everywhere. Three quarters of them survived, for a while, some of the goats chewing away unconcernedly while all hell was breaking out about them. Two celebrated survivors, Pig 311 and Goat 315, remained so healthy for so long that they were brought to Washington, DC, and put on display at the zoo.

The second reason for the Able shot's historic importance is more technical, and has whispers of the macabre. For the plutonium in the core of the weapon, which had been manufactured in August of the previous year, had already been involved in no fewer than two fatal accidents at the nuclear program headquarters at Los Alamos.

The components of the core, when kept apart from each other, were not especially dangerous—but when pressed together, and under certain circumstances, they could go "prompt critical," as the phrase has it, and release sudden immense amounts of radiation. This is what had happened to this particular core, twice.

---

* When the German crew finally left their ship at Panama, the American sailors discovered they couldn't work the *Prinz Eugen*'s boilers. Tugs had to be ordered, and the eighteen-thousand-ton ship had to be towed across the Pacific, bound for this vain attempt to destroy her.

First, on August 21, a physicist named Harry Daghlian dropped a tungsten carbide brick onto the core, causing it to go critical and douse Daghlian with enough radiation to kill him four weeks later.

The more notorious second incident took place in 1946, the following May. A flamboyant Los Alamos experimenter named Louis Slotin was carefully turning the blade of a screwdriver to lever the two nickel-plated sphere halves apart, and then move them closer to each other to measure the increasing radiation—"tickling the dragon's tail," as it was called. During this delicate process something startled him—in the feature film made of the event, it was the breaking of a dropped teacup—and he jerked the screwdriver, causing the hemispheres suddenly to close on each other. There was a blinding blue flash of Cherenkov radiation, and all the Geiger counters in the room went promptly off scale. Slotin stood up and shoved the top hemisphere onto the floor, ending the criticality and, with it, the radiation burst. But in doing so, he received a formidable dose of neutron and gamma radiation on his hand, and he calculated within hours that he was soon going to die. He was exactly right; and he did so, in intense pain, nine days later.

From then on, as a consequence of these two deaths, the twin half spheres of plutonium, with their shields of nickel and beryllium, became collectively known as the Demon Core. It was this very core that would become the operational heart of the Able shot. One could imagine the physicists just wanting to use it up, to explode it and take it out of their tiny inventory of plutonium bomb charges. But given its sorry history, the superstitious might well say that the use of the Demon Core guaranteed that the Able shot was doomed, either to cause more accidents or to be a failure.

In the end the only certain "casualty" of the first Crossroads bomb was the captain of the plane that dropped it, who banged his lip when the shock wave struck his departing B-29. What otherwise haunted Able was not disaster, but indeed, a certain sense of failure.

For it also largely failed to impress. Few of the UN observers sent to monitor the event were captivated. A Soviet professor, Simon Alexandrov, gave a very Russian shrug of his shoulders and, using a locution more modern than he knew, declared the bomb "not so much." A Brazilian said it was "so-so." A New York congressman said he felt the heat wave, but agreed with others aboard the observers' ship that the eighteen-mile distance to the drop zone had somewhat reduced the spectacle. Weathermen said the humid air had also deadened the sound and heat radiation. Newspapers in the United States photographed the mothers of the pilot and bombardier who'd dropped the weapon. These two bespectacled ladies, gathered in La Crosse, Wisconsin, appeared curiously unmoved. Only after the weapon had exploded did they say their boys must surely have enjoyed their adventure.

It was beginning to look as though the Bikinians had been turfed out of their home for nothing. But then came the second of the Crossroads tests, the Baker shot, on July 25. This was a far greater spectacle—at first a supposed success, in a military and a public relations sense. Yet it was also, in some other ways, a disaster—the first, as it happened, of several.

Able had been an air-dropped bomb. *Baker*, which was also designed to measure the effect of atomic weapons on the waiting congregation of capital ships, was by contrast to be exploded underwater. The official explanation for *Able*'s signal failure to sink as many ships as expected was that most of its damage had been done above the vessels' waterlines. Baker would, by contrast, do its damage to underwater hulls rather than above-water superstructures. As a ship killer, it should have been more effective.

Indeed, it was—both effective and spectacular-looking enough to be, for a while, the poster child, quite literally, of the atomic age. The bomb was suspended ninety feet down in the water, inside a concrete container held by steel cables from the underside of an old landing craft. So confident was Admiral Blandy that this shot would be a success that he summoned Juda, the Bikinian leader, over from Rongerik to view it. Blandy didn't imagine Juda would show up, but when he did, and boarded the navy observation vessel, he told his hosts he was looking forward to the explosion, and he hoped that once it was done with, he would be able to bring his people home. It was the wishful thinking of the truest naïf.

The weapon was duly exploded at 8:35 a.m. and provided the watchers with a spectacle they would never forget. With a gigantic whoosh, it suddenly created at first a mile-wide glass-bubble sphere of water and steam and condensate and crushed coral and mud that thundered out of the mirror-calm blue of the lagoon, and out of which erupted, at fantastic speed, a perfectly symmetrical hollow column, a mile high, of millions of tons of near-white water and seafloor sand topped by a ragged cloud of spray and coral debris—and which, caught by cameras as it fell slowly back into the lagoon, remains today one of the iconic images of the time. To those enthralled by matters atomic—and many young Americans especially were utterly captivated—it was to be the perfect wall poster, to be set alongside a pouting Brigitte Bardot and Marilyn Monroe laughing at her billowing dress. The mushroom cloud had already become something of a cartoonable cliché: that there was none produced after the Baker shot—subsea detonations produce much more of a crown-shaped, cauliflower-shaped arrangement, it was realized—made for an originality, a certain coolness, the bomb as a term of art.

The explosion entirely vaporized the landing craft, no measurable parts of which have ever been found. But it did far more than that. It sank ten ships, including two battleships, an aircraft carrier, and three submarines. Most potently, photographs taken a millisecond or two after the blast reached the surface show a dark stain rising vertically up along the side of the great water column. This stain is believed by analysts to be the entire

The plutonium bomb *Helen of Bikini*, used in the underwater Baker shot of Operation Crossroads, expelled disastrous quantities of fallout, ending the series. The dark stain at the column's lower right is said to be the entire battleship USS *Arkansas*.

battleship *Arkansas*, upended by the enormous blast and seemingly pasted onto the column's side before being hurled into the maelstrom that followed and then thrust back into the water upside down. This was a mighty battleship, with a displacement of twenty-six thousand tons. To be reduced to a mere stain, a midocean skid mark—with the whole starboard side of her hull, the side that had faced the bomb blast, crushed as if by some monumental hammer blow; and then her ruined self thrown backward into the Pacific mud, with her guns lolling out of their upended casemates like the tongues of the hanged Mussolinis—is a fate few would wish on any ship. Especially not a ship with so proud a heritage as the *Arkansas*, built in 1910, with service in both world wars, and with Normandy, Iwo Jima, and Okinawa among her battle honors. Her sudden annihilation as an almost casually picked-up and tossed-away victim of the supersonic subsea pressure wave from the bomb must have made many an admiral shake his grizzled head.

There were months of subsequent scientific fascination with

this one bomb—a whole conference was convened eight weeks later to deal with the vast amount of data that came from the explosion. Elaborate new atom bomb terms were created: the Wilson Cloud, the slick, the crack, the bubble, the base surge, the cauliflower.

Geophysicists, unexpectedly, learned something from the explosion that helped solve a near-Pacific problem of old: why the 1883 eruption of the volcano Krakatoa had caused a tsunami. It turned out that, unwittingly, the two detonations, the volcano and the atom bomb, somewhat mimicked each other. The A-bomb's explosion created a huge underwater bubble of fast-expanding gas; and the water displaced by the bubble formed a wave ninety feet high, which then rocketed toward Bikini Island, and was still fifteen feet high when it got there seconds later and picked up ships and tossed them onto the beach with cool impunity and then flooded the entire island.

Krakatoa's explosion did much the same thing: the island of the volcano was vaporized; seawater rushed into the white-hot void and then similarly flashed into bubbles of superheated steam, which triggered a surface wave. Big volcanoes are very much larger than anything even nuclear-armed mankind can manufacture. The Krakatoa tsunami killed forty thousand and then spread around the world, being seen and felt ten thousand miles away hours later. Bikini did no such thing.

But this second Bikini bomb also caused one terrible and entirely foreseeable wrong of which Krakatoa was manifestly not guilty. It spread abroad a vast and deadly amount of radiation. The military had been given due warning that this would happen. Admiral Blandy, who had once famously declared, "I am not an atomic playboy . . . exploding these bombs to satisfy my personal whim," was told that this bomb would be much more dangerous than its predecessor. Its plume of radioactive by-products would not be swept away by upper-atmosphere winds, but would be dumped directly into the lagoon, and would contaminate the waters and the shore and any ships that might survive the initial explosion. The scientists said that to go ahead would be foolhardy. But Blandy, who would later celebrate Operation Crossroads with a party whose centerpiece was a cake decorated with a large mushroom cloud, decided to go ahead with the test anyway—and the result was a catastrophe.

The cloud of falling debris itself produced a considerable amount of radiation, as expected; but as this column was falling back to the sea, a nine-hundred-foot-high wall of mist—the base surge, as it was later called—spread outward from the column and quickly enveloped the surviving ships as it rolled over them. This turned out to be the killer wave, and no one had known it would occur or how dangerous it would be. But it contained the majority of the fission products of the explosion, and though their total mass (three pounds or so, combined with about ten pounds of plutonium left over from the blast) might seem trivial, the substances were so toxic that an immense cleanup operation had to be undertaken, and very, very fast.

Yet the navy had made no advance contingency plans to do this. The result was an instant panic among the officers, and then sheeplike obedience by thousands of sailors who, wearing in most cases shorts and T-shirts, and using hoses, sprays, mops, and buckets of lye, were landed on each of the intensely radioactive vessels and ordered to clean away the residual material as quickly as they could. Fifty ships promptly set sail into the lagoon with fifteen thousand enlisted men, all soon bent on measuring and cleaning and hosing and decontaminating— and at the same time unwittingly absorbing, in their clothing, on their skin, in their hair, in their lungs, and on everything they subsequently touched, unimaginably excessive amounts of radiation. Plutonium debris was in any case not detectable by Geiger counters, so contamination with this most insidiously dangerous element went unnoticed at first.

Navy commanders on the spot had been given an impossible task, one that was incredibly perilous and that displayed the cruelest peacetime folly of having well-protected officers order-ing wholly unprotected servicemen to perform the most treach-erous labor. Pictures show groups of men swabbing the decks as they might have done after a topside dinner party, cheerful and vastly amused. One man said the ships were covered with sand and chunks of coral from the seabed, and he proudly displayed a chunk of rock he planned to take home, then put it in his pocket.

Though statistics relating to the later fates of these men—specifically, figures showing which of their number died of cancers that could reliably be put down to the Bikini bomb—are muddied, scientists quickly recognized, as the navy brass clearly did not, the terrible potential dangers. As a result, the next scheduled test, Crossroads Charlie, was canceled, and the Crossroads series formally terminated. Admiral Blandy was moved away from the Pacific to command the Atlantic Fleet, where he retired after three more years. He died in 1954.

But this was by no means the end of Bikini's nightmare. For one thing, the displaced islanders—by now largely overlooked in the drama of the weapons testing program—were in ever-worsening shape. When Chief Juda returned to Rongerik from the Baker test, and reported with his characteristic innocence that their islands still looked much the same and all the palm trees were still standing, he was addressing a community on the verge of starvation. The supply caches left behind by the Americans had run out; most islanders now survived on thin gruel and barely edible fish; a fire had devastated their main coconut plantation. A visiting Marshall Islander reported that the Bikinian exiles were emaciated, "just skin and bones," and an American doctor found compelling evidence of real malnutrition.

The islanders found an unanticipated champion. Harold L. Ickes, who had been Roosevelt's interior secretary for more than a dozen years, the man who desegregated the national parks and who dedicated Boulder Dam and who was in many ways the personification of the practical implementation of FDR's New Deal, got involved. By now retired, he was still a formidable champion of the underdog. In late 1947 he wrote a syndicated column decrying the treatment of the Bikini Islanders: "The natives," he declared, "are actually and literally starving to death."

All Washington read Ickes's essay, and it shocked Truman's administration into action. The government tried at first to deny responsibility—asserting, untruthfully, that the Bikinians were at fault: "[T]he natives selected Rongerik themselves," said a statement. "We built them houses, schools and watersheds on that island, and they were perfectly happy initially. Later it developed that the island was not as productive as originally expected, and we had to augment their food supply by bringing in food for them."

Few bought the lie. So boats and seaplanes were suddenly scrambled, and far away from the White House and the National Press Club, out on a sleepy mid-ocean atoll, an operation commenced that was born out of a sudden sense of national guilt. Scores of bewildered and unhappy Bikinians, most who by now had quite broken faith with the American government, were suddenly being moved again. This time they were shipped more than two hundred fifty miles to the south, to the great base atoll of Kwajalein, where they were put up in tents set up in lines along the huge airstrip.

It was noisy, busy, frightening, a world far removed from their isolated life up on a detached coral chain, far distant from a culture that had been based for hundreds of years on the stark simplicity of lagoon fishing. On Kwajalein, then as now a fully functioning American military base, all was stark, and little was simple. There was food and water in abundance. Too much abundance, many say today, since this was where the Bikinians began seriously and lethally to modify their diets, adding Spam, Coca-Cola, white sugar, and flour—and to change their working habits, to become what many regard them as today, participants in a handout culture. Few would dispute that from this moment on, the exiled Bikinians began to change, their native attitudes steadily eroded and diluted as the years away went on: Kwajalein is where the great alteration began to take hold.

Within months the U.S. government swiftly realized how unsuitable it was for the Bikinians, especially the growing number of newborn children, to be living in tents on a military airstrip. So in November 1948 they were moved for a third time, now to a tiny uninhabited speck in the southern Marshalls called Kili Island, a place that neither was an atoll nor had a lagoon. The island has no harbor, and during high seas a landing can be impossible. Airdrops from military cargo planes have to be arranged still, when sea conditions are too trying. A grass airstrip theoretically allows Air Marshall Islands access, but flights are few and very far between. Nonetheless, Kili is where the Bikinians, now transmuted from unwilling atomic exiles into perpetual atomic nomads, have been based ever since 1948. It now seems they may never go home.

If this proves to be so, it will be for many reasons—one being the obvious and long-lasting radiological contamination of their home in 1946, in the aftermath of the Crossroads Baker shot. But their exile is also a consequence of their atoll being massively polluted yet again, by the one most disastrous bomb for which Bikini has become most notorious, and which was exploded eight years later, on March 1, 1954: *Castle Bravo*.

By this time, the mid-1950s, there was no doubt that the Pacific was the place to test the truly big bombs of the future. On January 19, 1950, President Truman had made his decision. The superbomb, the thermonuclear fusion bomb, was to be made, and tried out—and it was to be employed as a bargaining chip with the Russians. The first prototype, *George*, had been tested in 1951; a bigger version in 1952. And now this one. The first potentially deliverable American thermonuclear weapon,* a classic hydrogen bomb, it was code-named *Castle Bravo*. It remains by far the biggest nuclear weapon ever exploded by the United States, and its enormous and little-anticipated explosive impact resulted from two big mistakes, a combination of a major technical miscalculation and mulish stupidity.

History has left someone to blame for the error: a brilliant physicist with a curiously interesting stake in the nuclear world. He was named Alvin Cushman Graves, and a previous mistake with fissile material in 1946—a mistake not his own but one that killed the man who made it—very nearly killed him, too. That Graves survived the accident, and then recovered sufficiently to preside over the disastrous 1954 Castle Bravo test, was probably not entirely unconnected with his cavalier, cocksure attitude toward radiation risks from fallout. Such risks, he once famously declared, were "concocted in the minds of weak malingerers."

The accident Graves survived was the second of the two lethal accidents that famously involved the Los Alamos lab's notorious Demon Core. Graves was the man standing just behind Louis Slotin when the pair of three-inch hemispheres of nickel-beryllium-plated plutonium briefly touched each other and a sudden surge of blue light and viciously dangerous radiation flooded the room. Graves was partly shielded by Slotin's body, but he nevertheless received a sufficiently scalding bath of gamma rays, X-rays, and neutrons to kill him. Few of his doctors

---

* The first true hydrogen bomb, code-named *Ivy Mike*, had been successfully detonated on the nearby Enewetak Atoll sixteen months before. But the hydrogen in that experiment had to be supercooled, making the combined bomb—it had to have a Nagasaki-like *Fat Man* bomb as a trigger—truly massive. It weighed sixty-two tons, so it was far too big to be used as a weapon. *Castle Bravo*, by contrast, used solid fuels and weighed in at only ten tons, and the success of the test convinced both the U.S. Navy and the U.S. Air Force that H-bombs could now be made in sizes that could be delivered by aircraft or missiles.

thought he would live. He was in the hospital for weeks, briefly lost all his hair, and developed serious neurological and vision problems. But to the amazement of all, he then slowly and steadily got better, ultimately recovering almost totally. Physically at least, there was little scarring, except one small spot of baldness, which he liked to display.

Infamously disdainful of the supposed dangers of atomic fallout, the nuclear accident survivor Alvin Graves ordered the fateful firing of the *Castle Bravo* thermonuclear weapon, the biggest of all American nuclear tests.

He eventually became well enough to be appointed—though at this remove, and considering what then happened, it is surely right to wonder at the wisdom of his appointment—scientific director of the Castle series of thermonuclear bomb tests. He swiftly became the most enthusiastic advocate of these new weapons—not least because he was well aware that the Soviet Union's nuclear program was catching up fast.* Once he arrived at the Enewetak headquarters of what were now called the Pacific Proving Grounds, he made one thing abundantly clear to his staff: since he had survived the very worst that the atom could throw at him, he would stop at nothing to detonate the Bikini-based fusion device that was now in his care.

The bomb designated for the Castle Bravo detonation was an

---

* The very first Soviet A-bomb had been exploded in 1949, more than four years after the first U.S. test in New Mexico. But Moscow's first thermonuclear H-bomb test came in August 1953, just *nine months* after the United States' *Ivy Mike* fusion bomb on Enewetak.

innocent-looking steel cylinder fifteen feet long, four feet in diameter. It looked rather like a large propane tank. It had been designed at Los Alamos, where, to suggest its innocence of purpose, it had been given the code-name the *Shrimp*. It had been shipped in great secrecy—lights off at night, aircraft and destroyers keeping pace with the cargo ship—to Enewetak in February, and was taken by barge to Bikini, with tarpaulin wraps to prevent the unauthorized curious glimpsing its size and shape. There it was suspended from the ceiling of a large shed, called the shot cab, that had been erected on an artificial island built on a reef off Nam Island, at the very northern tip of the atoll. A causeway connected the shot cab with dry land; the wires that would lead to the electronic firing bunker snaked across the sandbanks and coral reefs and past the Bikinians' now long-abandoned houses, to the tiny sliver of Enyu Island, twenty miles away.

At the end of February, all staff members were evacuated from Bikini and all ships were removed from the lagoon. Only the firing crew, nine men buried beneath concrete a dozen feet belowground, stayed behind.

Before the firing button was pressed, there were two serious uncertainties. The first was just how big this bomb would be. The *Ivy Mike* explosion of sixteen months before had been a thumping ten megatons, spectacular and memorable—and when that bomb blew up, it did so exactly as powerfully as the physicists had predicted. But *Castle Bravo* was using a solid rather than a liquid source of hydrogen—the hydrogen that would be compressed with such force and heat as to make it undergo fusion, and release the massive amount of energy that would cause the explosion. The solid compound in the new bomb was lithium deuteride, an amalgam of lithium and isotopic hydrogen. And no one knew exactly how much hydrogen it would release, or how big the detonation would be.

The testers would soon find out. And because of the other uncertainty—over the weather and, more specifically, the direction of the winds on detonation day—a great many others would find out as well.

For several days before the test date, the winds had been blowing in what was considered an acceptable direction: toward the west, where they would carry any radioactive fallout over an empty expanse of sea. The United States had declared a 57,000-square-mile "danger area" in an official Notice to Mariners, suggesting that craft keep away if possible, but without stating why. Had matters stayed as they were, the detonation would have caused little obvious harm.

However, on the night before the planned blast, February 28, the wind began to veer toward the east, away from this designated danger zone. Matters then got worse. As the sun inched up on the morning of the shot, meteorologists started reporting that at upper altitudes a powerful gale was now blowing directly from Bikini and toward the other populated atolls of the Marshalls, most notably in the direction of Rongelap, a hundred miles away, and forty miles farther on toward Rongerik, where the Bikinians had first been sent. On Rongerik there was still an American duty weatherman; he later told the newspapers that the wind, even at sea level, had been blowing directly at his island home from the west—from the direction, in other words, of Bikini, where they were counting down to firing the bomb.

Alvin Graves was aboard the command ship, the USS *Curtiss*, a venerable seaplane tender that was well accustomed to bombs, since she had been damaged by and had survived both the Pearl Harbor attacks and then a kamikaze strike in mid-Pacific. And though this bomb was a military device, Graves, the civilian chief of the project, had been given ultimate authority over the army general who was in command of the task force operating the weapon.

Graves was told of the wind direction and knew that radiation would spread downwind and contaminate, at the very least, Rongelap Atoll. But he had his orders, which were to proceed with the test without delay. Moreover, whatever the wind direction might be, no one had any idea how much radiation would be produced. Not that this was strictly relevant, of course, since Graves still cleaved robustly to his views about the malingerers who had concocted all this fuss about radiation being so terribly dangerous.

So he gave his orders to activate the automatic firing mechanisms. The *Castle Bravo* bomb should be allowed to explode. The men in the bunker took cover, and then pressed the brilliant red firing button.

At 6:45 a.m. on that clear, windy, blue-sky Pacific morning, it was as if the world had suddenly stopped, blinded by a vast white light of an intensity never before experienced. The iron gates guarding some terrible inferno seemed to clang wide open and unleash a ball of fire and shock waves and roarings of unimaginable speed, violence, and loudness. A white fireball four miles across was created in less than one second. A minute later, a cloud of debris ten miles tall and seven across rocketed into the sky. Ten minutes on, it was twenty-five miles tall and sixty miles across. The dawning sky lit up for hundreds of miles, and islanders from faraway atolls looked on in horror—for this was a secret test, unannounced, with no prior warnings—as a gigantic pillar of fire and smoke hurled itself into the air, a mushroom top boiling fuming orange and black miles above it, with rings of new-formed cloud expanding and coiling and writhing around it as it raced up through the layers of the atmosphere.

The shock wave tore across all the islands of the atoll, snapping blazing trees like twigs, razing almost all the hundreds of buildings and towers and sheds and docks and warehouses and barracks erected for management of the tests. Ships waiting beyond the islands were buffeted by giant waves as the shocks ricocheted across the sea.

A theoretical physicist, Marshall Rosenbluth, was on such a ship, thirty miles away. "There was a huge fireball with these turbulent rolls going in and out. The thing was glowing. It looked to me like a diseased brain up in the sky. It spread until it looked as if it was almost directly overhead. It was a much more awesome sight than a puny little atomic bomb. It was a pretty sobering and shattering experience."

Down in the bunker, the nine men of the firing party were rocked by what felt like a massive earthquake. Pipes broke, drenching them with water. The concrete walls swayed and cracked. Radiation swept in through the ventilation shafts. Radio contact with the command ship was degraded, ruinously—though the terrified men were able to understand that they would not be picked up by helicopter, as planned, as it was too dangerous for anyone to be on the atoll.

The men retreated into a single room deep in the bunker, where the radiation levels were a little lower, and there they stayed put—first turning off the air conditioners to stop radioactive air from entering the room, but then having everything else turned off for them when the outside diesel generators failed. There they waited in the sweltering darkness, until finally, late in the day, three helicopters arrived and ordered the team to come to the surface. They emerged draped in sheets, eyeholes cut out, looking like bizarre Halloween exhibits, eager only to get away from Bikini, and from the insistent chattering of the Geiger counters.

Bikini's *Castle Bravo* bomb was a quite extraordinary event, jaw-dropping, awesome, and, except for a few scientists who had advised caution, generally unexpected. It released a truly vast amount of radiation, and all of it was now spreading fast eastward across the Pacific in an enormous plume of dust and debris that for hours following the explosion was raining chunks of highly radioactive coral down from the sky and contaminating everything below. The explosion was greater than anyone had calculated: as a lawyer later told a court during arguments for compensation for the Bikinians, a train hauling the fifteen million tons of TNT that was *Castle Bravo*'s equivalent would have stretched in an unbroken line of freight cars from Maine to California, with hundreds of cars to spare.

The damage done by Alvin Graves and the bomb under his command was unprecedented. Within moments, everyone who was watching the blast column, and who knew the geography of the islands, realized that the islanders on Rongelap would probably be contaminated. A ship was ordered to speed across, and by midafternoon had landed a number of sailors in protective clothing to take Geiger counter readings from two of the village wells. They saw islanders who were clearly ill: staggering, vomiting, lying listlessly on the sand. But they said nothing to them, asked no questions, and left in a matter of minutes.

They were consequently unaware that the islanders had been startled that morning by what appeared to be a great sunrise in the western skies; and had then felt a sudden, jolting warm wind like a stuttering typhoon, followed by an unimaginably loud, thundering roar. They were also unaware that a fine mist had enveloped the island, that showers of grit and great gray flakes had fallen from the sky. Nor did they know that once the roaring had stopped, the islanders had immediately tried to resume their morning routines (breakfast, baking, fishing) and started to live a normal island day until, hours later, they began to show symptoms of some mysterious ailment.

The Geiger counters knew what had happened. The 236 people of Rongelap had received doses of radiation every bit as great as those suffered by the Japanese in Hiroshima, who had been just two miles from ground zero. But on Rongelap, no alarm had sounded. Instead, the bomb managers' first reaction was to think of employing the Rongelapese as case studies, as human guinea pigs. Radiation scientists at federal laboratories such as Brookhaven on Long Island expressed a kind of distant delight: "The habitation of these people on the island will afford most valuable ecological radiation data on human beings."

So, for the next fifty hours, the Rongelap islanders were left

to their own devices, to suffer in isolation until it became clear
that the radiation was so powerful it might actually kill them all,
whereupon official panic ensued, boats and planes arrived, and
the islanders were told to get out, quickly. They were hosed down
with water, ordered to wash, checked with Geiger counters, and
washed again, a routine repeated three times. They were told to
take nothing, to leave with only the clothes on their backs. Those
who looked fit enough were taken by ship down to the airbase
on Kwajalein. The old and frail went by seaplane. "We were like
animals," said an islander named Rokko Langinbelik, who was
twelve at the time. "It was no different from herding pigs into a
gate."

By now most were complaining of pain, burning, itching, hair
falling out, and skin lesions forming. But there was still no official concern for their condition—only an academic interest. They
might as well have been in cages. They were scared out of their
wits, having no idea what was happening to them, why they were
suddenly so ill, whether they were suffering from a fast-spreading
contagion. The doctors at the air base did little for them, other
than to advise them to wash and to subject them to constant monitoring with the ever-chattering radiation counters.

Six days later a secret investigation, to be known by the anodyne name Project 4.1, was initiated: "A Study of Response of
Human Beings Exposed to Significant Beta and Gamma Radiation Due to Fallout from High Yield Weapons."

No matter that these "human beings" had been the victims of
a monstrous and entirely avoidable accident, the consequence
of a decision made with casual, almost cynically calculated neg-
ligence. The subsequent racism of their treatment at the hands
of the authorities was obvious, or at least is amply recognizable
at this remove: had the islanders been Caucasians, then official
inquiries would have been instantly convened, congressional
committees would have been revved into high gear, presidential
apologies offered, compensation packages showered like rain.

But these were not Caucasians—they were mere Marshallese people, colored natives, members of a subject citizenry, a population now to be firmly contained and kept simply fed, watered, and, above all, docile. So there was never to be any inquiry of substance or value. The victims had worth not as members of any society, but as specimens—of importance principally to science. They might as well have been cadavers handed over to anatomists. They might as well have been branded with the term used by Japanese in their notorious human vivisection experiments—their human victims they called *maruta*, "logs of wood," a deliberately dehumanizing description, given to lessen the crime. These innocents from Rongelap were America's *maruta*, people rendered up as logs of wood. They were to become no more than the accidental subjects, serendipitously offered up to a group of faraway radiation scientists, of a detached, unemotional, and top-secret clinical study, a project of supposed significance for all in the ever more radioactive postnuclear world.

And for a while it seemed this project would remain top secret—except that an army corporal named Don Whitaker glimpsed a group of the evidently very sick islanders in their hastily built camp on Kwajalein and wrote to tell his relatives in Cincinnati, who were sufficiently horrified by his letter to pass it to the local paper, the *Cincinnati Enquirer*. The letter was published on March 9, a little more than a week after the blast. The news then spread rapidly, and it backed the U.S. government into a corner. It was forced to admit that, yes, there had been a nuclear test; that, yes, some islanders had been briefly exposed; but that they were being treated and that all was well.

The chairman of the Atomic Energy Commission, Lewis Strauss, angrily denied that the islanders were being regarded as guinea pigs, or that their evacuation had been deliberately delayed during those first two days so that their now unique biologies might be studied. Any suggestions otherwise were "utterly false, irresponsible and gravely unjust to the men engaged in this patriotic service," he declared. Moreover, he had taken the trouble to fly out to Kwajalein and see the islanders, and they "appeared to me to be well and happy."*

The people of Rongelap were not alone; there were other casualties. Most notably, a Japanese tuna fishing boat, the *Lucky Dragon Five*, happened to be innocently fishing in the waters near Rongelap that day; she was quite drenched in radiation.

Twenty-three men were aboard. The wooden hundred-footer had sailed from the southern Japanese port of Yaizu some five weeks previously, and after an expedition off Midway Island from which the pickings were extremely slim, the skipper decided to try his luck down in the Marshalls. He knew the dangers, he was well aware of the various Notices to Mariners about testing, and when the western sky lit up with a blinding white flash and then a huge orange fireball on that March 1 morning, he knew very well what had happened. Seven minutes later came the unmistakable Godzilla-rumble of the detonation; all aboard knew it was time to head north, to get as far away as possible.

But the men had to haul up their nets, and while they were engaged in this laborious task, the ash started falling. It was made up of great white flakes of scorched Bikini coral, quite tasteless— one crewman licked an especially large piece—odorless, cold. It fell incessantly, like snow mixed with cotton candy; after three hours, the men were covered with the stuff, their hair was matted, their bare brown shoulders were gray with grit. And very soon after these sea-weathered fisherman had stowed their gear and begun to chug away from the danger zone, they started to fall sick: nausea, burns, headaches, hair loss, stomach problems.

---

* Whether this was a deliberate employment of economy with the truth can never be known. But it is worth remembering that Strauss famously and wrongly predicted that nuclear fusion would allow for the generation of electricity "too cheap to meter," and that he was also largely responsible for destroying the postwar career of the Manhattan Project's leader, J. Robert Oppenheimer, suspecting him, also quite wrongly, of being a Soviet spy.

The irony is that these men, all victims of a hydrogen bomb, were Japanese, and were quickly diagnosed back at their home port as suffering from acute radiation sickness. The diagnosis was made so swiftly for the bleakest of reasons: after Hiroshima and Nagasaki, Japanese doctors knew all too well—by the way that this unique, newfound, and one might say *American-made* ailment presented itself—exactly what they were dealing with.

For weeks the men were terribly ill, bedridden, and dangerously vulnerable to infection. The American authorities did little to ease their medical misery, by declining, at least at first, to explain fully what isotopes had so contaminated them, since to do so might reveal something of the bomb's internal design.

Lewis Strauss, the AEC chairman who had already issued such trenchant denials about the alleged ill-treatment of the Rongelapese, now found himself performing similarly robust damage control over the *Lucky Dragon Five*. The boat, he suggested mendaciously, may well have been in the pay of the Soviet Union, and was spying. The burns on the men's skin were no more than a chemical reaction to the lime in the calcined coral. And their boat, in any case, had had no business fishing inside the danger zone. Mr. Strauss also suggested that the tuna caught both by this ship and others known to be in the danger zone was uncontaminated and harmless—though he said nothing when the U.S. Food and Drug Administration later placed severe limits on the importation of Japanese fish, which had the effect on the Yaizu fishing community of adding economic insult to radiation injury.

One of the crewmen, the ship's radio operator, died six months later,* leaving behind a note suggesting that he be regarded by history as the first fatality caused by a hydrogen bomb.

---

* The man who licked the falling dust lived into his eighties, and opened a dry-cleaning business, while another opened a tofu restaurant. All received the 2015 equivalent of five thousand dollars in compensation, once the United States formally took responsibility. The ship was hauled out of the water and now stands in a museum—not as a local monument in Yaizu, but in Tokyo, where she still gets national attention.

These are all episodes in a sad and shameful saga, and a story without a visible or imaginable end. Many Pacific peoples have suffered the unhappiest of fates, and to no obvious advantage. There is the fate of the contaminated Rongelapese, now all exiled, irradiated, sick, with sickly offspring and terminated pregnancies and tumors and mysterious growths and varying other legacies of florid illness and early death. There are the more casually forgotten islanders from the other test atoll of Enewetak, now home to the huge crater from the so-called Cactus test of 1958, which is currently entombed under a bizarre stadium-size dome of thick and leaking cement. There are the surviving crew members from the *Lucky Dragon Five*, most of them now living miserably far from home, self-scattered anonymously around Japan. Shame is still attached in Japan to the so-called *hibakusha*, "explosion victims," because some people are still scared that radiation sickness is contagious and can be spread, like leprosy. So the fishermen are exiled, too, victims until they die.

Underpinning all, most infamously, is the fate of the Bikinians. Though some remain on the congested islet of Kili, most of the 400 known members of the group (children, mostly, of the original 167 exiled inhabitants) are scattered, too, many now far afield. They are to be found all around the Pacific, their ancestral homes irradiated, their health compromised, their understandably querulous attitudes found tiresome by some—and, with their layers of lawyers, involved in interminable disputes about their compensation.

Unsurprisingly, Washington has dealt with its nuclear polluting of the modern Pacific mainly by paying out uncountable millions in taxpayer money and hoping the problem will go away. "Bombing Bikini Again," read the headline in a newspaper article in 1994: "This Time with Money." *Trust funds*, *compensation*, *claims*, *payouts*, *investments*—these days such words pepper the language of the Bikinians: "In all our meetings now," said a former Peace Corps volunteer who now acts as liaison with the U.S. government, "it's just money, money, money."

One means of gathering money for the islanders these days is by promoting the sunken ships of Bikini Atoll, catnip for the world's richest and most elite deep-sea divers. So even though the local Marshall Islands airline has only one plane, and it is almost always grounded, tourists who are willing to go by charter ship make their way to the atoll to dive down onto the superstructure of the USS *Saratoga* and to swim alongside sharks and to enjoy the bragging rights of having visited one of the best-known, least-seen places on earth. A place now declared by UNESCO to be worthy of designation on the list of World Heritage Sites, to be a place of "outstanding universal value," an outstanding example of a nuclear test site, associated with "ideas and beliefs . . . of international significance."

The divers who visit the lagoon occasionally do take their dinghies across to land, where they can poke around under the new-growing palm trees, stroll past the abandoned bunkers of rust-stained concrete, imagine much about the atoll's explosive recent past. But they will see precious little to remind them of Bikini's more ancient history, of the time before 1946, when the islanders were asked, ever so gently, to clear themselves out and to allow the American forces to begin their conduct of God's work, for the good of all mankind. The houses of these people are long gone, their memorials vanished, their fishing boats long decayed, their island traditions long since assimilated into other, alien ways.

In August 1968, there seemed a chance that matters might come back to normal. President Lyndon Johnson ordered that the people of Bikini be allowed to enjoy the comfort of their own homes once again. His scientists had told him, and he was now telling the world, that it was safe for everyone to return. Everyone, he said, should go do so.

On the night of the president's announcement, the Bikinians who still lived in shacks down south on Kili, the tiny, prison-like speck that had been their exile home for the previous twenty years, rejoiced. At last, they thought, their great national sacrifice was over and they could resume the peaceful rhythms of their former lives of fishing and copra making, and of voyaging in their outriggers to spend time with island neighbors of the western Pacific seas. So more than a hundred of them went off home, exuberant, relieved. An image from the time shows a group of island elders disembarking onto the coral shore, wearing shirts and ties, and so turning their homecoming into a formal event, an episode suffused with the proper dignity.

But the scientists had been wrong. "We goofed," one of the AEC officials said, with that breezy detachment of language that has marked so much of the official accounting of the saga. "The radioactive intake in the plant food chain had been significantly miscalculated." It turned out there was still a great deal of radiation deep down in the Bikini soil. The vegetables the islanders grew were contaminated, lethally so.

Congress then had to be asked for a further fifteen million dollars to take the islanders away again. They all left in 1978 and are now back on Kili, or have spread themselves around to other places in the world that will have them. "We were so heartbroken," an islander named Pero Joel told an interviewer in 1989. "We were so heartbroken we didn't know what to do."

Where they and their ancestors had once lived had, during the twelve years from 1946 to 1958, seen the explosion of twenty-three atomic bombs, with the combined force of forty-two million tons of conventional explosives. Everything the islanders had known had been obliterated: their homes and boats destroyed, their soil and the seawater contaminated, and their lives changed and spoiled forever. And for what purpose? To what end?

The blue Pacific now churns ceaselessly each present day along Bikini Atoll's quite deserted coral beaches. The palm trees lean into the breeze, unclimbed. There are no sails out in the lagoon, no sounds of chanting as the fishermen pull in their nets, no villagers gathering to chatter under the coconut groves. Bikini is today a place of a strangely deadened silence—a terrible, unnatural emptiness that compels any visitor to turn somewhere, to try to face the eternally invisible perpetrators of all this, and demand of no one and of everyone: *just why?*

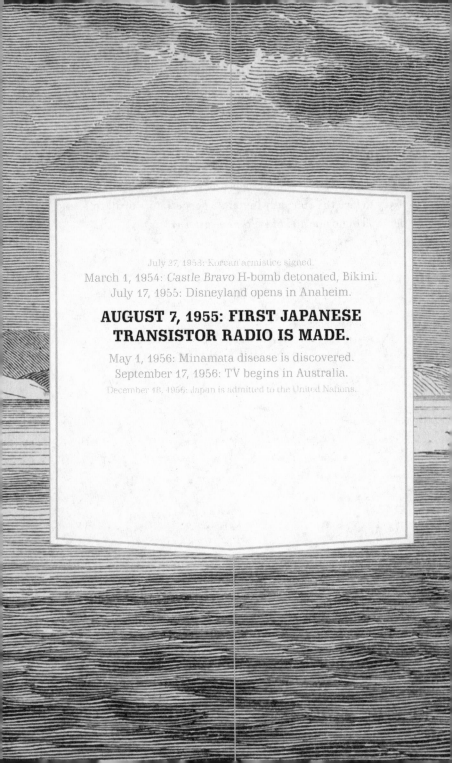

July 27, 1953: Korean armistice signed.
March 1, 1954: *Castle Bravo* H-bomb detonated, Bikini.
July 17, 1955: Disneyland opens in Anaheim.

AUGUST 7, 1955: FIRST JAPANESE TRANSISTOR RADIO IS MADE.

May 1, 1956: Minamata disease is discovered.
September 17, 1956: TV begins in Australia.
December 18, 1956: Japan is admitted to the United Nations.

Chapter 2

MR. IBUKA'S
RADIO REVOLUTION

———◦◦◦∪∪∪◦◦◦———

This is not the age of pamphleteers. It is the age of the engineers.
The spark-gap is mightier than the pen.
—LANCELOT HOGBEN, *Science for the Citizen*, 1938

It was piercingly hot in Canada in the late summer of 1955—so hot, the newspapers said, that apples in Ontario were baking on the trees. Indoors it was sweltering, and those who came home from work and wished to listen to the evening news or learn how their local lacrosse teams were faring found it necessary to keep their windows open, crank up the radio's volume, sit out on the stoop or the lawn, and hope the passing traffic didn't drown out the broadcast.

But those few who had passed by electrical stores in downtown Winnipeg and Edmonton, in Toronto and Montréal and Vancouver—most especially Vancouver, which at the time had a sizable Japanese population, who had some prior knowledge about such things—would have noticed on sale that month a small boxlike device made from greenish-brown plastic that, all who saw it swiftly realized, brought an answer to their summertime prayers.

It was a radio set no bigger than your hand, with no wires connecting it to anything. Until August 8, when this device first went on sale, most radio sets had been pieces of furniture. They were, by and large, behemoths made of walnut veneer that needed to be dusted and polished, and that more often than not provided a resting plinth for potted plants. But this little box was different. It wasn't furniture at all. It ran off batteries and didn't have to be connected to the wall. It was lightweight, didn't need time to warm up, and in fact didn't get warm; it emitted sound the moment you turned it on, and it could go anywhere—certainly well beyond the oppressive heat of an August living room. You could use it outside, under the shade of a tree, in the cool beside the fine spray from the sprinkler. It was so neat and tidy, with its tiny plastic feetlets, that you could set it down on a table in the yard or on the lawn itself, or on a table on the porch—or it could be carried to and fro as you wished, perhaps as you went to the icebox to fetch another bottle of Molson.

It was a pretty little thing, very modern, very midcentury. Most of it seemed to be a loudspeaker, with a grille and scores of tiny perforations. There was a small red and black wheel on the side that turned it on and off, and another to adjust the volume. On the front right-hand side was a dial and a knurled, revolvable disk that allowed you to change from one CBC station to another. On this disk were words, most probably unrecognizable to all but the immigrant Japanese: "Tokyo Tsushin Kogyo," which in English meant "Tokyo Telecommunications Engineering Company."

There were two other words on the front of the little device. In raised plastic lettering across its top was embossed a lately created but suddenly quite familiar description of the electronic organelle that lay in the beating heart of this radio, and that essentially, if incomprehensibly, made it work: TRANSISTORIZED.

Then, in a minuscule oblong space above the tuning wheel, a space easily overlooked, there was the second word, which was destined to become one of the best-known brand names in the world. That word was SONY.

The tentacles of what would become a giant global corporation, whose inventions would affect the ways many millions of people took pleasure from their lives, had started extending their way east across the Pacific. The Japanese electronic century, as some would call it, had officially slithered into existence.

It is tempting to suppose that Sony sent its first radios to Canada rather than to the United States for reasons having much to do with the Second World War, then only a decade past. But the truth is more mundane. A Canadian businessman named Albert Cohen, wandering through Japan in search of opportunities, happened to spot an advertisement in a Tokyo newspaper seeking a distributor for a new kind of radio. He arranged an appointment, sought and offered terms that were mutually agreeable, closed a deal on a handshake, and lugged a crate filled with fifty radios back to his company headquarters in Winnipeg. That Sony's first beachhead beyond Japan was thereby established among the grain elevators and green expanses of the North American prairies, a world away from Asia and the great blue expanses of the Pacific Ocean, speaks volumes: an early hint that the Pacific's economic and cultural reach was to be unimaginably vast, almost limitless.

The man with whom Albert Cohen had his first dealings in the summer of 1955 was Akio Morita, the better-known and most public face of what would become the Sony Corporation. In 1955, Morita was well on his way to becoming the elegant silver fox of Japanese electronics. He was worldly, sophisticated, patrician (the silkily affluent heir to a Nagoya sake-brewing dynasty), and a visionary. He had a degree in physics, but he was not, strictly speaking, an engineer. And the beginnings of Sony—indeed, the beginnings of everything that would underpin Japan's remarkable revival after the ruin and humiliations of a roundly lost war—were rooted solidly in the world of engineering and in the grimy hands of its practitioners.

Happily the man who was to be Morita's cofounder of Sony, the much less well-remembered Masaru Ibuka, was a true engineer, a classic of the breed. By the time the pair first met, in 1944—when Ibuka was already thirty-six, a great crag of a man; bluff, myopic, and shambling; ursine and untidy; and towering over the twenty-three-year-old Morita—Ibuka was already known as a tinkerer, a maker, an inventor. In 1933, when he was still at university, he had been handed an early award (a Gold Prize from the Paris World's Fair) for devising a way to make the glow in neon lamps appear to flow—so-called dancing neon, created by attaching a high-frequency power supply to one end of a neon tube and varying its output, a technique still much used today in advertising signs.

He was captivated by all things mechanical. He was a ham radio operator, had made his own gramophone, and had built a pair of stupendously large loudspeakers for use in a local sports stadium. He collected music boxes, player pianos, and organs, and to amuse himself, he had a remote-controlled helium balloon. He was also entranced from childhood by model railway trains, and would in time become president of the Japanese Association of Microtrains. More often than not he could be found on his knees on the tatami, reconnecting a length of miniature railroad track or tightening the screws on a steam locomotive.

It was the summer of 1944, and one desperate, last-ditch effort by the Japanese military to reverse the tide of what was now clearly an unwinnable war, that first brought Ibuka and Morita together.

The munitions ministry wanted to devise a new kind of antiaircraft missile, to harry and maybe bring down some of the American B-29s that, with their relentlessly lethal firebombing campaigns, were so devastating Japanese cities. They turned to a Tokyo maker of naval radar, the Japanese Instrument Measurements Company (JIMCO), and ordered the young naval lieutenant Morita, who had a degree in physics, to act as liaison.

Ibuka was the managing director of JIMCO, and from their first meeting, his inventive genius entirely captivated Morita. He was evidently a true lateral thinker, years before the concept was born: to solve a problem concerning the oscillations of a new radio transmitter, for instance, he had hired a score of young women, all of them music students from a nearby college, to employ their perfect pitch to help him adjust the radio to the exact frequency of a tuning fork. Such ingenuity! Morita thought. Such imagination! Ibuka-san, rough around the edges though he might be, and with a Tokyo workingman's accent, was one memorably creative individual!

Either the heat-seeking missile was never made or else it did no good, for eight months later the war was over. But the professional collegiality that had sprung up between Morita and Ibuka developed swiftly into an inseparable and lifelong friendship. The first stirrings of creative energy soon began to display themselves.

Japan in the immediate aftermath of the war was steeped in a miasma of misery. The population—throughout Japan, though most particularly in the capital, Tokyo—was afflicted by a hitherto unknown condition that had been given a new name,

plucked from psychiatrists' manuals: it was *kyodatsu*, a phenomenon that mixed exhaustion with despair in equal measure.

Hardly surprising: Barely half the inhabitants had a roof over
their heads. One in five had tuberculosis. On all sides in the
capital were ruined buildings, broken water mains and sewage
drains, shattered schools. There was no public transportation: all
the buses were destroyed; the trolley lines and their cars had been
obliterated. There were the daily degradations and humiliations
of the American occupation; there was a pervasive lack of work and
its kin: a want of money and widespread beggary and destitution.
There was also, or so it seemed, a collapse in society's moral fiber,
with gang warfare, prostitution, thievery, and black marketeering
pasted onto a national sense of remorse, guilt, resentment, and a
deeply felt, unfocused, and chance-directed bitterness.

Yet, for all that, as 1946 got shakily under way, something curious happened: the Japanese people began to ready themselves,
though they knew it not, to rise up and display a mettle quite unimaginable in its scope, heft, and range. And the Pacific Ocean
was the theater in which this display was to be most vigorously
mounted.

In those first months after the surrender, the country was
gripped by a spasm of self-repair, of make-do and mending, of
precipitous institutional about-faces and adaptations. Factories that had weeks before been making war materials switched
their production lines to start making items needed not by generals and admirals, but by the bone-tired civilians and by the
ragged menfolk returning from the battlefields. So bomb cas-
ings became charcoal burners, sitting neatly upright on their
tail fins and helping households get through that first bitter
winter. Large-caliber brass shell cases were modified as rice
containers, while tea caddies were fashioned from their smaller
shiny cousins. A searchlight mirror maker turned out flat glass
panes to repair thousands of smashed Tokyo windows; and for
country dwellers, a fighter plane engine piston maker turned

his factory to building water pumps. A piston ring fabricator named Soichiro Honda took small engines used during the war as radio generators and strapped them onto the frames of Tokyo's bicycles—the resulting Bata-Bata motorcycles, the name being onomatopoeic, later evolved into a brand of bike still famed from 1950s Japan as the Dream. Its popularity and commercial success heralded the birth of today's automobile giant, the Honda Motor Company.

As with Honda, so with the company that would soon be founded by Ibuka and Morita. It was Ibuka himself who first set matters in train. Within moments of the emperor's broadcast to his nation, announcing the surrender, Ibuka told his radar-making colleagues that he was returning to Tokyo, immediately. He had divined, with what now seems almost messianic clarity, that the country's future depended on engineers and on their ardent use of technology. He also believed that only in the country's capital was such progress possible.

As he packed his bags, he dared others to go with him. Six men did—one of them, Akira Higuchi, remarking later that he made his decision "in two shakes, and left without a second thought. It was as if we were communicating telepathically. I followed him then, and I have never left him."

Higuchi, who eventually became Sony's head of personnel, was much like Ibuka: a memorable figure. He was a formidable mountaineer, for example, and in later life kept a globe in his office studded with tiny flags indicating the more than one hundred peaks he had scaled. He was still employed by Sony into his eighties and celebrated his eighty-fifth birthday on the top of a ski slope near Lake Tahoe, in the California Sierra.

So Ibuka and Higuchi and their five colleagues took themselves down to Tokyo and promptly set up shop. They managed to rent for a pittance a cramped third-floor room in a near-derelict department store building, and bought desks and worktables. They first agreed on a name, Tokyo Telecommunications

Research Institute, though they then changed it, twice, before finally agreeing on Tokyo Telecommunications Engineering Company—in Japanese, Tokyo Tsushin Kogyo, familiarly to be called Totsuko.

Despite having no business or any real idea of what the company might do, make, or even dream about, Masaru Ibuka next wrote out a formal company prospectus. The handwritten ten-page document—written vertically on horizontally lined paper, with blots and crossings-out, and the uncertain look of a school-boy essay—is preserved now in a specially made glass display case in the Sony archives in Tokyo. It still offers a model for what a company, anywhere in the world, might aspire to be.

According to an equally lovingly preserved English trans-lation of the prospectus, the purpose of Ibuka's firm was "to establish an ideal factory that stresses a spirit of freedom and open-mindedness, and where engineers with sincere motiva-tion can exercise their technological skills to the highest level."

We shall, he pledged, "eliminate any unfair profit-seeking exercises" and "seek expansion not only for the sake of size." Further, "we shall carefully select employees . . . we shall avoid to have [sic] formal positions for the mere sake of having them, and shall place emphasis on a person's ability, performance and character, so that each individual can fully exercise his or her abilities and skills.

"We shall distribute the company's surplus earnings to all employees in an appropriate manner, and we shall assist them in a practical manner to secure a stable life. In return, all em-ployees shall exert their utmost effort into their job."

Finally, his new company would *help his country*. Its formally stated national intent was to help "reconstruct Japan, and to ele-vate the nation's culture through dynamic cultural and techno-logical activities."

Yet this high-flown language—in truth more Grandisonian than grandiloquent, as the firm's later progress would show—

masked many early difficulties. Neither Ibuka the man nor Totsuko his company had any real idea what to make. The first invention was a crude electric rice cooker, no more than a wooden tub with a flat aluminum element at its bottom. You poured in rice and water, plugged in the device, and the mixture's conducting wetness triggered the switch that powered up the heating element. The rice was cooked and duly dried, and the nonconducting dryness broke the circuit and switched the device off. It was in theory a fine and clever idea—except, the vagaries of the year's crop made it almost impossible to cook the rice properly. Sometimes it was fully cooked, sometimes not. Sometimes the machine switched itself off while the rice was still wet, like porridge. At other times the cooked rice was quite dry but had the consistency of a fistful of shotgun pellets. As a result, Totsuko's first foray into the mercantile world was a complete dud, and the hundreds of rice cookers languished unbought on the office shelves, for years.

But before long the firm did in fact find its feet, once Ibuka had insisted that instead of flailing around with some truly eccentric ideas (building miniature golf courses on bomb sites, selling sweetened miso soup), his engineers stick to their core pursuit: electronics. So, by the end of 1946, when Akio Morita, newly released from the navy, joined his new friend's firm, the business model swiftly focused the minds of all the employees on one particular and widespread electronic need: the repair of radio sets.

All Japanese households owned radios, but during the war some had been damaged by the bombings, and others had been destroyed by the much-feared Japanese secret police, the Kempeitai, who had campaigned to stop civilians from listening to American shortwave propaganda. Now, with peace returned, households wanted cheerful music; they wanted to hear such news as the American censors allowed (official announcements, information), and radio was the obvious best means of disseminating it. So Ibuka and his team—now swollen to more than twenty, outgrowing their one-room premises to fill one floor of the old department store—began doing real business. A reporter from *Asahi Shimbun* stopped by. Ibuka, with his shrewd sense of what made good PR, must have delighted in the resulting article, which described his radio repair business and reported that the work was being performed "quite apart from any commercial motive." The door was soon thronged by customers bearing broken radios.

Steadily the inventive energies accelerated—as did the quality of the technical work. The firm first made a voltmeter that was mundane enough in its own right, but cleverly enough designed and built that it got the attention of occupation forces' quartermasters, who sent samples back to America to be used as benchmarks of technical excellence. Suddenly a Japanese machine was winning kudos beyond Japan. Then Ibuka, swelling with pride, made the first fully functioning and complex electrical device that would perform the kind of task for which the firm would eventually win worldwide fame. He made a tape recorder.

This machine was expensive both to develop and to construct, but it eventually sold in respectable numbers. He was able to build it because Akio Morita and his old, rich, and highly traditional family decided to put money, serious money, into the young company. It was the firm's first investment, at the now-legendary sum of 190,000 yen. The Morita Company, run at the time by the fourteenth generation of farmer-dynasts, had for centuries concentrated on businesses sacred to the spirit of the nation: on the growing, harvesting, and storage of rice and soybeans and on the delicate brewing of sake, miso, and soy sauce. Yet now, and with remarkable prescience, the clan elders were able to discern a future of a very different kind. Helped in addition by an abiding faith in the artless genius of Masaru Ibuka, the family chiefs felt a stirring of commercial possibility—and instructed their son and presumed heir* to join the new firm as partner, and go back down to Tokyo and utterly transform the Pacific world.

Ibuka was fascinated by the idea that the human voice, music, the sounds of daily life, could all be mysteriously transplanted

Masaru Ibuka, a lifelong collector of model trains, ham radios, and helium balloons, was the engineering genius behind the first Japanese transistor radio, and later the Trinitron and the Walkman—and the cofounder of Sony Corporation.

onto a length of thin brown tape and played back through a loud-speaker. He first saw a tape recorder—a concept that had been born in Germany a decade before—at the American censors' office at the main Tokyo radio station. He did not know how it had been done—maybe the tape was plastic; probably it was in some way magnetic; maybe magnets were employed in some fashion to spread the sound onto the tape. However it was made,

---

* Akio Morita would remain connected to the family firm throughout his career at Sony, and returned without fail to the village to chair the Morita Company's annual board meetings. He also saw to it that the family's ancestral home and Buddhist temple were fully restored, to the delight of local villagers.

though, and whatever the magic of the tape itself, he could easily imagine the possibilities of such a device. It would be ideal for education, for training, even for what was then so keenly needed in Japan: sheer *entertainment*. He vowed that the firm would build and sell such a machine, whatever the cost, whatever the likelihood of immediate profit.

It took a while for the company's accounting chief, a dour man sent down from Nagoya to look after the family's investment, to sign off on the project. Morita and Ibuka infamously took him to a black market restaurant and got him drunk enough to agree. Once the money was available, the team sat down to solve the technical challenges.

Obtaining tape was the greatest problem, and from the outset the company decided it should manufacture the tape, rightly anticipating that owners of the recorders would need to buy ever more reels of the stuff. The plastic that was used in the American recorders was simply not available in Japan. Cellophane, which could be found, stretched, and so was useless. The only other available substance that could be magnetized was paper. So a specialized papermaking company was found in Osaka, thousands of sheets of the smoothest available craft paper were ordered, and Morita and Ibuka settled down to cut them into countless narrow strips.

The strips were then glued together and laid out on the factory floor—hundreds of feet of them, weighted to stop them from blowing about. All thirty-six of the company employees—the voltmeter business had nearly doubled the staff count—now armed with brushes made of fine raccoon belly hair, fell to their knees and, their heads bowed like monks in a scriptorium, applied with infinite care a magnetic paste concocted from a mixture of ferric oxide and geisha-quality white face powder. There was a down-home aspect to the business: the ferric oxide had been cooked up in frying pans; the powder had been bought wholesale from a cosmetics company.

The resulting pasted tape was left overnight to dry, and then tested the next morning by being run over magnets connected to speakers. The result was the so-called talking paper—fragile, scratchily imperfect in the first tests, but increasingly more workable as the cutters and the gluers and the brush wielders got better at their tasks. The acceptable batches were then wound onto reels, and these were placed on a hefty machine that had been cobbled together from motors and magnets, and was equipped with an external microphone and a built-in loudspeaker.

Finally, here was the prototype of the Totsuko Company's G-type tape recorder, a bulky hundred-pound confection of steel and copper and glass and raccoon-hair-pasted paper tape. It worked, quite reliably. It would both record and play back whatever the microphone picked up. So the firm painstakingly hand-built fifty recorders, priced at 160,000 yen each—more than twice what was then the annual Japanese salary—and then crossed its corporate fingers. The dour Morita Company accounting chief, back again from the countryside and by now wise enough to remain sober, waited nervously to see how the market would react and whether his masters' investment was secure.

If it was, it was more by luck than judgment. Sales were painfully slow. Everyone who saw and heard the device was impressed. A noodle shop bought the very first and tried to encourage a primitive form of karaoke, which drew in crowds of diners. But few others wanted something so heavy or so costly.

The company then began doing what it subsequently became famous for, something that the Japanese people had been doing for centuries: shrinking things. The old notions—the neatly nested lacquered box, the tightly concertina'd fan, the foldable-to-nothing room screen—were for the first time translated into this electronic corner of the Japanese corporate world. The first giant tape recorder—those few of the original fifty that did sell went to the government and into the courts, for transcription—

was cunningly distilled into something that was neater, lighter, and very much smaller. This second version was called the Model H, for "home." It weighed just thirty pounds and cost eight thousand yen. It was followed by the Model M, which was designed to suit the fledgling movie industry; and finally, by the truly popular and successful Model P, a cheap and miraculously how-do-they-do-it? lightweight portable tape recorder, with a shoulder strap and an appearance of near-chic modernity—which started selling at the rate of six thousand units each year.

With figures like these, and the firm's newfound ability to come up with new and smaller models and then swivel its production lines to satisfy public demand at what seemed a moment's notice, the little company could afford to rent more space and hire more people. By the end of the 1940s, Totsuko had a staff of almost five hundred and had expanded offices in a former barrack block in a hilly western suburb, where the company is still based six decades later.

Shrinking the product seemed to have been the key. The engineers who mastered the mysteries of squeezing more and more features into smaller and smaller volumes were the early heroes of the story. But in later years, they were to be greatly assisted by an invention from the late 1940s—an American invention, as it happens—that would allow the small to be made tiny, the tiny minuscule, and for a real electronic revolution to get itself properly under way.

This was the transistor. This small, simple, and now all too easily made electronic amplifying device is widely accepted as one of the greatest of all modern inventions. It is an essential in the making of all today's computers, is key to the birth of the Pacific coast technologies of Microsoft and Apple and more generally of Silicon Valley (so named, since 1974, as silicon is the transistor's core material), and helped light the fuse of Japan's postwar success. It was invented, all agree, on December 23, 1947. A trio of electronics engineers, who would later win the

1956 Nobel Prize in Physics for their discovery, made the first working transistor where they were employed, at Bell Labs, in Murray Hill (a New Jersey suburb founded by a spritzer maker who had migrated there from Murray Hill, in Manhattan). It was one of the last gasps of Atlantic coast inventiveness in a field of technology that would become increasingly dominated by the much greater ocean to the west.

It took Drs. Bardeen, Brattain, and Shockley* years of intense application to maneuver the tiny slivers of semiconducting germanium and the even tinier conducting electrodes of pure gold leaf into performing their magical feats of amplification. But once they had achieved this world-changing miracle, the vacuum tube (that fragile, hot, cumbersome, and slow-to-warm-up *valve* that had managed to switch and amplify electrical signals before) was effectively retired, to be replaced by the semiconductor and the new-made transistor. Once such transistor-based circuitry could be integrated onto single pieces of silicon, eventually allowing thousands and then millions of transistors to be etched onto a slice of semiconductor no larger than a fingernail, the modern high-technology world, or at least a substantial part of it, began to assume the complexion it still has today.

Masaru Ibuka became immediately intrigued by what he learned of the transistor. News of its invention trickled into the Japanese papers, though initially the only suggested use most could imagine—a use that took advantage of its tininess—was in the making of hearing aids, which were seldom worn in Japan. So in 1952, when Ibuka went on his first-ever journey to the United States—"[A] stunning country!" he reported. "Really fantastic.

---

* Beyond the electronic world, only Shockley's name is now widely remembered, and that mainly because of his preternatural enthusiasm for the supposed benefits (much debated and derided today) of eugenics. He also performed, for the wartime U.S. government, cold-blooded assessments of the number of likely casualties in any invasion of Japan. His official estimate—that ten million Japanese might have to be killed and that eight hundred thousand Americans might die accomplishing this—is said to have influenced the decision to drop the two atomic bombs instead.

Buildings brightly lit. Streets jammed with automobiles"—he was not initially bound for Murray Hill. The sole official purpose of his expedition was to see how tape recorders, then still the company's core (and really, only) business, were being used.

He worked hard. He discovered many new uses for the recorder, and each time, he sent telegrams back to Tokyo demanding action, and Morita would invariably comply. One suggestion was to start making recordings in stereo. Within days, Morita's engineers had solved some trivial technical challenges, whereupon Morita himself, exhibiting his soon to be legendary marketing acumen, cleverly approached NHK, Japan's national broadcaster, and offered it equipment that would allow it to present a thirty-minute radio broadcast in this newfangled stereophonic manner. On December 4, 1952, NHK introduced its stereo experiment, "produced" as the continuity announcer solemnly intoned, "by Tokyo Telecommunications Engineering Corporation and NHK." It was a stunning success. Thousands heard it and responded with unalloyed enthusiasm. "Our cat, who had been sleeping on the foot-warming table," wrote one listener, "was shocked by the sound-effects and jumped out of the room." The reports were telegraphed to America, and back to a clearly elated Ibuka.

By chance he was now coming toward the end of his expedition. He was staying at the Taft Hotel in New York—and on a night now famous in company lore, was being kept relentlessly awake by loud music from the Roxy Theatre nearby. Lying sleepless in his hotel bed, he suddenly connected two ideas that were floating through his mind.

First, Western Electric, parent company of Bell Labs, the inventor of the transistor, had just announced that it was looking to license outside companies to produce transistors in bulk. Second, Ibuka knew he had more than forty newly hired scientists of exceptional brainpower still tweaking the finer points of the company's tape recorders, but with not a great deal else to occupy their minds. So, without asking for anyone's agreement back home, he spent his final hours in America applying for the necessary production license—characteristically dismissing the idea that anyone at company HQ might balk at the twenty-five-thousand-dollar fee being demanded for it.

He must have pondered the matter more deeply as his Northwest Airlines DC-6 thundered westward toward and then across the Pacific, and whenever he got out to stretch his legs in the increasing chill of the airports at Minneapolis and Edmonton, Anchorage and Shemya, a lonely and gale-swept U.S. Air Force outpost in the Aleutian Islands. When finally he arrived back at Haneda Airport two days later, he was convinced: "Radios," he declared to the assembled senior staff. "We are going to make this transistor. And we are going to use it to make *radios*—radios that are small enough so that each individual will be able to carry one around for his own use, with a power that will enable civilization to reach even those areas that have no electric power yet."

A stunned silence greeted his announcement. "Too wild, too risky" was how one of the company managers summed up their reactions. Maybe the established, big-time companies (Mitsubishi, Toshiba, Hitachi) could make them. They already had licensed agreements with Bell to make their own transistors, and they had the resources to do so. But not tiny Totsuko, the new kid on the block.

There were financial and bureaucratic problems—getting twenty-five thousand dollars out of the company coffers was trying enough; getting these dollars wired out of currency-starved Japan was at first well-nigh impossible. There were also technical problems, which Ibuka solved through determination and prescience, but also with the help of the figure who would become the third member of the triumvirate of titans of this story, a brilliant young geophysicist and volcanologist named Kazuo Iwama. He came from the government's main seismological observatory, and like everyone else at Totsuko, he knew next to nothing about semiconductors.

But Iwama proved to be a phenomenally quick study. He and Ibuka flew back to America in the summer of 1954, to learn more about transistor technology, the revenue from tape recorder sales covering their hotel bills and food. The Sony Archives today hold the fruits of that three-month visit: four fat file folders crammed with hundreds upon hundreds of blue one-page onionskin paper air mail letters that Kazuo Iwama sent back, often several at a time, every single day.

The letters are crammed with detail, tissue-thin Rosetta stones of jumbled numbers, arcane formulas, Chinese ideographs, Japanese phonetic scripts, and a scattering of English words and phrases, together with fine filigree drawings of crucibles and diagrams of oscillators and depictions of circuitry that give the letters the appearance of some strange new art form, the designs for the future of an exotic new world. "Zone leveling single crystal," one letter reads. "Pure paraffin wax," another. "Detexile paper—no sulfur." The assembled papers constitute a small encyclopedia of transistor wisdom, a distillate of all that was then known in America about this magical new device. And in 1954, all of it headed westward to Japan, there to help create an economic, eventually transpacific revolution.

There were other challenges. Even in the early 1950s, the Japanese still felt something of a sense of cultural cringe, a pervasive lack of self-confidence. Years of hard work and dedication had improved the appearance of most Japanese cities, but the shame and humiliation of the war still exerted a powerful drag on progress. Morita recalls being in Germany—noting how rapidly it had rebuilt its own ruined cities—and having a shopkeeper in Düsseldorf offer him an ice cream with a miniature paper parasol stuck in it, remarking kindly that it came from his country. Is this all we are good for? he asked himself. Is this what the world thinks of us?

Yet it was rather more complicated than this. I am sure I am not alone in believing that many East Asian sciences, in particular, have long suffered, have long been held back, by the basic Asian concept of "face," of what the Japanese term *mentsu*. This (which, very broadly, relates to the giving of respect and the protection of one's own dignity and regard) plays a profoundly important role in the social exchanges of many countries in the northwestern Pacific. The socially lethal consequences of losing face or, more dangerously, of causing others to lose it, may well have inhibited certain kinds of scientific progress, in large part because such consequences militate against *experimentation*, which invariably embraces failure, even public failure. Picking oneself up and beginning again, making the experiment subtly different, and performing many experiments until finally one works—such is the essence of scientific advance. And this was not always an easy concept for Asian scientists to accept.

This is not to say that failure plays no part in Japanese society—far from it, indeed. To watch a sushi chef, for example, compelling his apprentice to cook *tamago* (the egg-and-vinegar-and-soy-sauce omelet that is a key component of a full-blown *nigiri* dinner) is to watch the pursuit of perfection through the repetition of countless attempts, most of which initially fail. Time and again the youngster falls short of making satisfactory *tamago*, and each time, the master contemptuously throws it away. Yet no shame is attached to the apprentice's failures, even though they seem to happen day after day and day. For, eventually, one hopes, the boy succeeds in this crucial task, is ultimately inducted into the corps of the minimally accomplished, and then slowly, painstakingly, makes his way toward becoming an acceptable sushi chef. Failure is just part of the process—in this and many other callings in Japanese life.

But science is very different from sushi making. Japanese cui-

sine is a time-honored craft, with teachers (sensei) who will cajole
and berate an apprentice along the hard road to success. A scientist, on the other hand, has to engage alone, in a quest for the
undiscovered and the unknown. He has to trust himself to do so
without a sensei at his elbow, with only his own curiosity to compel
him. This would be a formidably difficult challenge for any scientist. For one who might be further burdened by the concept of
"face," by the abhorrence of public failure, even more so.

The great empiricists, from Bacon and Galileo through to
Watson and Crick, all failed, but a mark of their greatness was
that they never abandoned their quest for scientific truth. The
same cannot easily be said of those early East Asian scientists,
particularly those who worked during the years of the Enlight-
enment in the West. Such advances as were made in Europe of
the time were simply not happening in the East, no matter the
centuries of progress (most especially Chinese progress) in the
years before. Puzzlement over just why this was has generated
interminable debate over the years. The so-called Needham
Question*—why, after so much earlier progress, was there so
little advance in China after the fifteenth century?—distills this,
and has never been satisfactorily answered. Face is suggested as
a component, one among many.

It was clearly a component in those early Totsuko days. One
member of the research team working on the licensed transistors remarked that "the voice of Bell Labs is like the voice of
God"—implying that for his Japanese colleagues to try to do anything different from the way the Americans were doing things
back in Murray Hill would be to court failure, disaster, and the
consequent loss of *mentsu*—and if not that, then perhaps also to
humiliate the generosity of the licensees at Bell Labs, to cause
them to lose face also. Respect for others, for elders, for per-

---

* Joseph Needham (1900–1995) was a Cambridge biochemist who spent much of his life
studying the origins of Chinese science. His story is told in my book *The Man Who Loved
China*.

ceived betters—these were concepts similarly central to Chinese and Japanese thinking: while it was dangerously uncomfortable to lose face yourself, it was unforgivably shameful to cause another to lose face. So, at first, timidity ruled in the Totsuko laboratories on the floors high above the tape recorder production line. Everyone was nervous, and for many weeks during 1953 and the first months of 1954, nothing very much was done, and less was accomplished.

Such hesitancy sorely tested Ibuka and his team of leaders, all of whom were doing their best to spur the scientists upstairs to do their best. Months after their successful purchase of the Western Electric transistor license, it was starting to seem as if they might never create anything better than the American model.

They seemed unable in particular to take the radical steps necessary to achieve the one technically risky but most commercially vital thing: to create a unique kind of transistor that was powerful enough and would work at a high enough frequency to allow the miniature radio set that Masaru Ibuka demanded his company manufacture to work. And this was causing major problems for the accountants. The income from the tape recorder business might still be healthy, but the burn rate (the payment of salaries to all these scientists and engineers who for all these complicated reasons were achieving rather little) was getting out of hand.

Months ensued, of cajolery and chemistry, of patience and physics. Ibuka and Iwama continued to write home from America, cabling their more urgent instructions for making the needed transistors. BUY HEAVY DUTY DIFFUSION FURNACE, one cable read. ACQUIRE DIAMOND GRINDER FOR SLICING GERMANIUM CRYSTALS, read another. Then, slowly, beating against the undertow of traditional thinking, the team in Japan started to nudge its way toward success. The timid became the tentative. Hesitancy morphed into determination, and the dragging weight of *mentsu* began to evaporate. Progress started, and through the mist the vision of the true Japanese transistor started to solidify.

The first device was completed late in the summer of 1954, while Ibuka and Iwama were still in America. It was in essence just a fair copy of that made at Bell Labs—a so-called point-contact transistor, primitive and not so small. But the principle was established: the needle on the detecting oscillator swung, with all watching nervously, indicating that the gadget was indeed creating an amplified output. By the time Iwama arrived back home, the team already had a more sophisticated model, a junction-type transistor with a perfectly cut germanium crystal—sliced with a rusty old cutting machine that had been found out in the rain in a Tokyo suburb—that was making the oscillator swing its needle even more vehemently. The little company that could was finally on its way to perfecting an invention.

The technology behind what is now an entirely routine procedure—even if we don't entirely understand what they are doing, we are well accustomed now to seeing images of workers in protective suits in brilliantly lit, clean rooms, directing the etching of tiny circuits onto minuscule slices of semiconducting material—was, in the 1950s, dauntingly complex. But using a procedure that Bell Labs had tried and discarded, a technique known as phosphorous doping, Totsuko eventually made the breakthrough it had long sought.

In June 1955, six months after the Iwama expedition to America, the company set up its first grown-crystal transistor production line. In the first weeks, maybe only five in a hundred of them worked; Ibuka's sanguine view was that so long as a single transistor worked, then perfecting the production technique could be accomplished at the very same time that production was under way. So the button was pressed, the factory started producing, and hundreds of tiny radio-frequency, high-powered, grown-crystal, phosphorous-doped Japanese-made transistors began cascading off the line.

Now all the company had to do was make a radio to put them in; to establish a brand name under which to market and sell this radio; and then proceed to change the lives of millions. This is what Ibuka demanded, and this is what he, Kazuo Iwama, and Akio Morita achieved.

There were hiccups, of course. An American company based in Indianapolis, named Regency, launched the first-ever transistor radio, the TR-1, in October 1954. "See it! Hear it! Get it!" blared the advertisements. Jewelry stores in New York and Los Angeles sold the sleek little sets for $49.95. The TR-1 sold well initially, but performed poorly: radio reception was often scratchy, and the set ran out of power too quickly to be of much use.

The first-ever Totsuko radio rolled off the production line in the spring of 1955. Called the TR-52, it was a tall rectangle, the size of a large cigarette packet. The four hundred square holes of its white plastic speaker grille looked like tiny windows, leading critics to say it resembled Le Corbusier and Oscar Niemeyer's UN headquarters, opened in New York two years before, and causing the radio to be called "the UN Building." Totsuko made a hundred of them, but the TR-52 never went on sale, because the grille bent and peeled off in hot weather. It was, or could have been, a major embarrassment.

However, the Bulova Watch Company saw the prototype—in cool weather, presumably—and very much liked the concept. Buoyed by the news of Regency's very modest success, Bulova reached out to Morita and ordered one hundred thousand of his radios, a staggering number.

Yet, to the dismayed astonishment of all back in Japan, Morita balked. He refused to take the order as offered. He did so because the American firm declared that it wanted to sell the radio in America under the Bulova name—and to that, Morita, a proud man, simply would not and could not agree. Especially since, just a few days prior to receiving the order, he and his colleagues had decided to rename their company, to call it Sony.

The employment of the name Sony came about entirely because of the American market. Morita had found that almost no one in the United States could pronounce either Tokyo Tsushin Kogyo, the company's formal name, or its diminutive, Totsuko. Something easier was needed, he wrote in a company memo. Something short; four-lettered, if possible. Something memorable, like "Ford."

The Totsuko principals explored only modestly, searching either for an existing word or for an arbitrary word—Kodak, created at the whim of George Eastman fifty years before, seemed an ideal. They thought of two-letter words, with which the Japanese language abounds, but to which English mainly consigns prepositions. They considered three-letter combinations (NBC, CBS, NHK). Perhaps their own existing initials, TTK, might work. But then they began to think of four-letter combinations. The name of Ford kept striking Morita as ideal, as being brand perfection—so he and Ibuka combed through their various dictionaries. As to whether they had a Latin dictionary to hand, corporate history is silent; but somehow or other they eventually came across the Latin word for sound (the ultimate product of all their engineering), and liked what they found: the Latin word *sonus*. Five letters, true, but very nearly perfect.

Since 1928, when Al Jolson had sung, "Climb upon my knee, Sonny boy, / Though you're only three, Sonny Boy," the term *Sonny* had won widespread affection, especially in America. Occupation forces, now three years gone, would throw sticks of Wrigley's gum to children, calling out, "There you are, sonny!" The word had pleasing connotations. It echoed the Latin word. It was easy to pronounce. It had universal appeal. And to make it into a Ford-like quadrilateral just a small modification in pronunciation and spelling was needed. Thus, in 1955, the word *Sony* was born. The word. The company. And history.

Stubborn to the end, Bulova refused to use the name on its products. "Who ever heard of Sony?" asked the president. Akio Morita politely replied, "Half a century ago people would have asked—who ever heard of Bulova?" But no ice was cut. The American's heart did not melt. And so Morita, with exaggerated courtesy, left the office—without the precious order for a hundred thousand radios. If Japan's first transistor radio was going to sell in America, then it would be called a Sony—and the Sony team would have to do their best to sell it themselves. The company's future was now very much on the line.

A concatenation of curious events then got under way. The melting plastic grille prevented the firm from ever producing a significant number of the TR-52s that Bulova had wanted. Instead, the more modish and functional TR-55 was the radio that made its debut in Winnipeg in the late summer of 1955. This was the radio bought by a lucky few blisteringly hot Canadians, and which allowed them to listen to the CBC while under their garden shade trees. And if any of the sets found their way to the United States, it was more by luck than adroit corporate judgment.

Whether the American makers of the Regency TR-1 ever saw an example of the Sony radio remains unknown. But something frightened them or their backers. For, suddenly, the firm announced that it would stop manufacturing the sets and would withdraw from the marketplace. It was a decision (still quite inexplicable, even at this remove) that left a gaping hole in the radio marketplace, and one that the newly named Sony Corporation[*] was poised, and happy, to exploit.

---

[*] For a while, products were branded "Sony—made by Totsuko," but in January 1958, Morita formally renamed the company Sony Corporation, despite opposition from Mitsui, the firm's very conservative bankers. For many years afterward, Sony fought legal battles with the newly formed Tokyo-based makers of Sony Chocolate. It won.

The device that Sony then made in an effort to fill this gap was designated the TR-63, the so-called pocketable radio. Company lore has it that Sony created the word *pocketable*, but the word made its first recorded appearance in the English language as far back as 1699. The same internal histories suggest also that this radio wasn't exactly as pocketable as the brochures had it. It certainly didn't fit into the breast pockets of most Japanese shirts. The wily Mr. Morita, it is said, had his salesmen's shirts modified with a slightly bigger pocket, so their demonstrations of pocketability could invariably progress without mishap.

Such claims may well have been buffed by time and expensive PR firms, and perhaps understandably so. The event that truly made this elegant little radio famous, and that made Sony a familiar name into the bargain, was entirely true, and involved a robbery.

The tale appeared on page 17 of the *New York Times* of Friday, January 17, 1958. Most of the other news items close by were quite routine. Noël Coward had a cold, and so could not go on for his matinee performance of *Nude with Violin*. Winston Churchill's actress-daughter Sarah, who had already been fined fifty dollars for disorderly conduct in Malibu, was now in the hospital suffering from exhaustion and emotional upset. A twenty-five-year-old prostitute named Sally Mae Quinn had squeezed her evidently rather slender self through an eight-inch window to become the first person ever to break out of a prison for women in Greenwich Village—though the trail of blood on the roadway thirty-five feet below the window suggested to police she might have something of a limp.

But the lead story on page 17 was of somewhat greater moment: "4,000 Tiny Radios Stolen in Queens," read the headline. The story was a sensation. A manager named Vincent Ciliberti, turning up for his morning shift at Delmonico International, an import-export company based across from the Sunnyside rail freight station in Long Island City, had discovered to his alarmed dismay that, during the night, a posse of thieves had broken in through a second-floor window and taken "400 cartons of green, red, black and lemon-colored radios."

The men had then, apparently displaying great fortitude and eagerness, broken no fewer than four locks to get into a freight elevator, backed a truck up to a loading bay, moved the radios in their boxes onto a pair of skids, and then hauled them onto the back of the truck, and vanished into the darkness.

Each carton held ten of these tiny radios, which Delmonico had been holding before sending them off to the stores to sell at $40 apiece. Some $160,000 worth of high-tech merchandise had just disappeared into the wilderness of outer New York City. It was the lead story on the city's radio stations throughout the day. Detectives were investigating what was said to be the biggest heist of electronics equipment in American history. More than fifty potential witnesses were questioned at length. No one, of course, had seen a thing.

Then, confirming the adage about ill winds and the doing of good, came a crucial piece of information. Delmonico, reported the *Times*, "is the sole importer and distributor of Sony Radio, built in Japan. Each of the $40 radios is 1¼ inches thick, 2¾ inches wide and 4½ inches high." A search suggests that this was the first time the name Sony had ever appeared in the *New York Times*.

Most crucially of all—and most delightfully, so far as Tokyo was concerned—it was only these Sony-brand radios that had been taken. Left behind, unclaimed and disdained, were twenty cases of other radios, and several tons of other electronics equipment. Since only the Sony devices were taken, it suggested to most readers of the paper that Sony radios were the highest-value items, the only radios worth stealing. If the thieves thought they were good and valuable, then they probably were.

That truly set the market afire. The little radio promptly became an essential. To this day, most Americans of a certain age remember their first transistor radio: a small plastic box, with a tinny loudspeaker and perhaps an earphone, that could be smuggled into high school, perhaps so that a baseball game could be listened to during algebra; or taken in the Impala at night to provide soft music while one was parked on a clifftop, hoping for rather more than the view.

All of a sudden an entire new industry swept into being, an industry bent on employing electronics, and devices with electronics at their heart, for the sole purpose of entertaining, amusing, and informing the public—either en masse or, more often, in person. Other manufacturers might continue to satisfy other, more traditional demands of heating, lighting, clothing, feeding, and moving the public about. Others might build cars or ships, mine coal, make stoves or washing machines or razor blades. But this new industry skillfully blended technology with the humanities, married the machine to the artist; and by doing so, its leaders were seeking to improve the daily lives of the average person by amusing and interesting him, by playing on his emotions and to his sentiments. It did so by the employment of transistors, semiconductors, and printed circuit boards.

The term *consumer electronics* was instantly coined[*] to describe this new business—backed by an industry that was born on the Pacific Rim, and has in one form or another come to play a sustaining central role in the betterment of human life, in most corners of the world.

And Sony, in Tokyo, one of the first entrants into the business that it had invented, promptly did its best to satisfy the market it had created. Factories expanded and hummed with energy,

---

[*] Curiously, a term not to be found, at the time of writing, in the *Oxford English Dictionary*, the usually omniscient accumulator of the language. *Consumer society* is listed, along with *consumer goods*, *consumer research*, and an ugly phenomenon, *consumer terrorism*, a phrase first noticed in the Pacific in 1984 after Manila police found deliberately poisoned pineapples. *Consumer durables* is also listed, but an editor regretting (and vowing to reverse) the absence of *consumer electronics* noted that *consumer durables* was already sounding a somewhat dated combination.

and hired thousands; and more plants were built, some hastily, most in more considered fashion, and with both investors and company bosses now cleaving to a firm belief in the firm's ever-more-settled future. Smokestacks belched, machines roared, heavily laden trucks lumbered off to the airport—entire cargo planes had to be chartered from the newly formed Japan Air Lines to meet Christmas demand—and containers, containers, containers were packed with boxes, bound in those early days for Seattle, Long Beach, and San Francisco Bay, and later for just about every major maritime port.

The containers were eventually to be crammed with much more than cartons of simple radio sets. The inventions that would be dreamed up by Masaru Ibuka and his swiftly expanding teams of engineers included microphones and videocassettes, computers and video cameras, games and storage devices, and a thousand other essentially inessential gadgets for the improvement of the daily lot of lots of people. The Walkman—a tape player that didn't record, seen initially as a heresy for a company that had made its name by recording sounds and not simply playing them—was a worldwide success.* The Trinitron—which Ibuka said later was the creation of which he was most proud—made full-color high-quality television inexpensively available to all.

A change in perception also started to occur as this steady stream of new products began to emerge from the Sony engineering benches. In the immediate postwar years, Asian countries were seen largely as peddlers of the shoddy, the gimcrack, and the second-rate. But now, with the inventions being shipped eastward by Sony and its like, Japan was swiftly winning quite another reputation, a name for itself such as it had never enjoyed

---

* With almost no marketing budget, the backers of the Walkman had to employ guerrilla tactics. Pretty young secretaries from the New York Sony offices were asked to stroll around Central Park, or to roller-skate through Union Square, while listening to their music, and wait for passersby to ask what they were doing, what they were listening to. The success is part of marketing legend.

before: for being a past master of the precise, the particular, and the highly accurate. All these devices, at least in the firm's early days (and this applies to the products of most of the other Japanese firms as well), were made with the kind of precision that was more readily associated with products made in Europe, especially in Switzerland and Germany.

Japan was a society built on traditions born largely of nature: of working in bamboo and water; of tatami and silk; of ceramics and flower arranging and the presentation of tea and the hammering of razor-sharp sweeps of steel; of adapting a natural world that, ipso facto, existed utterly without mathematical perfection, without straight lines. Now, all of a sudden, and thanks to men such as Ibuka, Morita, and Iwama, this country was becoming known for its masters of precision, for its ability to work with germanium and titanium and the pitiless certainties of the micrometer, the caliper, and the vernier scale—and yet never for a moment abandoning its intimacy with nature and the charming, spiritually important imprecisions of the natural world. The dexterity with which the Japanese bridged that gap—to employ and revere both titanium and bamboo, the die-straight and the gently curved—says much about the Pacific Ocean more generally.

For the Pacific had become some kind of cultural meeting place, for a certain kind of marriage—whether permanent or temporary, it was then too early to say, and with details and conclusions to be teased out. It would be a marriage of, on the one hand, a congeries of ancient natural cultures, most of them animist in origin, that permeate and define those countries that make up what the West likes to call the East; and on the other, the more numerically based, more ruthlessly practical capitalist and Judeo-Christian cultures that tend to dominate America, the American West, and what indeed Western peoples more roundly like to call the West.

As far as Sony was concerned, the company seemed first to enjoy and exult in its success, but then later to pay and suffer the often inevitable price of the pioneer. It first rose in public esteem on what seemed an impossibly steep trajectory. It soared with seeming effortlessness through the twentieth century. Its founders died and were honored and memorialized. Akio Morita is widely remembered; Masaru Ibuka, the true creator, rather less so. Then the company began to stutter, to lose velocity and altitude; and it commenced, in the first years of the new century, a long and painfully public decline, with assets sold, management changed, unwise ventures attempted. There came an all-too-regular litany of apologies, meetings dominated by the deep bowing of abject sorrow offered in silence by sad and dignified men who felt they had let everyone down. No excuses, though. No blame attached to others. Just acceptance and endurance, as is the Japanese way.

Sony was hardly alone in its sufferings. The consumer electronics business turned out to be a field of extraordinary competitive brutality. The Japanese companies—Sony, of course, but also Matsushita, Sanyo, Sharp, Toshiba, Panasonic, and a host of others—had at first vanquished the Americans. Their impeccable Japanese-made products and adroit marketing campaigns had reduced firms such as RCA, Magnavox, Zenith, and Sylvania to quivering wrecks, and eventually made them curl up, wither, and perish. The Japanese then assumed lead position and, from the western Pacific, commanded the heights of this new world order.

Then, as is the way of things today, the Japanese began to cede ground as well. First they lost ground in the making of products, and companies based in Korea and Taiwan, firms such as Samsung and Foxconn, began to create devices quite as impeccable and revolutionary as those the Japanese had made, but for much less money. Second, the new Japanese firms weren't generating as many innovative products, a development that presented an opportunity for American companies, of which Apple had to be the most vivid example, to step into that space. If the transistor radio was the electronic icon of the sixties, and the Trinitron its counterpart in the eighties, then the twenty-first-century equivalents were the iPod, iPad, and iPhone—all conceived in and around the Pacific Ocean, but now on the other side, in America.

I first saw Tokyo Bay at close hand in the late 1980s. It is a remarkable sight at any time, but on this windy and cold blue-sky day in early spring, it was memorably so. I had taken the rattling Japan Railways train out to the town of Futtsu, on the bay's eastern side, on the inner limb of the Boso Peninsula and the old volcanic hills that protect Tokyo from in-sweeping ocean storms. There is a spit of land, now Futtsu Park, where it is said the first American servicemen landed after the 1945 surrender. Its breakwater was slippery that morning, with water blown up by the stiff westerly breeze, and the entrance to Tokyo Bay was a confusion of dark blue water and white horses, spume lifting off the wave crests. In the distance, shockingly familiar, was the symmetric exactness of the cone of Fuji, the summit slopes blinding white with the sun reflected on new-fallen snow. The wind had blown the city smoke and fumes well away, and the view was perfect, rare, and serene, the mountain majestic.

The view of the bay, helped by binoculars, had a majesty about it also. This was the world's commerce writ large. To my left was the Uraga Channel, the six-mile-wide entrance to the Tokyo approaches. The biggest of the vessels inbound that day were all destined for one of the many ports inside the bay (Yokohama, Yokosuka, Kawasaki, Chiba, and Tokyo), each of them at the time among the busiest in the Pacific, and together, just about the busiest in the world. By the entrance to the fairway, I could see a jumble of the vessels stopped, or slowed, waiting at the pilot station, the small white shuttle boats hastening to and fro bringing yet more and more channel pilots to ease the logjam.

The navigable channels stretched clear across my field of vision, maybe two miles from where I was standing. The inbound ships, general-cargo vessels and ore carriers and tankers and liquefied natural gas carriers rust-stained from days in mid-ocean, bringing raw materials and food and oil, were sliding home to port from left to right, from south to north, along the shipping lane closer to me. But it was the outbound vessels that caught my eye that morning—because they were nearly all of the same type, heavily laden NYK container ships piled high with twenty-foot-long and forty-foot-long boxes, all of them loaded solidly with the goods of the Japanese industry, almost all of them headed for the American West Coast.

Once in a while a ship even larger than the last, an immense and featureless wall of green steel, eased into the shipping lane from the junction fairway from Yokohama: a car carrier laden with Toyotas, I assumed, and also en route for the United States. Some of the cars would be out on the interstates within the month, I imagined, and the stereos and cameras and televisions in the container vessels would be in the chain stores even more quickly.

Then, maybe twice that day, and often on my subsequent visits, I saw slip out from behind the large warehouse structures across the bay in Yokosuka a warship—one time it was a destroyer; on the second occasion, an aircraft carrier. They were American naval vessels, part of the immense forces that operate from bases that have been dotted across Japan since the end of the last war—in Okinawa; at Sasebo, near Nagasaki; at Misawa and Atsugi; and here, the biggest conglomeration of all, at Yokosuka.

The American ships, the aircraft, and the thousands of U.S. Marines based near here are said to be perpetually ready for action anywhere in the western Pacific, at a moment's notice.

Admirals in Hawaii could order them to battle stations. Their simple existence here, in such strength, has consistently deterred any potential enemy—North Korea and China being the most likely candidates—from behaving in an unfriendly manner. And the corollary is the message that is spelled out repeatedly by the Hawaii navy headquarters: that the prosperity of Japan, and of all the allied nations nearby, had been won and sustained and underwritten and secured, in large and undeniable part, by the presence of this protective ring of battleship gray steel provided in places like this by America.

So the cargo ships eased their way back and forth through the spume and spray, unmoved by the gathering swell. The skyscrapers of Tokyo rose to the distant north; the towers of Yokohama stood directly ahead, across the miles of sea; the smokestacks of Chiba and Kawasaki and Setagaya and Ichihara ranged all around—and all this immense enginework of commerce and prosperity was protected, so all are led to believe, by the comforting presence of the slim gray American warships that regularly emerged from their lairs, stealthy and unannounced, and sailed off to exercise, to remind, and to warn.

I came back to Tokyo Bay almost thirty-five years later, in the summer of 2014, and this time I came by passenger ship. We were inbound from Russia and the Kuril Islands, and we had spent two days winding our way steadily south along the coast of Honshu. Our captain, in a fit of misplaced anxiety, gave the irradiated shoreline near the nuclear power plants at Fukushima a wide berth. He later said that an American liability lawyer aboard had warned him not to go too close, since any subsequent illness among the passengers could be technically ascribed to radiation, however unlikely. "You know lawyers," the man remarked, sardonically.

We hove to beneath the slopes of Mount Daisen and boarded the channel pilot, who looked far too young to take command of so vast a tonnage as ours. The moment we turned to starboard, and headed north toward the narrows and into Tokyo Bay, the radar on the bridge came alive with contacts, hundreds of them, ships crowding into the bay heading this way and that and at a variety of speeds: lumbering freighters, tiny gnatlike hydrofoil ferries, spinnakered yachts out for a dangerous day sail, crude-oil carriers and ore ships and a few inbound Chinese, Korean, and Evergreen container ships (the last from Taiwan) pressing against the ebbing tide.

We got into line ahead with a few other vessels and passed the buoys into the fairway, then settled down at the regulated maximum of twelve knots, heading toward the city. There are same-way separation lanes in the Tokyo Bay fairway, so that we, a relatively fast ship, could overtake the more ponderous or the small or the underpowered. We soon found ourselves passing, rather too close for comfort, a flotilla of liquefied natural gas carriers, each with her three spherical tanks filled with highly flammable gases won from the fields off the Russian coast, north of Hokkaido.

But in the outbound lane that hot morning—with the air too thick to see Fuji this time, the waveless waters still and greasy-looking, and Tokyo a blur through the hot brown haze—there was precious little traffic. There were ships, to be sure, but in the two hours it took us to weave our way to the turning basin and the dockside, no container ships at all. The only container vessels were inbound, and they peeled off for Yokohama.

The statistics tell the story. Tokyo has been falling from grace as a container cargo export hub for many years. At the last count, the Japanese capital was thirty-second in the world table, just behind Colombo, just ahead of Mumbai. Yokohama was the only other port listed in the top fifty, ten places below. And a further statistic applies, giving some clue to the realization that the only container vessels I saw that summer's day were inbound, bringing imports to Japan, and not exports bound for the rest of the world: most probably the imports were televisions and laptop computers, because, incredibly, Japan in 2014 became a net importer of such devices, forty years after essentially creating the industry and then dominating the field.

Most probably these devices had come from China, because China has taken over most of the industries that Japan once dominated. First this dominance traveled a trifle westward, to Korea, jumping—with many Japanese engineers granted work visas to help—the sea between, on whose name neither Japan nor Korea has been able to agree for the past half century. (Korea wants the East Sea; Japan, the Sea of Japan. Neither side will budge.) Then the production of consumer electronics fled farther eastward, still across the unarguably named East China Sea, and to factories that are dotted along the eastern Chinese coast between Dalian and Hong Kong. With the result being that the largest container ports in the world, outstripping the export figures of Japanese ports by almost an order of magnitude, are now Chinese.

Shanghai is by far the largest, eight times the capacity of Tokyo. Seven of the world's top ten container ports are currently in China. And all the Chinese products in these great shipping boxes—aside from a small number that, like those piled high on the vessel we overtook on its slow passage into Tokyo Bay—are heading out daily, almost hourly, across the Pacific Ocean. The transoceanic momentum, a phenomenon started by the products of Masaru Ibuka's inventive mind, is still very much in process. But the axes through which it gathers speed today are pinioned in China and the United States, and are run between ports such as Shanghai and San Francisco, Hong Kong and Los Angeles, Shenzhen and Seattle.

Hand in hand with this evolution, a change has occurred that is potent with irony. In the days when America did so much business with Japan, it was trading with a nation that had been an enemy but had later turned into an ally. With China, it is the reverse: the nation with whom America once dealt as an ally is now considered a potential foe, a country that is presently regarded with varying degrees of suspicion, wariness, fear, and alarm.

This could explain why, as we passed Yokosuka Naval Base off our port side, we didn't see a single American warship. There was a lone destroyer far out at sea, but it turned out to be Japanese, part of Japan's own newly revitalized self-defense forces. The American vessels (the carrier group of the Seventh Fleet and a host of other ancillary warships) were all far from home that day.

Where? Though the admirals seldom tell, it is most probable that on that summer's day they were patrolling as near to the coast of China as they comfortably could, keeping watch.

The U.S. Navy is still paying close attention to a part of the world that evolved so radically and quickly into an economic powerhouse—an evolution that was set off when Masaru Ibuka began selling his tiny radio set in August 1955. But the eyes of the U.S. military are no longer so firmly fixed as they once were, on Japan and Korea, on the defense of Taiwan and the Philippines. Instead they are now focused, near-obsessively, on China.

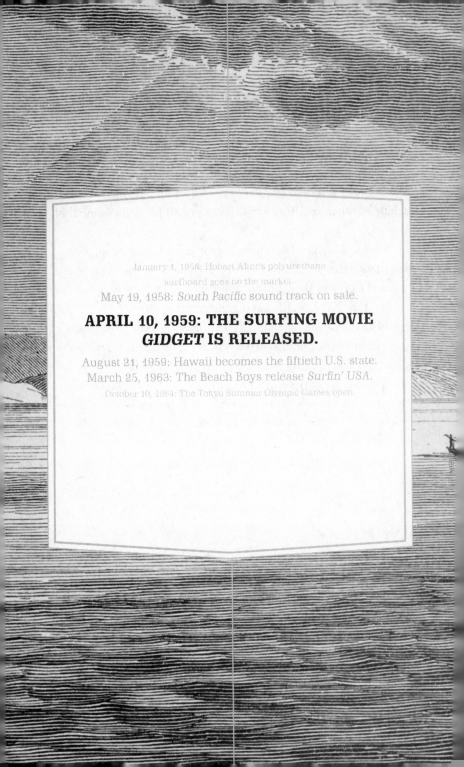

January 1, 1958: Hobart Alter's polyurethane
surfboard goes on the market.

May 19, 1958: *South Pacific* sound track on sale.

APRIL 10, 1959: THE SURFING MOVIE *GIDGET* IS RELEASED.

August 21, 1959: Hawaii becomes the fiftieth U.S. state.

March 25, 1963: The Beach Boys release *Surfin' USA*.

October 10, 1964: The Tokyo Summer Olympic Games open.

Chapter 3

THE ECSTASIES OF
WAVE RIDING

He is a Mercury, a brown Mercury. His heels are winged, and in them is the swiftness of the sea.
—Jack London, "A Royal Sport: Riding the South Sea Surf," in *Woman's Home Companion*, 1907

The Pacific is a liquid place, and on most of its inhabited coastlines this liquid is warm and ultramarine and inviting. It is also by its very nature ceaselessly in motion. For centuries native peoples who lived on many of the islands of the ocean's tropical interior have made great use of all this motion in ways that provided them with the purest joy imaginable. They rode out on long wooden boards through the beachside surf and spume and waited, floating, for a wave to lumber in from the ocean, and then stood up on the boards, toes gripping the leading edge, and from the wave's summit crest, rode the boards down its steep green face, all the way back into shore.

This wave riding, as it was initially called, seemed to be practiced only on the Pacific. On other oceans, the shore dwellers merely swam, or dived, or else ventured out in boats, and very little more.

At the end of the nineteenth century, European travelers came to the Pacific, most notably to the islands of Hawaii, and many of them observed, mostly with amazement, some few with shocked horror, this strange and unfamiliar kind of ocean-derived pleasure. A few enthusiasts then attempted to emulate what they had seen, and dabbled: some made crude facsimiles of the boards and tried to do what the islanders did, on wave-battered beaches in California, in Australia, in South Africa.

For half a century, this wave-riding, wave-gliding phenomenon was no more than an arcane pursuit, a mysteriously joyous pastime pursued only by a *corps d'élite* of a very few. But in the late

The original Gidget was Kathy Kohner, whose teenage exploits in Malibu prompted her father to write his blockbuster novel. Sandra Dee and Sally Field starred in screen adaptations, spawning huge worldwide interest in surfing.

spring of 1959, everything changed. The world beyond became fully aware of this curious ocean-side magic—and it did so because of a cheerfully unexceptional Hollywood movie named *Gidget*.

Few imagined that this little film (based on a true story of a youngster named Kathy Kohner, and a novel written by her father) had much of a future. This lack of confidence on the part of its makers likely explains why Columbia Pictures didn't open it in any major American cities, but only in their suburbs. It was released during the cool prelude to the last summer of the decade.

There it might well have languished, largely unviewed and entirely unremembered—except that Howard Thompson, then film critic at the *New York Times*, was savvy enough and curious enough to venture out to see it. On April 23, 1959, in the Thursday edition of the newspaper read then as now by most East Coast Americans who mattered, there appeared one of his characteristically snappy and quick-fire top-of-the-page reviews—and a positive one, to boot.

To the surprise and undoubted relief of all involved, Thompson very much liked what he had seen. Moreover, with what passed for wild enthusiasm in the generally sobersided writings of the *Times* of the day, he declared that *Gidget* was, in his view, really quite enough "to make anybody light out for Long Island Sound." Go out there, he was urging his readers—stimulated by the film to learn of something quite new that was, in his view, "the ideal way to usher in the beach season"—get out of the sweltering city and to the ocean!

The movie that so excited Thompson was unashamedly slight; this was no *Citizen Kane*, to be sure. To compensate, it was highly color-saturated, and shot for grand effect in CinemaScope. It was frothy and faintly sexy, a love story of sorts, with a defined plot and a firmly fifties moral stance. Its cast was made up of handsome young men and women in swimming costumes—worn as brief as was allowed at the time, with female navels concealed to the point of nonexistence. Almost all in the cast were under thirty. In the title role was a pert and wholesome teenage newcomer named Sandra Dee.

Gidget's limited artistic merits are of little matter, since today's historians of the field regard the film as having an impact well beyond the merely cinematic. It is, by their general agreement, the single greatest influence on introducing surfing, the Pacific Ocean's most sublime and lasting gift to playtime, into the mainstream of life in America.

So the little shrimp of a girl, a pint-size emblem of determination, courage, and adventure, was seen by millions as she learned the sport. They watched as Gidget paddled out to sea on a surfboard as large as she was and then knelt on the slick flatness of its upper surface. After this, the audiences learned as Gidget learned: how to find an incoming wave and recognize it for what it offered; how to stand up on the board and, probably as with all beginners, wobble and hesitate and almost fall; but then, in one moment, how to keep standing and to balance upright; and then how to lean and steer, with arms spread-eagle like balancing wings, and then be blissfully captured by and swept up into the previously invisible onrush from behind. Then there were the moments when the board and its tiny rider began to gather and rise with the wave's own rising waters, and then to slip onto and over the lip curl of its breaking top*—and the moment when the board and rider started to slide downward and forward with the wave as if they were becoming a part of it, heading toward the shore edge at an ever-

---

* A wave *breaks* when the depth of the shore water is less than one-seventh of the distance between adjacent wave crests. The drag at the base of the column of water slows this lower part, leaving the top of the wave speeding along—but unsupported from below. So it *breaks*, with the onrushing top curling over and beginning to fall down the wave's leading face. The white mess into which all this eventually disintegrates is the *surf*—from the Anglo-Saxon term *suff*, indicating the inrush of water toward the shore.

increasing speed and with a swelling, joy-filled grace quite
unimaginable in all prior experience, before finally crashing
exhilarated into the ragged mass of white foam at the wave's
end.

Of course, this is not the end. Instead, Gidget recovers,
catches her breath, turns back, and paddles happily out into
the water once more. There she and other like-minded devotees
wait, bobbing idle and cold on the sea surface, gazing to the near
horizon for that next curling green uplift of water, for the next
wave that might allow them to do it all over again, only better.
For next time—and there will be a score of next times in the film
as in the life of this particular calling—promises a new wave with
still more grace and power than before, which they can ride in
and slide in faster than before.

Thus was seen here in this little movie, with its lithe little
heroine, the birth of a young girl's love affair with pleasure and
purity and the possibility of serene wave-riding grace. In the
weeks that followed the film's release, the thousands and then
millions who saw it were inspired to do the very same, to ride
and slide the waves with Gidget, to adopt a sport that seemed so
uncomplicated and joyous and so very *democratic*—perhaps this
last being the greatest joy of all, because the water and the waves
were always free, were always there, and were available to all who
ever went in search of them.

If Thor Heyerdahl was right, and if his famous voyage in the deliberately unsteerable balsa wood raft *Kon-Tiki*[*] ever had serious

---

[*] Heyerdahl wanted to show that Polynesia could have been settled by South American
boatmen who drifted with the currents, and that the cultural basis for the Pacific islands
is thus all incontrovertibly Incan. Later research showed that Polynesians knew very well
how to navigate without instruments, and had long sailed the often considerable distances
between the ocean's islands. DNA results disprove Heyerdahl's theories, and show that
Polynesia was settled from the west, from Asia. His 1947 expedition is seen now as little
more than an amusing, though courageous, stunt.

scientific meaning, it might be possible to believe that surfing began in Peru, and that it spread west across the Pacific from there.

Surfboard-like creations are to be found in Peru, for sure. When the anchovy fishermen of the pretty little coastal town of Huanchaco venture out each misty morning to examine their nets, they ride through the surf not in boats, but by sitting astride small and very graceful one-man rafts called *caballitos de totora* and paddling with oars of bamboo. Their craft (which in a dim light and with a powerful imagination resemble the bulkier old-fashioned surfboards) are hand-built, made by lashing together hundreds of bundled bulrush reeds. Locals like to say that these light, buoyant, and rather awkwardly mobile little craft have been built in Huanchaco for more than three thousand years. If so, they are fully a part of Incan culture, and are almost as ancient as the strikingly similar papyrus reed rafts that were once sailed on the ancient Nile.

But just as Heyerdahl's belief that the Pacific was settled from the east was eventually debunked, so the Peruvians' claim to being the originators of surfing has also been dismissed. Genetics and linguistics did in both theories, showing that the Polynesians were born, essentially, from an original human stock in what is now Taiwan, coming eastward to their islands by way of the Bismarck Archipelago in Papua New Guinea; and that Peruvian Incas have no proven genetic or cultural connections with Polynesia, and are descendants of those Native Americans who came from the Arctic north, across the Bering Land Bridge.

This is a shame for Peru's impoverished anchovy fishermen, who had long clung to this claim, their one amulet of worth. The anchovy men are in any case a vanishing breed, as their skills and their will to continue an economically unrewarding pursuit are fast evaporating with time. A poignant sight to be seen often these days on cold Huanchaco mornings is that of an elderly fisherman making his way back from his net inspection,

gamely half-standing on his *caballito* as it lumbers home slug-
gishly through the waves. Beside him, and zipping past at great
speed, are groups of young men, out playing before school starts,
weaving and curling through the waves on proper modern polyurethane surfboards, taking advantage of the waves in a manner
that the old *caballitero* beside them never can achieve, nor probably ever could.

No, not Peruvian. The surfboard and the craft of surfing are, by
all modern accounts, purely Polynesian inventions, and surfing
is a sport born in Tahiti and the islands of southern Polynesia—
perhaps as far south as New Zealand. The written record is in-
complete; all certainty is fragile. Given that caveat, "wave riding,"
which is the literal translation of a pan-Polynesian phrase *he'e
nalu*, is said to have begun as a children's pastime, maybe as long
as four thousand years ago. It became more of a pan-Pacific phenomenon only when settlers moved north to the Hawaiian Islands perhaps twelve hundred years ago. It was probably one of
the many skills brought in by these newcomers, along with all the
food and utensils and other necessities of colonization brought
when they navigated their way across the ocean in the great long-
sea canoes in which they explored its outer reaches.

When surfing reached Hawaii, it flourished, and fast became
a popular obsession. The combination of latitude, weather, and
topography made these islands a natural place for the calling to
mature—and to become, as it soon did, the island's principal recreation. The northeast trade winds that blow at a steady fifteen
knots onto the cliffs and reefs of the islands' lee shores produce
endless trains of eminently glidable waves. Depending on the
scale of the barrier reefs; on the depths of shoreline sea; on the
presence of rocks, spits, and headlands—all these by-products
of the geology of the hot-spot volcanoes that first formed the
islands—the Hawaiian waves can be either grand or modest,
challenging or lethal. They can present long sweeping swells or
short, sharp, massive ship-destroying monsters. They can form

high walls that break with curling overtops; or they can present themselves as long hollow tubes of water, roofed with millions of tons of green and translucent sea. In the season of storms (January and February, in particular) they roar ceaselessly in line toward the shore before breaking in vast explosions of foam and spume. This is the liquid Pacific personified.

To all precontact Hawaiians, a people pious and shamanistic by nature, such wildly roiling seas were the playgrounds of the gods—but gods with whom, if conditions were right, the seas also could be shared and enjoyed by those humans who lived under their divine care. The actual process of surf sharing, however, was intricate and scrupulously regulated.

This being Hawaii, a place that in antiquity was bound by ceremony and rigid class distinctions, a complex surfing hierarchy was soon established. There were strict rituals relating to the selection and felling of board-suitable trees—a red kumu fish was offered to the spirit gods once the timber had been hauled away—and then other rituals concerned with the making, blessing, and storage of the boards, and with precise directives for the rubbing of coconut oil and the wrapping of tapa cloth. The methods of surfing—standing up on the liquid or lying prone as it swept inshore, surfing with clothes on or not, surfing silently or reciting chants and letting out exultant cries—all these were subject to age-old codes and precedents. And social hierarchies dictated the use of different kinds and sizes of boards, and by different kinds and classes of users; they also led to beach restrictions familiar as a kind of apartheid, with this section of a surfer beach reserved for commoners, but the best of the wave-rich breaks reserved for the ruling aristocrats.

These last, often gigantically fleshy men whose physiques had been stoked on the consumption of the well-pounded taro paste colloquially called three-finger poi, traditionally surfed quite naked, and used the longest of all surfboards, known as the *olo*. Nineteenth-century British explorer Isabella Bird, that most epic Victorian wanderer, passed through Hawaii: "I saw fat men with their hair streaked grey riding with as much enjoyment as if they were in their first youth."

Some of these long boards were improbably twenty feet tall and two feet wide, and though made of the lightweight wiliwili wood normally used for canoes, they still weighed as much as a good-size man. On these absurdly large planks, the Hawaiian royals—fat maybe, but strong as oxen, their physiques more sumo-chubby than beer-blubbery—would sluice through the surf as if shot on cannonballs. Their boards were keelless, and so impossible to steer; this made the ride all the more exciting and dangerous. Today, ceremonial *olo* are much prized and copied. There is a trade, akin to that in antique Hawaiian shirts, with five-figure prices happily paid by true believers.

The Hawaiian commoners—and surfing involved everyone, "farmers, warriors, weavers, healers, fishermen, children, grandparents, chiefs and regents," according to one historian of the sport—would engage in their passion on more manageable but rather less magisterial boards. These were the so-called *alaias*, ten feet long and made of the wood of the breadfruit tree; most modern boards are sized and styled after these. Children used tiny round-nosed boards known as *paipos*, which were often a little longer than the child, and which provided space for hordes of them to scoot around on the water with enviable speed and nimble-footedness.

The swashbuckling adventure writer Jack London first encountered a swarm of these peskily cheerful Hawaiian surfer children in Waikiki on a hot summer's day in 1907. He was swimming along in shallow water, blissfully minding his own business, when he was suddenly overtaken by an infestation of youngsters streaming past him on their boards. He was astonished, and on reaching the beach, London persuaded one of the boys to let him try his *paipo*. The result was less than distinguished, no matter how many attempts he made. "We would all leap on our boards in front of a good breaker. Away our feet would churn like the stern wheels of a river steamer, and away the little rascals would scoot while I remained in disgrace behind."

London's main legacy is his literature, of course: *White Fang*, *The Call of the Wild*, and dozens of short stories, with "To Build a Fire" eminent among them. But his discovery of the pleasures of surfing, and his subsequent preaching of its glories and virtues, exerts a powerful pull still today, more than a century later. London, with the help of two other formidable, half-forgotten figures he met in Hawaii, first managed by his writings and his enthusiasm to resurrect this most kingly of Hawaiian sports from the threat of near extinction.

When Jack London arrived on his forty-five-foot ketch, the *Snark*, two fast-metastasizing assaults were being leveled against surfing. First, diseases brought into Hawaii by foreigners had reduced the islands' population in the nineteenth century to a mere tenth of what it had been in the days before Cook's first encounters in the eighteenth; and such islanders as remained had more on their minds than beachfront frolicking. Second, Protestant missionaries, whose pious hypocrisies condemned any public displays of the human body, had compelled previously contentedly near-naked islanders to don all-enveloping Mother Hubbard modesty dresses, to stop performing hula dances and rituals, and to maintain a dignified and sackcloth reverence while bathing in the sea. This had forced surfing deep into the shadows. London had to confront a century of views like those that had been expressed years before by the preacher-father of modern Hawaii, Hiram Bingham,* a man who was convinced he was bringing Christian civilization to an utterly savage people.

---

* The first of three, whose grandson, Hiram Bingham III, was an early claimant to finding the ruins of the Incan citadel of Machu Picchu, in the Peruvian Andes.

"Can these be human beings?" Bingham infamously asked on his first arrival, as his church-sponsored boat was mobbed by swimmers and surfers and men in canoes. "The appearance of destitution, degradation and barbarism, among the chattering and almost naked savages was appalling. Some of our number, with gushing tears, turned away from the spectacle." To Calvinists, surfing was a sinful exercise, leading only to unbridled licentiousness and godless impiety. Go surfing, they pronounced from their pulpits, and eternal flames awaited.

But Jack London saw firsthand the unbridled delight of those *kanaka*—"native locals," the word being used also for all Polynesians—who still embraced the surfing life. He realized how, as one Los Angeles newspaper had it, life in the rainbow days on the Hawaiian seas "seemed a-flutter with enjoyment." So he vowed to learn how to become a surfer himself. "The *Snark* shall not sail from Honolulu until I, too, wing my heels with the swiftness of the sea." He was quite determined: he would win any future races with those sea-skimming jackanapes who had so bested him on their wretched *paipos*, and then, if he became halfway competent, he would write about it.

Two remarkable men bent him to the task.

The first was the man who taught him: a moneyed flâneur named Alexander Hume Ford, a Honolulu social butterfly and promoter, the son of a Dixie plantation owner who had drifted into Hawaii society on his way to Asia, had discovered surfing, discovered he was good at it, and had stayed, a confirmed addict at the age of thirty-nine. Ford's mission was unashamedly one of public relations: he wanted to sell the idea of Hawaii to the public, and surfing was the vehicle by which he would achieve it. Jack London's wife, Charmian, said of Ford that "he swears he is going to make this island's pastime one of the most popular in the world."

Ford was no sun-browned hulk, in no sense the surfing archetype. He was small and slight, bespectacled and donnish, with a tiny goatee and a vegetative mustache, and pale and perpetually sun-afflicted skin. But his own experience in learning the craft of high-speed surf riding (which he would pass on to Jack London, and so to millions of London's readers in the outside world) speaks to the sport's most significant aspect in those heady prewar days: it was poised to cross the color line.

For, up until around 1906, surfing had been an almost wholly Polynesian craft; the skill was one believed to be peculiar to what were then known as the Pacific races, and so a skill utterly unattainable by whites. Mark Twain had said so when he visited Hawaii in 1866, declaring that "none but the natives ever master the art." Yet Ford, though a prime-type specimen of frail white gentility, had indeed managed to master it. He spent four hours daily out in the waves, waiting for the breaks, waiting for his opportunity to try to stand up on the board, to try to get all the way into shore. He did this every single day for three long months until, finally, he mastered it. And he did so with the help of the second of the team who would eventually propel London on his way to become surfing's first great champion.

This was George Freeth, the half-Hawaiian grandson of an Irish* shipowner. A swimmer and Waikiki lifeguard, Freeth is thought by many to be the true godfather of the art. He was unusually adept at surfing standing up, for one thing. Vertical wave riding was a skill that barely survived the late nineteenth-century decline in surfing, when it fell victim to the twin scourges of fanaticism and fever: those few Hawaiians who still surfed employed a technique that involved merely speeding inward on their boards while lazily lying prone on top. The erotic, arching elegance of a standing surfer that had so entranced and shocked

---

* Ireland has tried hard, though with scant success, to claim George Freeth. A film, *Waveriders*, was made in 2008, to press the case; in the end it was the great waves off Ireland's west coast, most notably at Mullaghmore in County Sligo, that caught most viewers' attention, and helped place Ireland prominently among the world's great surfing centers.

the first visitors a century before had now all but disappeared—until Freeth revived it, astonishing all who witnessed him. "I saw him tearing in," wrote London, "standing upright on his board, carelessly poised, a young god bronzed with sunburn." Young indeed—Freeth was just nineteen years old, champion of all islanders.

So Freeth taught Ford, Ford taught London, and London made the first attempt to teach the world. And surfing's transit across the color line can be seen vividly in this trinity's progression, because Freeth was bronzed and half-Hawaiian, Ford was white and wellborn, and London was a caricature of a white man's man—to such a degree that many think of him today as a casual racist, maybe even a deliberate one. "The White Man must be rescued," he once wrote, demanding that a white heavyweight boxer come out of retirement to defeat the black man who then held the title, which London saw as an affront to his kind.

London wrote about the surfing lesson that finally snared him. "Ah, delicious moment when I first felt that breaker grip and fling me. On I dashed, a hundred and fifty feet, and subsided with the breaker on the sand. From that moment, I was lost."

He wrote his essay for the October 1907 edition of the *Woman's Home Companion*—then a vastly popular monthly, which had agreed to take all of London's dispatches from the *Snark*. It appeared under the heading "A Royal Sport: Riding the South Sea Surf." The essay created enough of a stir for it to experience the Edwardian equivalent of going viral, appearing in slightly amended form and titled "Joys of the Surf Rider" in England's equally popular *Pall Mall Magazine*. Eventually a fleshed-out version appeared in London's long-planned book *The Cruise of the Snark*, giving unprecedented book-length credibility to a sport that had hitherto been little more than marginal. Now it was potentially everybody's—and lest there be any doubt about where London stood on the racial spectrum, "[W]hat a sport," he declared it, "for white men."

Freeth was the "brown Mercury" of this chapter's epigraph, the inspirational vision that set London's mind racing, and his pen dipped in its customary purple. Freeth was properly flattered by all the oily references to him in the *Woman's Home Companion* article, and asked London to write him a letter of introduction, so that he might try to peddle his wares (only his surfing skills were of value) on the American mainland. London obliged, and Freeth promptly took off for California, with a one-way ticket to plant a seed in what would turn out to be the most fertile of soils.

Los Angeles in 1907 was a modest city of some 275,000 people, three-quarters of them newcomers. It was a place of eccentricity and abundance, a place of half-made mansions and modest migrant houses, of churches catering to a bewildering mix of beliefs, of sanatoriums and asylums, of endless orange groves, of forests of derricks and drilling rigs from a just-beginning oil boom, of small factories and stores, of a fledgling movie industry, and with a vastness of beaches that were little more than wind-swept and half-deserted dunes. Local residents didn't go to the beaches. For recreation, they preferred to head inland; a popular amusement was the potting of rabbits from streetcar windows. Few took much interest in the pleasures of the sea, which was seen as a wild and dangerous place, with enormous waves that could break a man to pieces. Moreover, there was no easy way to get to those dunes, nowhere in beach country where one might stay, no base for the beginning of any kind of beach culture.

But there was money about, and there were local moguls—Henry Huntington, who ran streetcar lines; and Abbott Kinney, who made a fortune from his vast tobacco company*—who decided they could manufacture a beach culture. So they concocted plans to develop separate and competing portions of the coast-

---

* Sweet Caporal was Kinney's best-known brand, with his factory in New York's Chelsea making eighteen million a week until 1892, when a fire started by a suspended gasolier destroyed the giant building and its entire stock. The formidably wealthy Kinney was asthmatic, and eventually left for Southern California and clearer air.

line, each tycoon vying for the leisured pursuits of the new leisured classes.

The undeveloped and untamed coast in question ran for
twenty miles, a pristine stretch of sunset-facing Pacific ocean-
front between the massif of the Santa Monica Mountains to the
north and the old displaced schists of the crags of the Palos Verdes
Peninsula to the south. Kinney claimed beaches up north, and
at a site two miles from the small city of Santa Monica, he constructed a bizarre and very costly network of canals. There were
six of them, gaslit, plied by gondoliers, and with arch-bridged
stretches of freshwater reservoirs that ran along the shoreline,
just in from the sea. With more imagination than commercial
good sense, perhaps, Kinney placed antique-looking stores
along them and then named his creation for the Venice that he
had so admired in Italy. The canals were eventually filled in and
replaced by roads, but Venice Beach remains.

Henry Huntington, nephew of one of the backers of the trans-
continental railroad, who was in amiable competition with this
same uncle for the laying of rail links within Los Angeles, built
his coastal experiment rather more modestly than Kinney, and
well to the south. He arranged, presciently, that one of the lines
for the red-enameled streetcars of his Pacific Electric railway
company would terminate at his stretch of beachfront, bringing customers out of the city and to the ocean. And to give
them a destination, a place to while away the hours, he began
to build an immense hotel, with a saltwater plunge pool and a
Moroccan-themed ballroom and innumerable restaurants. He
named it for the previously little-known mile of coast that he
had purchased: the Redondo Hotel, in Redondo Beach.

City dwellers would come, in steadily swelling numbers,
either to sit and take their ease beside the big Moorish arched
windows of Huntington's Redondo, or else at window-side tables
in the chic faux-Italian espresso houses of Kinney's Venice.
Their eyes would invariably be drawn westward, and they would

gaze, mesmerized by the view and the colors, out at the ceaseless rolling thunder of the wild Pacific waves.

Until one day, in the late summer of 1907, there appeared in the waves off Venice Beach the figure of a man, who was doing what no rational Angeleno thought possible: *he was walking on water.*

Or at least, he *seemed* to be walking, until the waves brought him in and crashed him down onto the sand directly in front of the astonished watchers—whereupon this vision of muscular manhood stepped lightly off the eight-foot-long board that had seemingly carried him, standing, all the way in from the breaking surf. He gave a cheery wave to the watchers, picked up his redwood board, walked it back out into the sea, leaped aboard, paddled it right out toward the horizon, to where the inrushing waves originated, and then stood and seemingly walked back in again on top of one of them, at great speed and with unimaginable grace and elegance; and after landing, he went back and did it again.

This man was Freeth, the bronzed Mercury of Waikiki, putting on a show, a show that, just by chance, a local reporter happened to witness. On July 22, 1907, Henry Huntington read the resulting essay, "Surf Riders Have Drawn Attention," and in an instant of commercial genius, he realized how best to promote his hotel, and have visitors flocking to Redondo Beach by rail from downtown.

By the end of the year, Freeth found himself an official employee of the Pacific Electric railway, charged with a duty that would be the envy of all. Twice each weekend day, at 2:00 p.m. and again at 4:00 p.m., clad in a tight green woolen singlet and close-fitting green woolen shorts, he would paddle his board out to the surf line and, on cue to an announcement boomed from a Redondo Hotel megaphone, catch a wave, stand, and then soar effortlessly back onto the beach.

For the surfers of the American mainland, Freeth is the

sport's acknowledged godfather; yet he is now almost forgotten. He was a quiet, patient, rather solemn and unremarkable man—in pictures, suffused with an air of melancholy, never quite the bronzed Mercury of Jack London's overheated first essay. He taught surfing skills to local youngsters, and he demonstrated the sport to thousands. He worked as a lifeguard, drove a motorcycle to the more distant rescues, and invented the paddleboard as a means of quickly reaching the distressed and the drowning.

He lived on his own, frugally, quite without friends, and he never married. He died, alone, in the influenza epidemic of 1918–19. He was thirty-five years old. A bronze bust of him was erected as a memorial on the Redondo Beach boardwalk, with his back to the sea, his gaze directed toward a multistory parking garage. The plaque reads, THE FIRST SURFER IN THE UNITED STATES. (The original one was stolen, but it was replaced.) Few beyond the beach know his name. He is probably better known in Ireland.

One final figure then set the capstone on what would come to be coastal America's new obsession—but this man was charismatic and noble where Freeth was sober and unprepossessing. Moreover, this man was truly bronzed, because he was truly Hawaiian, albeit with a first name that was very much of the white man's world. He was Duke Kahanamoku, a swimmer to beat all and a surfer to crown all, and if not the father of surfing, then the first true surfing icon, its greatest of ambassadors, to America and beyond.

His father was a clerk with the Honolulu police who had also been named Duke, in honor of the British naval officer and son of Queen Victoria the Duke of Edinburgh.* He passed on the name to

---

* Prince Alfred was given the resurrected title to the dukedom of Edinburgh by his mother, Queen Victoria. His life was colorful: on the same transpacific journey that brought him to Hawaii, he was shot in the back in Sydney by a would-be assassin (who was hanged), and then one of his own sons shot himself during the duke's twenty-fifth wedding anniversary

his oldest child of nine, who promptly rose to fill the ducal image by being regally tall, utterly charming, and eternally polite, with good looks that drew crowds of admirers—"the most magnificent human male God ever put on this earth," according to one of his fans. He dropped out of school to become a creature of the beach, endlessly swimming, playing beach volleyball and water polo, and taking surfing to a new level of graceful competence.

The Hawaiian Olympic swimming champion Duke Kahanamoku, named by his policeman father for the Duke of Edinburgh, was the first true ambassador of surfing, touring the world to give exhibitions of graceful wave riding.

Kahanamoku's swimming first took him around the world. He won Olympic medals: a gold and a silver in 1912 in Stock-

celebrations. The tiny capital of Tristan da Cunha, still a relic of British colonialism in the South Atlantic, is called Edinburgh, named after this most peripatetic of British royals.

holm, two golds in 1920 in Antwerp, and a silver in 1924 in Paris (where his younger brother Sam won the bronze, and the actor who would later play Tarzan, Johnny Weissmuller, won the gold). And if swimming took him places, then it was to these places that Duke Kahanamoku spread the message of surfing, far and wide.

On his way back from Stockholm, for instance, this gentle giant of a man gave demonstrations of wave riding in Atlantic City, in Rockaway Beach, and in Sea Gate, at the Manhattan end of Coney Island. He did the same in California, at Long Beach—which had already somewhat caught the craze, thanks to the efforts of Freeth a little way to the north—and then set out to come full circle back across the Pacific to Hawaii.

Two years later he was demonstrating his prowess in Australia. This world-famous Olympic swimmer, the best in the world in the 100-meter, showed off before thousands of swimming-obsessed Sydney-siders. He staged his performance at Freshwater Beach, displaying how he could ride effortlessly and for hours on a board that he had planed down from a hastily hewn slab of sugar pine bought from a local lumberyard. He had designed the massive sixteen-foot-long board to resemble the old *olo* once used by Hawaiian royalty; and since there was room for more than one on its aircraft carrier–scale flight deck, at one point he dazzled his Australian audience by taking along a fifteen-year-old schoolgirl named Isabel Letham, who stood just in front of him on the board as he veered and twisted and flew and spun around on the wave face, picking her up each time she tipped or stumbled, and never losing sight of bringing her safely back to shore. Miss Letham, who went on to teach swimming in northern California, lived until she was ninety-five years old, surfing well into old age and generally regarded as Australia's First Lady of the sport.

And by the end of the First World War it was indeed now becoming a proper, recognized sport. As it spread around the world—slowly, fitfully, initially in pockets and very much on the margins—it developed rules and standards, and there were contests and championships. Australia was the first country to hold regular competitions: the Surf Life Saving Association organized races, to see how fast a surfboard-mounted paddler could reach an imagined swimmer in difficulties—a fine demonstration of how a plank board could be used for the common good, if not perhaps quite the exercise in wave-bound joy that had been inspired in Hawaii and was now sweeping California.

The first proper contests were held in Southern California, in 1928—and these squadrons of once car-crazy youngsters, infused with the cash and good cheer that marked the beginnings of the Jazz Age, had scouted the entire coastline for the best possible wave breaks. Redondo Beach and Venice may have looked good in the early days, but now Corona del Mar, south of Los Angeles, on the road to San Diego and Mexico, was decreed the Valhalla of the moment; it was there, with the dropping of an "aerial bomb" firework to kick things off, that the Pacific Coast Surf Riding Championships got under way—with a "thrilling rough-water surf-board race" culminating the day's water sports and setting in train the whole complicated business of judging who might be the best of the surfers around.

Judging was never an easy business, and remains so. With the wind and water endlessly capricious, with differences of opinion over whether speed trumps style, whether grace trumps courage, the grading of surfing took many years to evolve. But contest rules were laid down, and a system of award points was established among those organizing in Corona del Mar. As the word spread, surfers from around the world arrived (from Peru, from France, from the Basque country) and, in later years, would transmute the Pacific coast of the United States into the premier international gathering place, the rule-setting headquarters, and the ultimate arbiter of a fast-growing pastime that was now on its way to becoming a very popular and profitable business.

At first the money was being made by the ancillaries—by the

owners of hot dog stands and the peddlers of beer, of suntan oil and beachwear. Then serious commerce started to take hold, and with it came commercial competition, and with that came the constant handmaiden of *technology*, the kind of technology that offered improvements to a surfer's speed, and ways of augmenting his water-borne grace and elegance—and thus his chances of winning.

As the century progressed, neither commerce nor technology ever lessened its grip. The core discoveries, which started to be made in the mid-1930s, had all to do with materials. How could a surfboard be made lighter, stronger, more flexible? Was wood—which floated, true, and thus had a natural advantage—the only suitable material? In 1935, *Popular Science* offered readers diagrams of how to build the ideal surfboard. It was made entirely of wood. It was eleven feet long, barely two feet at its widest point, with a core of lightweight balsa, ribs of spruce, and dowels and molding of redwood.

Refinements were added: different shapes for the nose and tail, a fin here, a keel there, convexities, hollows, concavities, taperings, crowns. Some boards were covered with elaborate marquetry, fancy mahogany on the outside, quite empty inside. Then came the breakthrough discovery of resins, and of fiberglass, and Bakelite. The traditional woods were replaced or augmented by different kinds of resin shells, and their noses were armored with lightweight fiber, to prevent cracking and bruising from the constant beach sand hammering of a typical summer's afternoon.

Girl boards—the term would never be used today—were made and became vastly popular in the town of Malibu, where girl surfing truly took off on these smaller, nippier, and very fast inventions. The emergence of young and highly competent female surfers (no longer just decorative figureheads for the boys' amusement) brought with it an erotically charged energy that enfolds Malibu still. Men claimed that competition with the crews of very good surfer women affected their own style, improved the water performance of all, and fueled the need for better and better equipment.

There was a certain 1950s fickleness to the sport, with sudden trends, sudden fashions. For a while everyone searched for boards of a down feather lightness, until technological advances allowed some to be made as light as twenty pounds, which resulted in some being tossed about on the sea like chaff, to their riders' evident peril. Then (in reaction, no doubt) there came a new trend for heavier boards, with more stability, and with greater heft and purchase in the ever bigger waves to which these Californians were now being increasingly drawn.

The quintessentially American desire for steadily more superlative experiences, in surfing as in just about everything, was fast pushing the sport to its limits—which is why technology was and still is so crucial. The languorous pleasure that had satisfied the Hawaiians for so many centuries was simply not good enough for mainland Americans—and whether that spoiled or enhanced the experience of surfing is a matter for debate to this day.

For Hobie Alter, there was no question. To this quiet and sober young Californian, the son of a wealthy orange farmer, improvement and innovation were everything. He had started his teenage surfing life as a shaper—the word has little meaning beyond surfing—working in a tiny beachfront store, financed by his father, sculpting lengths of balsa wood imported from Ecuador into configurations suitable for the waves and riders on and near Manhattan Beach. He set up a shingle: Hobie Surfboards. He set up his board-making business just a mile or so south of where George Freeth had displayed his skills half a century before. Now the beaches in front of his shop were thronged by surfers, and his shop, which this twenty-one-year-old ran with scrupulous care, did well.

Then, one day in 1957, a salesman called in with a small chunk of a newly invented by-product of the petroleum industry: polyurethane. German chemical plants had first created this light, almost woodlike plastic magic, which could be foamed into size, then sawed and cut and milled to the perfect shape—which Alter realized in an instant would be the undisputed future of surfboards. Together with one of his employee laminators, an engineer named George Clark, who was a late convert to surfing from a half-formed career in soldiering and in the oil fields (and who became so passionate about surfing that he would go days without bathing, hence his nickname, Grubby Clark), Alter started experimenting with foam.

Hobart "Hobie" Alter, creator of the inexpensive learner yacht the Hobie Cat, also invented the polyurethane surfboard, and along with his onetime friend George "Grubby" Clark turned surfing into an entirely affordable sport.

"It was dirty, messy and smelly, nothing you'd dream of doing for a career," Clark once said. In the early days, it involved mixing a dark and oily-looking substance, toluene diisocyanate, with a polyol and a blowing agent: within seconds the two combine to form a foam that expands to a frightening volume, up to twenty-five times that of the liquid components, and then sets, hard. Alter and Clark (Hobie and Grubby in the lore of surfing, since they promptly almost monopolized the business) took two years to work out how best to capture the foam inside a surfboard mold, and how to persuade it to set without trapped air bubbles or any other distortions. Their patience paid off, because in the winter of 1959, the company began producing polyurethane boards, placed a small advertisement in the *Los Angeles Times*, and began to reap the beginnings of a fortune.

That was a scant eight months after *Gidget* made the sport popular. Soon after, the five-man singing group the Beach Boys came along and eventually provided the anthems and the mood music; and then came more movies, Bruce Brown's 1960s paean, *The Endless Summer*, most justly famous of all. A craze that had been straining at the leash since the 1920s suddenly broke free: an industry was born, fortunes were gathered in, turf wars were begun and ended, and millions became involved in a sport that was uniquely American, but that, unlike baseball or gridiron football, had as its playground the entire coastal world, where warm waters and big waves and white beaches provided a sandpit of inexhaustible size.

I was in Hawaii in the early spring of 2014, when Hobie Alter died at the age of eighty. Hawaiian public radio stations offered what seemed like wall-to-wall encomiums—Alter also created skateboards and the famously inexpensive learner yacht, the Hobie Cat, so he was revered by many. By chance, I had the radio on while driving along the north shore of Oahu, along the stretch of coast south of Pupukea, the site of the Banzai Pipeline, one of the best-known wave breaks in the world.

On this blissfully warm afternoon, the swells were long and powerful, the tubes were rolling in over the three reefs, and under the press of the northeasterly trade winds, they were curling in from the right, a configuration that prime surfers say they like best. So the sea was dotted with the tiny black silhouettes of men and women floating, gazing, looking over their shoulders into the blue distance for the faraway sight of the oncoming waves. Leaving the engine running, I got out of the car so that I could hear the tributes to Hobie—the phrases "legend," the "shaper of a culture," references to the man who "never wanted to wear hard-soled shoes or work anywhere east of the Pacific Coast Highway," drifting from the speaker.

His legacy was on this great beach in front of me. His boards were there in abundance. Either they were standing up in the sand in multicolored ranks; or they were being brought in to shore under the arms of droves of exhausted, happy people who had just spent hours riding, floating, wave gliding; or they were still out on the ocean, either waiting with their riders aboard or else plunging, wheeling, soaring, gliding on the great green wave faces or through the Banzai tunnels that swept in, minute by minute, to the enthralled ecstasy of all the watchers and waiters in the sun. Few legacies can be wedded to so singular a pursuit of human happiness than this ancient Polynesian tradition, born of the Pacific, and now found everywhere, made available to so many by inventors such as Hobie Alter.

George Clark's legacy is the more enigmatic. He parted ways with Alter soon after the board company became a success, and set up his own company, Clark Foam, to make polyurethane surfboard blanks, to be shaped according to need. He proved a formidable businessman, and by the turn of the millennium had a hammerlock on the industry, controlling nine-tenths of the market for surfboard manufacture in America, 60 percent of the business across the world. And by 2005, twenty million people worldwide were defined as surfers. Clark and his firm were powerful beyond imagination, and he won a reputation for ruthlessness, and worse. He seldom spoke to the press, lived far away from the surfing scene in near-monastic seclusion. He was known for a vindictive streak, and a pitiless and ruthless attitude to competition, quite out of character with the genially relaxed sport that his ability and expertise had long allowed him to dominate. Unlike Hobie Alter, George Clark was decidedly not a well-loved figure.

Not least because of one moment in his career, when, on December 5, 2005, he suddenly shut down Clark Foam, destroyed all his blanks and concrete master molds, and locked away from prying eyes all his proprietary chemical formulas—effectively denying to everyone else the immense storehouse of technical knowledge and expertise he had accumulated during his four decades in the business.

The surfing industry was left aghast. There seemed no reason for the closure. Power-madness, some said. Plain meanness, suggested others. Clark himself sent a lengthy and barely coherent fax to his dealerships, suggesting that there was a lawsuit pending against him because of the suggested carcinogenic nature of one of the hardening agents used in his manufacturing process. But no suit was ever filed, and no known complaints of a chemical kind had ever been leveled against his firm.

An image remains of that day, long since known to surfers as "Blank Monday." Jeff Divine, one of the sport's best-regarded photographers, abandoned his customary universe of sunshine and blue waves for a day, and took off for the cloudy inland of the industrial world, where he found and captured an enduringly melancholy moment: it was of a man standing disconsolate in a concrete recycling yard, gazing down at a pile of broken and discarded surfboard molds, dumped there on Grubby Clark's orders, amid a tangle of rusting rebar and mountains of cement dust.

The industry took a deep breath, and wondered hard about its future. Why so many board makers had relied on one supplier remains a mystery. It took the industry months to recover, until others built new molds, learned the tricks of the trade, and the spigot was turned back on. Clark never properly explained why he'd acted as he had. He moved to an immense ranch in Oregon,

Mosaic Books

411 Bernard Ave, Kelowna, BC
250-763-4418 or 800-663-1225
Fax 250-763-5211 GST#115387854
orders@mosaicbooks.ca

Open 7 days a week, 363 days a year

Sun Feb11-18 2:43pm
Acct: 63863 Inv: C45655 P 00
Gwyn Prokop

Qty	Price Disc	Total Tax

978006231542R Pacific: Silicon Chips and
| 1 | 8.99 | 8.99 a |
| | Points | -8.99 |

| Items | 1 Total | 0.00 |

Total Discount Savings: $8.99

======== Frequent Buyer Status =========
Credit earned with this purchase $ 0.45
Total credit on your account $ 11.37
Credit is available for future purchases

Refunds are accepted 7 days after
purchase. Exchanges or store credit
are accepted 14 days after purchase
Special orders are non-returnable
Non-Book items in their original
package are valid for exchange only
www.mosaicbooks.ca

a reclusive octogenarian, the puzzling centerpiece of a swirl of surfing rumors.

Inevitably, one must suppose, sorry tales like this go hand in glove with the evolution of an industry—for *industry* is the word today, rather than *sport, pastime,* or *ancient Hawaiian tradition*—into such a monster of wealth and power. There are corporate giants who cater to surfing now; and television shows, and magazines, advertising and sponsorships, and all the other necessary evils of juggernautery have tinted and tainted a thing once so pure and innocent. George Freeth and Duke Kahanamoku and Alexander Hume Ford and even Jack London would hardly recognize and would scarcely welcome what has happened, I suspect.

Nor would Kathy Kohner, the five-foot, ninety-five-pound fifteen-year-old who traded stacks of homemade peanut butter sandwiches for a few moments' use of a Malibu surfboard—and became skilled enough that she would be the Gidget of a novel written by her father, and then in due course the Gidget made even more adorable by Sandra Dee, of the movie that a little-known Malibu director of war movies, Paul Wendkos, was persuaded to make.

The world of surfing they all knew bore little relation to the industry it had now become.

The ethos of the surfers themselves—the detached, otherworldly, laid-back, relaxed, pleasure-addicted world of sun, sea, sand, and endless pure enjoyment—has left its own legacy, and in unexpected places.

Industry, most particularly. A handful of company founders—most of their companies based on the American West Coast, and in places close enough to the Pacific wave breaks to allow employees to surf in their spare time—decided that it might be prudent to hire people who surfed, and then to accommodate the firm's working practices to those employees' peculiar needs, and hope for good and fair treatment by their workers in return.

A pioneer of such an idea was Yvon Chouinard, the son of a French Canadian plumber, who became an avowed environmentalist, bird-watcher, and climber—and surfer—and who decided in 1957 to teach himself blacksmithing so that he could make his own climbing equipment, since he found American-made pitons and crampons so wanting. A decade later he similarly found that American-made foul-weather clothing could not adequately protect him and his other wilderness lovers, so in 1970 he established a company, Patagonia, that he said would produce the kind of outdoor clothing he would use. He also declared that he would follow as ethical a path as possible. He explained to me how his own enthusiasm for surfing colored his business practice and his approach to his firm:

> Because of the unpredictability of weather and wave forecasting in the past, a serious surfer was required to choose a lifestyle and a job which allowed one to go surfing whenever the conditions were favorable. One could not plan a schedule to go surfing precisely at 3:00 p.m. next Thursday afternoon. Surfing really took off in the mid-fifties to mid-sixties, which I believe was the height of the fossil fuel "party." Gasoline was twenty-five cents a gallon, a used car could be bought for twenty dollars, and they were so simple any teenager could work on them. Camping was free, and you could always find a part-time job. There was a lot of fat on the land, so it was a perfect time for a counterculture to exist on the fringes of society.
>
> My Patagonia business philosophy is based on a principle which I might call "Let my people go surfing." I named it after our flex-time policy here at the company. None of us ever dreamed of working in a normal business environment, so we hired self-motivated and independent folks: surfers and climbers. We leave them to decide when and how to do their jobs. So far it has worked out very well.

More than a few West Coast companies have since followed suit. The idea of flextime, the informal approach to office wear, the provision of a working environment that one might imagine someone with a penchant for the outdoor life might actually *like*—such features have inspired enormous firms such as Google, Facebook, Twitter, and Apple to consider, even to appropriate, some of Patagonia's management techniques.

Such company practices may yet be spreading slowly elsewhere. For now, though, they are a principally Pacific phenomenon, born in part in tandem with this sport of the old Hawaiian kings, the Polynesians' great seaside calling, and alongside what one might these days fairly term the Pacific Ocean's greatest gift to the outside world beyond.

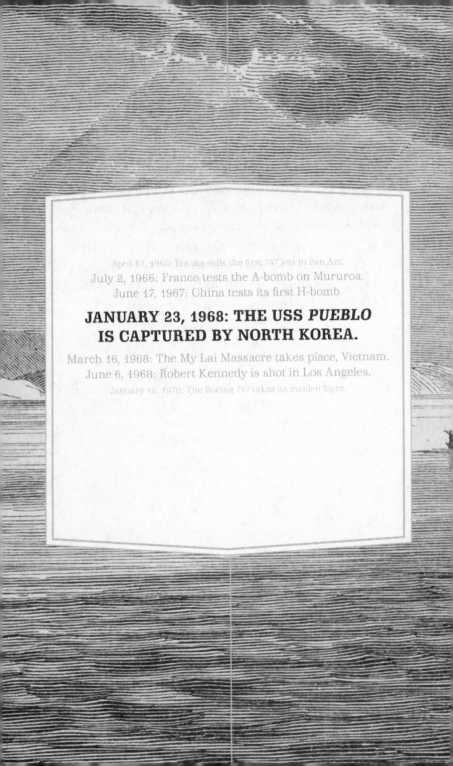

April 13, 1966: Boeing sells the first 747 jets to Pan Am.
July 2, 1966: France tests the A-bomb on Mururoa.
June 17, 1967: China tests its first H-bomb.

JANUARY 23, 1968: THE USS *PUEBLO* IS CAPTURED BY NORTH KOREA.

March 16, 1968: The My Lai Massacre takes place, Vietnam.
June 6, 1968: Robert Kennedy is shot in Los Angeles.
January 12, 1970: The Boeing 747 takes its maiden flight.

Chapter 4

A DIRE AND DANGEROUS
IRRITATION

O it's broken the lock and splintered the door,
O it's the gate where they're turning, turning;
Their boots are heavy on the floor
And their eyes are burning.
—W. H. AUDEN, "O WHAT IS THAT SOUND?"

One crisp, cold winter's day in 1968, men with burning eyes fired their heavy machine guns on a small, dilapidated, and barely armed American warship and managed, by doing so, to bring Washington swiftly to its knees.

The men were North Koreans, and the actions they commenced shortly after noon on Tuesday, January 23, culminated in the capture of the ship, a 335-ton army freighter turned float-

ing electronic eavesdropper named USS *Pueblo*. The ship's seizure resulted in one of the most calamitous intelligence debacles
the United States had ever suffered, and it was the first time an
American naval vessel had been seized on the high seas since
the British captured the frigate USS *President* off the coast of
New York City in 1815, more than a century and a half before. But
whereas the *President*'s crew was merely imprisoned in Bermuda
and then sent home, the *Pueblo*'s eighty-two surviving sailors
were starved and tortured for the better part of a year, until the
American government was compelled to make a craven apology
in order to have the half-broken men set free.

The humiliation caused by this incident was total, deliberate,
and public. It was also entirely in keeping with the long-standing
behavior patterns of the Democratic People's Republic of Korea
(North Korea, in its formal English-language rendering) toward
the United States and its allies in this corner of the western Pacific. The *Pueblo* incident was just one of hundreds of taunting
provocations, some of them lethal, some profoundly dangerous, others ludicrously extravagant, but most of them merely
annoying, that have taken place since the ill-tempered signing
of an armistice agreement marking the end of the Korean War's
formal hostilities in July 1953. No peace treaty has ever been enacted between the various sides in that brutal three-year brawl:
the 1953 agreement merely established rules meant to prevent
the resumption of armed hostilities. Considering that today the
principal parties have atomic weapons, these rules seem more
necessary than ever.

The roots of the nuisance—and although harsher words can
easily be used to describe the pariah state of North Korea, its
purely military significance in the Pacific theater, even with its
atomic warheads, is reckoned to be little more than a wretched
annoyance—lie deeper still than the 1953 armistice. And they
are entirely the result of a single hurried decision, by just one
man, with just one map, that brought about the division of the

Korean Peninsula. And both the man and the map were decidedly American.

The map was a large, wall-mounted, small-scale (552 miles
to the inch at the equator), National Geographic Society map of
the Pacific Ocean. It had been designed by the society's legendary
chief cartographer, Albert Bumstead, and its photo lettering was
designed by the equally famed typographer Charles Riddiford.
The map was pinned up in the Pentagon office of George Marshall, who was then the army chief of staff, and who had used it to
plan the movements of American forces in and around the ocean.

In Washington it was nighttime on August 14, 1945. Five days
before, the second atomic bomb had been dropped, with devastating effect, on Nagasaki, and Emperor Hirohito, in his thin,
reedy, and hitherto publicly unheard voice, was on Japanese
radio telling his people that the fighting was now over and that
Japan would surrender, bringing the war in the Pacific to an end.

Those listening in the Pentagon were relieved and satisfied,
though it was a satisfaction muted by the sober realization that
their supposed allies and friends, the Soviets, were now scyth-
ing their way fast and furiously past the defeated Japanese units
in Manchuria, on Sakhalin Island, and into Korea. And as the
Communist Russians did so, they were claiming immense new
tracts of territory. Fear about the long-term implications of this
land grab dampened any celebration inside the Pentagon.

The Cold War had not yet begun,* but already strategists in
Washington were supposing, correctly, that the Communists
had ambitions to mop up territories for themselves along the Pacific periphery.

Marshall's staff officers decided that night that they must
come up with immediate recommendations—for the White
House, for the Joint Chiefs of Staff—for halting the Soviets' ad-

---

* George Orwell would be the first to introduce the term *cold war*, in an essay in the
magazine *Tribune*, eight weeks later.

vance. They first decided to establish a series of checkpoints around the region where the onrushing expansion might be stopped. One of these young officers then walked over to the wall map and pointed to the scantily known and, of late, Japanese-governed peninsula of Korea.

The officer was thirty-six-year-old Colonel Charles Hartwell Bonesteel III, a third-generation soldier, a West Point graduate, and a Rhodes Scholar. Standing before the blue-washed expanses of the National Geographic map, he traced his index finger east and west across the entire breadth of the ocean, along a line of latitude that almost precisely joined the cities of San Francisco and the Korean capital, Seoul. Both of them lay some thirty-seven and a half degrees north of the equator, he observed—an uncanny coincidence.

He promptly declared that it was important for the city of Seoul, Korea's capital, to remain in American hands. He told his two colleagues in the room—one of whom, Dean Rusk, would become President Kennedy's secretary of state—that the Soviet

Born into an esteemed American soldiering family, Colonel Charles Bonesteel III—seen here after eye surgery—famously drew the line in August 1945 that divided the Korean Peninsula along the Thirty-Eighth Parallel, creating North Korea.

army should be requested to stop its advance just north of the capital. Since Seoul lay at thirty-seven and a half degrees, then "the Thirty-Eighth Parallel should do it," Bonesteel remarked with studied casualness. Then, using a grease pencil, he promptly drew a line clear across the map, from Asia to California, along the latitude line thirty-eight degrees north of the equator. The officers told General Marshall what they recommended.

Marshall thought the latitude line was perfectly reasonable and logical. Diplomats were told, and within hours, Moscow, to the astonishment of all, Teletyped back the Soviets' concurrence. The Red Army would continue its sweep down into Korea, the message said, and would accept the surrender of all those Japanese forces on the peninsula stationed to the north of the Thirty-Eighth Parallel. The United States could land its troops elsewhere, and confront the Japanese occupiers to this line's immediate south.

This was all written swiftly into General Order No. 1, the surrender demand that was handed formally to the Japanese the very next day. The two relevant paragraphs of this historic document were the second ("The senior Japanese commanders and all ground, sea, air and auxiliary forces within Manchuria, Korea north of 38 north latitude and Karafuto shall surrender to the Commander in Chief of Soviet Forces in the Far East") and the fifth ("The Imperial General Headquarters, its senior commanders, and all ground, sea, air and auxiliary forces in the main islands of Japan, minor islands adjacent thereto, Korea south of 38 north latitude, and the Philippines shall surrender to the Commander in Chief, U. S. Army Forces in the Pacific").

Immediately, and with this succinct command, two quite new and separate spheres of influence were created in the world. One would in time come to be called South Korea, and would fall under the general influence of the Americans; and the second, known as North Korea, would come under the influence and direction, first, of the Soviets and, later, of the Chinese Communists.

In due course the latter would become the oxymoronically named Democratic People's Republic of Korea. Within a decade, these two separate, distinct, and mutually hostile states would be riven by a disastrous three-year war, which would reach an uneasy conclusion in 1953 that has separated them ever since by the most dangerous and heavily armed international border on the planet.

Many military strategists have speculated that the world might have been a far safer place if postwar Korea had been divided four ways, among the United States, the Soviet Union, the Republic of China, and the United Kingdom, as was first proposed. Or if the Soviets had been given free rein to invade all of Korea, and be done with it. In this latter instance, there would have been no Korean War, for certain—merely a Leninist satrapy in the Far East that, most probably, would have withered and died, as did other Soviet satellite states.

Instead, the Pacific Ocean's most volatile choke point was created, innocently and unthinkingly, by one swift sweep of a grease pencil held by Charles Hartwell Bonesteel III, a man who had been known since childhood by the nickname Tick. Like so many other all-too-swift border creations—the one between India and Pakistan comes to mind—his would leave a legacy both unimagined and unimaginable. Bonesteel would live until 1977 and would see the many consequences of his casual defacement of that National Geographic map, not least the horrendously lethal war that erupted five years later, and then the thousands of subsequent cross-border shootings, kidnappings, incursions, tunneling, and myriad other nuisances for which the Thirty-Eighth Parallel is still known today—one of the most egregious being the capture in 1968 of the U.S. Navy spy ship USS *Pueblo*.

This shabby little ship, a vessel akin more to the USS *Caine* than to the USS *Constitution*, was part of a secret navy spying program of the 1950s named Operation Clickbeetle. Initially the plan was supposed to involve the recruitment of as many as seventy similarly elderly, clapped-out, and inexpensive vessels, most of them deemed useless for anything more than being melted down and made into razor blades. They were to be stuffed to the gills with electronic espionage apparatus and ordered to hug the coastlines of countries that Washington thought were troublesome, told to run out their aerials and to listen to radio chatter. But the budget couldn't stretch to seventy ships, and in the end the Pentagon got only three, the USS *Palm Beach*, the USS *Banner*, and the USS *Pueblo*.

This last was the mothballed runt of the litter, and she was handed the trickiest assignment of the trio. She was to be the eyes and ears in the far western Pacific for the National Security Agency (NSA), then as now the most secret of all major American espionage organizations.

Though nothing like the omnivorous Gargantua of electronic spying it is today, the NSA in 1968 was nonetheless an immense, powerful, and expensive agency. I remember all too well being shooed away by guards when I drove too close to its triple-fenced compound in Fort Meade, Maryland. It also had secret and well-guarded outstations around the globe, from which I was also turned away whenever I tried to visit. Some were static—in Britain, in Hong Kong, on Ascension Island, in Gibraltar, on Hawaii, on Diego Garcia. Others were mobile: airborne or seaborne; some of the latter were submarines. Still others, such as the three starveling ships of Operation Clickbeetle, rode the surface waters, always isolated from other vessels, their role in espionage publicly denied, their official capacity spoken about as softly as their budget requests were hidden, even though their tasks were seen as crucial to the fast-growing empire of intelligence gathering.

The tugboat-size *Pueblo*, which had a couple of small tarp-covered, downward-pointing machine guns on deck under a canvas awning that made her look more like a Pacific islands tramp steamer than a fighting vessel, was supposedly, ostensibly, officially conducting innocent oceanographic research. Improbably, and in an attempt to bolster the vessel's cover story, a college professor of nutrition was on hand at the commissioning ceremony in the Port of Bremerton, in Washington State, to tell reporters that the *Pueblo*'s work might involve extracting new high-tech foodstuffs from oceanic waters. This was still, after all, the time of TV dinners, and people liked to believe such things.

The *Pueblo*'s chosen commander was to be a hard-drinking, fast-talking, loud, chess-playing, Shakespeare-loving muscleman and former submariner named Lloyd Bucher (Pete

Built on Lake Michigan in 1944 as an army freighter and transferred to the U.S. Navy as an environmental research vessel, the USS *Pueblo* was captured by North Korea in 1968 while on a spy mission. She remains captive, as a museum.

to his friends). Bucher was initially disappointed that he was being given command of such an undistinguished vessel. But he changed his mind after ten days of briefings at the Pentagon and at Fort Meade. This was because his little ship, which would be homeported in Japan, was not going to work along the stoutly defended coasts of the Soviet Union or China. Instead, she would venture out northwestward, across the six hundred miles of the Sea of Japan, and patrol along the little-known coastline of North Korea, a ragged line never less than thirteen miles from shore—twelve miles being the territorial limit declared by most Communist nations of the time. From there she would intercept, listen to, and decipher all the radio transmissions and other electronic chatter that emanated from this most secretive of Asian states.

It was all very hush-hush, considered by Washington to be of critical importance. This, coupled with Bucher's understanding that he would be performing his work all alone, and that the full might of the U.S. Navy in the Pacific might well not be able to reach him should he ever get himself into trouble, thrilled the young captain's gung-ho side. He was a man eternally eager for adventure, a show-off notoriously given to braggadocio and fond of derring-do.

There was at the time a rough-and-ready code applying to the Cold War's spy ships, born partly of the belief among naval officers that the sea is the only true enemy, and that human arguments are subordinate to the sea's dangers and demands. So there was at the time, for instance, a gentleman's agreement that governed spying on the Soviet Union, each side accepting that espionage was a necessary evil, and therefore tolerating the fact that ships involved in such business were deployed by the other's navy. But Bucher was rather more fascinated to be told, and most forcefully, that no such agreement covered espionage against North Korea. True, Pyongyang was perhaps less of a threat. Its navy was small, with just a number of small coastal craft, and it had no blue-water ships of the kind operated by Moscow. But the North Koreans could be nasty and dangerous; moreover, they played outside the rules. This also thrilled Bucher, who was entirely undaunted by the dangers. So he flew off to his ship's fitting-out in Washington State wholly excited by his coming duties.

It was in Bremerton that the highly sensitive interception equipment (to be manned by technicians who worked in code-locked cabins) was installed on the ship and tested by NSA engineers. Dozens of radio aerials and radar dishes, some familiar, others ungainly and of strange shape, were mounted all over the ship, most of them sprouting from two high masts built amidships. Once this work was completed, Bucher was ordered to sail his technologically festooned little craft (which rolled mercilessly on the ocean swells, to the crew's discomfort and chagrin, causing Bucher to wonder if all the added gear might make the boat dangerously top-heavy) across to Hawaii. There, he would be given further secret and more specific briefings by the chiefs of the Pacific Fleet. At the end of 1967, the *Pueblo* finally set off for the naval station at the Japanese port of Yokosuka, on the western side of Tokyo Bay, where she would be based and from where she would eventually commence her duties.

On January 5, 1968, after spending the Christmas holidays of 1967 in port, the tiny vessel, a plague of mechanical problems all half-resolved, headed out for what would be her first real-world operation. The forecasters said the temperature would be cold, the sea state moderate.* A number of North Korean trawlers were noted to be operating in the region where the *Pueblo* would be patrolling, but Bucher was reassured by his commanders that they would be likely to do no more than pester a sovereign ship of the U.S. Navy.

---

* In fact, the *Pueblo* encountered a near-crippling storm on her first day at sea and, half blind, almost collided with a reef; she had to put in to a second U.S. base in Japan, Sasebo, for two days of repairs.

The presailing orders had been quite specific: the *Pueblo* was to spend two weeks sailing up and down the North Korean coast from the frontier with the Soviet Union down to the DMZ that marked the border with South Korea. She was to be on the lookout for two things: a series of newly built coastal antiaircraft batteries that America believed the North Koreans had been constructing, and any of four new submarines that Moscow had recently handed over to the North's fledgling navy. After two weeks of this, the ship was to turn about and head back to Japan. For the entire journey, the *Pueblo* was to keep strict radio silence, except in the event of a real emergency.

Bucher clearly relished the instructions. He eased away from the dock and out into the fairway, and as he set sail on a southbound course he stood ostentatiously out on the *Pueblo*'s flying bridge. He was in fine, exuberant form: as his men toasted him with eggnog, he decided to show off his singing talents, booming down to them, and to all the mystified crews of passing Japanese fishing boats, a jazz number popular at the time, "The Lonely Bull." It turned out to be an apposite choice.

After a storm-tossed but otherwise uneventful crossing, the *Pueblo* arrived off the North Korean port of Wonsan on January 13, and began her ferreting duties. As ordered, she kept well away from the coast, and to minimize her chances of being seen, she sailed dangerously through the nights with all her lights dimmed or doused. As instructed, she also maintained her radio silence, and with religious determination. Bucher would not even transmit the brief coded messages that all other American vessels sent daily, allowing Pacific Fleet headquarters to know where all its ships were, all the time.

The crew, alone and unleashed from authority, then fell into a simple patrol routine. With North Korea lying gray and cold and mountainous off to their west, and the Sea of Japan lying gray and cold and empty off to their east, there commenced a plodding sameness to every day: a lumbering, swell-tossed progress up and down the coast, the ship's bristle of aerials listening out for any transmissions, the sailors' watching eyes trained for anything unusual that might be observed. Other than that, a good deal of time was spent in the hammock, eating three plain meals a day, watching movies such as *Twelve Angry Men* and *In Like Flint*, and doing secret work for the codebreakers and the monitoring teams locked behind their blast-proof door with all their mysterious NSA machinery. The only work that kept the men physically active was chipping away ice from the superstructure. Bucher was worried that his already unstable ship could well founder if too much ice accumulated. This was work that all of them loathed.

Nine days of this zigzagging back and forth off the frigid Korean coast, and they were off the port of Wonsan once again on January 22, a Monday. Bucher and his crew had every reason to suppose that this day and the next would be as routine and as dull as the nine before. What they did not realize, because no one had seen fit to tell them, was that a series of events had unfolded on land the previous weekend that made their presence so close to North Korean waters particularly dangerous.

A heavily armed North Korean commando unit had been smuggled south of the border a week previously, with orders to mount a daring raid on the presidential palace in Seoul. The plan had been to execute the South Korean president, Park Chung Hee, to kill his family and personal staff, and to flee back across the border. Thirty-one tough young members of the 124th Army Unit, a North Korean special forces team, had cut their way through the fences, and then managed to get all the way into Seoul and to within half a mile of the palace, before being intercepted.

A massive weekend firefight had taken place on the capital's streets, watched with amazement and horror by civilian city dwellers. The surviving North Korean commandos fled into the surrounding hills, and were picked off one by one by some

of the six thousand South Korean troops who had been mobi-
lized to find them. Come Monday morning the team from the
North had all been killed, and the episode was all but over. But
the atmosphere was still electric, and incredibly tense—and it
is almost impossible to believe that no one at U.S. Pacific Fleet
headquarters bothered to tell the *Pueblo* that its presence so
close to North Korean waters was likely to be even less welcome
than normal.

That Monday morning was uneventful and peaceful for the
ship. Just after lunch, however, Captain Bucher was summoned
to the bridge. Two North Korean trawlers had been spotted
steaming toward them, and they were closing in fast. The craft
halted less than twenty-five yards from the American ship, and
their crew members, armed with more cameras than ordinary
fishermen might need, took scores of photographs. The men had
extremely fierce expressions. One *Pueblo* sailor remarked that
"they looked like they wanted to eat our livers."

An alarmed Bucher, deciding to break radio silence, told Japanese Fleet HQ that his ship had been identified. It took the better
part of half a day to compose the heavily encrypted message and
then send it on the one secret channel available to the ship.

Nothing more happened that day: the two trawlers left and
headed back to shore. The next morning, only an unusual
amount of North Korean coastal radio chatter suggested anything abnormal; as on Monday, the morning seas were quite
empty, and the *Pueblo*, stopped dead in the water fifteen miles off
the coast, seemed to be alone in the ocean. Then, around lunchtime, a small North Korean naval vessel, a lightly armed submarine chaser, suddenly appeared out of nowhere, heading toward
the *Pueblo* at what American sailors call flank speed, the highest
its engines could command. Its crew, all wearing helmets, raised
signal flags, one after another: "What nationality?" demanded
the first. "Heave to or I will fire" was the second.

The oncoming ship then radioed a flotilla of other fast patrol boats that were appearing over the horizon. "It is U.S. Did you get it? It looks like it's armed now. It looks like it's a radar ship. It also has radio antennae. It has a lot of antennae and, looking at the wavelength, I think it's a ship used for detecting something." This particular message—intelligent, articulate, savvy, but unheard aboard the *Pueblo*—happened to be intercepted by an American C-130 Hercules aircraft that was operating high above the North Korean coast as a direct consequence of the failed attack on the palace in Seoul three days before. The U.S. Air Force signals officer, alarmed by a drama that suddenly seemed to be enveloping an unwitting American ship far below, radioed promptly back to Japan, urging the staff there to warn the ship's captain of what now seemed about to happen.

But if a signal was then sent to the *Pueblo*, it went unheard. A number of North Korean ships—six in total, summoned by the subchaser's first broadcast, and now appearing from all corners of the compass—were closing in on the *Pueblo*, fast. Only twenty minutes had passed since the first ship was seen, and now heavily armed enemy vessels were everywhere. They had clear intentions. "We will close down the radio, tie up the personnel, tow it and enter the port at Wonsan," said the subchaser to one of her sister ships. "We are on the way to boarding."

Bucher, aware only of the presence of a flotilla closing around him, quickly considered his limited options. He might try to outrun his pursuers, or else he might scuttle his ship, destroying secret documents. Two MiG jets then flew low overhead, adding to the gathering drama.

The U.S. captain decided on defiance. He first ran up signal flags indicating he was staying put—he was, after all, officially an oceanographic ship, collecting water samples. But within moments, a number of North Korean sailors appeared on the deck of the subchaser, all of them armed with rifles and clearly readying themselves to board the American vessel. So, beginning to sweat, Bucher ordered his signalman to put out more flags, this time spelling out a more conciliatory message: "Thanks for your consideration—I am leaving the area." He pointed his prow eastward, spooled up his engines, first to two-thirds ahead and then to full speed, and lit out for the open sea. He was trying to leave with dispatch and dignity, he later said.

Up above, the American aircraft heard, quite distinctly, an alarming radio call from the North Korean subchaser: "They saw us, and they keep running away. Shall I shoot them?" Orders were evidently given from an unseen controller ashore, because Bucher then spotted a fresh set of signal pennants going up on the ship that was now half a mile behind him: it was the same warning—"Heave to or I will fire." But he decided to keep on running. He ordered his crew to destroy as much secret material as possible, to smash the electronics, to burn the paperwork. He doubted he'd have much time.

He heard the jets thunder past overhead once again, this time terrifyingly low, terrifyingly loud. The smaller North Korean patrol vessels broke away, leaving the *Pueblo* alone, firmly in the gun sights of the pursuing capital ship. Then, with a terrible fusillade of sound and smoke and flying lead, the North Korean navy finally opened fire on the fleeing American warship. *Piracy? War? Dangerous lunacy?* Whatever the official term, the matter had swiftly become fraught with peril.

It took no more than one frantic hour for it all to play out. Topside, the crew did their best to avoid or to mitigate the withering bursts of gunfire—but shrapnel was flying everywhere; fires were breaking out; glass was shattering and the shards flying; bullet and shell holes were being punched into the hull, the smokestack, the bulkheads. Men were injured—one young man, a fireman named Hodges, was hit in the abdomen and leg and died from an immense loss of blood.

Belowdecks the wreckage was even worse, though of the crew's making, since standing orders demanded that everything classified now be reduced to garbage, useless and unreadable by any captors. So technicians with axes and sledgehammers hacked away at the electronics—finding them remarkably robust, resistant even to the cruelest of blows by the strongest of men. There were tons of papers to burn or shred, and only a small incinerator and a one-piece-at-a-time shredder with which to do the job. So, below, scores of small fires were set, the smoke filling the classified radio room and forcing its ax-wielding technicians to leave, choking, in search of fresh air—only to be forced back into the inferno as yet more shellfire erupted from the shadowing Korean boat.

The radio operator opened a link with the navy operations room in Japan, and the whole sorry progress of the one-sided firefight was then broadcast to a team of presumably open-mouthed operators back in the safety of headquarters. Bucher was at first still trying to get away, still heading east, but his full speed was a lame and limping thirteen knots, which the enemy vessel, now alongside, could easily match. More shells were lobbed at the *Pueblo*, prompting the single mutinous moment of the incident: the ship's chief engineer, viewing the situation as hopeless, demanded that the vessel come to a stop and, seeing Bucher's momentary indecision, wrenched the pilothouse annunciator to stop all engines, which the engine room unquestioningly obeyed. The American ship decelerated quickly and sighed to a stop, finally bobbing sulkily on the half-calm swells. Black smoke billowed from every porthole and every ventilation shaft as more and more paper fires were set, the frantic continuance of a vain attempt to destroy everything.

Then the gunfire stopped, and there was sudden silence—just the slap of cold waves against the hulls of the near-touching boats. No one aboard the Korean vessel knew enough English to shout orders, so the Koreans raised yet another string of flags, with new orders. "Follow me—I have a pilot aboard." Bucher ignored them. The sailors then angrily pointed up at the flapping strings of colors, internationally recognized and requiring no language skills, demanding the American ship turn around and do as bidden. AK-47s were raised to shoulder height. The fore-deck cannon was trained and leveled. Korean sailors, unsmil-ing, put fingers to triggers, awaited orders to annihilate the Americans, so close, so vulnerable, so *theirs*.

Bucher could do little more than breach the cardinal rule of all naval officers worldwide: *never give up your ship without a fight*. James Lawrence's battle cry, his dying command "Don't give up the ship," is a mantra of consummate importance to every American sailor. But Bucher would do the opposite: he would make the entirely contrary decision, and one that would haunt him for years to come. For not only was he unable to destroy all classified documents and machinery (to the eternal chagrin of the intelligence community), but also he simply acquiesced when asked to turn over his ship to an enemy. He did what he was told and didn't fire a single shot in defiance.

Commentators in America would long be unforgiving: one could hardly imagine, they said, that John Paul Jones or Admiral Farragut or even Lord Nelson or Rodney would ever have done such a thing. They would have traded shot for shot, lead for lead, crewman for crewman, and they would have gone down if neces-sary with a crippled and burning ship, her ensign sinking into the depths just as the captain's hat floated off to join it. That was the way navies did things. To do otherwise was unworthy, unac-ceptable, un-American.

But Lloyd Bucher did it anyway. This forty-one-year-old rake-hell from the arid plains of southern Idaho, a man brought up as an orphan in the Nebraskan prairies, who went to college on a football scholarship, and who could in no sense ever be said to have had seawater in his veins, promptly acted in the western Pacific Ocean as no true-blue sailor would ever have done, they said. He ordered his helmsman to turn about and to limp slowly toward land, to slide westward (at a paltry four knots, to enable yet more papers to be added to the pyre). He sailed his sad little ship morosely toward the enemy shore, to acquiescence, and to be henceforward inevitably thought of in connection with such words as *surrender, captivity, humiliation*, and *shame*.

"Now hear this!" the *Pueblo*'s public-address system barked. "Now hear this! All hands are reminded of our Code of Conduct. Say nothing to the enemy besides your name, rank, and serial number!" Then, within seconds, there was the stamp of heavy-soled boots on the iron deck, and ten North Korean soldiers with automatic rifles and fixed bayonets boarded the American vessel. *Pueblo*'s maiden voyage as a spy ship was officially over, little more than two weeks after it had begun. She was under arrest, and so were her captain and all her crew.

The ensuing fate of the men and their ship is recorded either in the memories of eleven months of beatings, interrogations, starvation, and humiliations or else in the black-and-white photographic images taken by their captors. First there were Captain

The *Pueblo*'s eighty-two surviving crewmen, led by Captain Lloyd Bucher, were held in prison in North Korea for eleven months, before being freed at Christmas 1968. They were greeted at the DMZ by the then general Charles Bonesteel.

Bucher and his men, the officers in black leather jackets and with naval caps displaying their official insignia, the enlisted men in fatigues and woolen beanies, walking in single file through the night with their hands held high, abject, defeated, captive. Next there is a group of sailors seated in a prison cell, potted plants placed to suggest home comfort;* some of the Americans have their middle fingers extended, in what they later explained to the North Koreans was a Hawaiian good luck sign. (When the North Koreans found out from a tactless caption in *Time* magazine what the gesture actually signified, the responsible men were beaten.) There are images of one sailor, Stephen Woelk, in a reenactment of a surgery performed on him without anesthetic, and then of him smiling at having survived the surgery. His own recollections are somewhat more vivid:

> *I believe it was the evening of our tenth day of captivity, I was removed from our room and taken to what appeared to be a medical examining room, just down the hall. Up until then, no major medical help had been provided to any of us. In this examining room, I was placed on a metal examining table, my hands were bound and tied down to the table. My legs were spread, my feet bound and tied down so I was unable to move. Their so-called NK doctors commenced operating on me without any form of anesthetic whatsoever. I can still recall the scissors cutting away flesh and being sewn up with sutures that looked like kite string. A small handful of shrapnel was removed in the operation that seemed to last an eternity, but probably did not last more than twenty to thirty minutes. I was told later my screams could be heard throughout the building and many crewmembers thought one of us was being tortured. I was then returned to my room and fellow wounded crewmen.*

---

* The plants were taken away as soon as the picture sessions were over; any left were urinated on by the prisoners, in an effort to make them look sickly for any future photo opportunities.

Aside from the crewman killed in the initial raid—Duane Hodges, a firefighter colleague of Mr. Woelk, whose own injuries were sustained in the same attack—all the *Pueblo* crew members survived their months in prison. And Mr. Woelk, despite the exiguous medical care, recovered:

> *My medical treatment would consist of the doctor taking a pair of forceps and shoving long strips of gauze saturated with a type of ointment down into my open wounds as far as it would go. Each day the healing process would not allow him to shove it in quite as far as the last time. One day something came out with the gauze that drew the attention of the doctor and his staff. It appeared that a bed bug had found refuge and a warm place to sleep inside of me. This did not seem to be a big deal to the staff. This was probably a normal occurrence in the everyday life in the Democratic People's Republic of Korea. I also received several injections daily and one transfusion of a clear liquid in my leg. Although my leg swelled to twice its normal size, whatever medicine it was seemed to work since I finally started making headway towards a recovery.*

All the men, fit or injured, officers or enlisted men, had similar tales to tell, of brutality, isolation, hunger, sadism. They knew little of the negotiations going on between Washington and Pyongyang to allow their release, or of the utter unblinking intransigence of their captors and those who represented North Korea at the bargaining table. It took the better part of a year for a deal to be brokered.

The *Pueblo* talks were conducted across a plain baize-covered table at the DMZ crossing point of Panmunjom—the site, still notorious and much visited by tourists, is where the original armistice talks were held that brought the Korean War's fighting to an end fifteen years before. Like most negotiations with the North Koreans, these talks had the affect of a parallel universe, a strange dystopian *Alice in Wonderland* world where little was as it seemed, where there was much shouting, spluttering, and fist waving, and where truth was more fugitive than in other, more rationally organized places. The opening of the discussions—officially the 261st meeting of the UN Command Military Armistice Commission, to which the *Pueblo* incident had been formally added—began with a statement by the North's granite-faced and unsmiling chief negotiator, Major General Pak Chung Kuk. His opening remarks to the lone American negotiator more than amply set the tone:

> *Our saying goes, "A mad dog barks at the moon." I cannot but pity you who are compelled to behave like a hooligan, disregarding even your age and honor to accomplish the crazy intentions of the war maniac [President Lyndon] Johnson for the sake of bread and dollars to keep your life. In order to sustain your life, you probably served Kennedy who is already sent to hell. If you want to escape from the same fate of Kennedy, who is now a putrid corpse, don't indulge yourself desperately in invectives. . . . Around 1215 hours on January 23 your side committed the crude, aggressive act of illegally infiltrating the armed spy ship* Pueblo *of the US imperialist aggressor navy equipped with various weapons and all kinds of equipment for espionage into the coastal waters of our side. Our naval vessels returned the fire of the piratical group. . . . At the two hundred and sixtieth meeting of this commission held four days ago, I again registered a strong protest with your side against having infiltrated into our coastal waters a number of armed spy boats . . . and demanded you immediately stop such criminal acts . . . this most overt act of the US imperialist aggressor forces was designed to aggravate tension in Korea and precipitate another war of aggression. . . . The United States must admit that* Pueblo *had entered North Korean waters, must apologize for this intrusion, must assure the Democratic People's Republic of Korea that it would never happen again.*

In the end, the Americans did as they were told. Or they went through the motions of doing so. They agreed, in essence, to the three A's (to making an admission, giving an apology, pledging an assurance), only to repudiate all three within moments of signing the document. This tactic, this form of words, the use of bizarre new terms for what would happen—*prerepudiation*, a word absent from most dictionaries, was one; the term *prior refutation* another—was fully accepted beforehand by the Communists. And this merely added to the perception of these talks, as with most of the Panmunjom talks to have taken place in the sixty years since the end of the war, as being part of a witch's brew of near insanity, with many of the sessions unreal and frightening by turn.

The result of these negotiations was the eventual freedom of the eighty-two men and the return of the body of the hapless Duane Hodges. As President Johnson had wished, the men were released in time for Christmas 1968. They were brought in buses to the Joint Security Area (JSA), that small bubble of hope and horror that straddles the dividing line between the two Koreas. The MDL, the military demarcation line, runs through the bubble, with buildings of North Korea on one side, of South Korea on the other, and a few heavily guarded structures built across the line itself, and is where such talks as take place do so across tables that have the line passing through their exact midpoint.

There is a shallow and muddy river, the Sachon, that dribbles through the western half of the JSA bubble. The demarcation line runs through the midpoint of the river, so any bridge crossing it would, ipso facto, cross the MDL as well. There is in fact such a bridge, an unattractive concrete affair two hundred fifty feet long, with a North Korean guardhouse at the western end, a guardhouse manned by American troops at the other. It has long been known as the Bridge of No Return, since Korean War prisoners who elected to cross it for postwar repatriation were told that, once across, they could never come back. Captain Bucher and his men were brought to the western end of the structure on the chill morning of Tuesday, December 23. It was nine o'clock. They had had turnips for breakfast. It was snowing lightly.

In another building nearby, the negotiators were signing their sheaves of meaningless documents. It had been agreed that two hours after the final signatures, the men would be freed— but then, at the last minute, the North Koreans decided on another tiny torture, delaying matters by a further, quite pointless thirty minutes. Since the prisoners in their buses had no idea what was going on, they were unaffected. But for the negotiators and the political leaders back in Seoul and Washington, it must have been agonizing.

Finally word came down the telephone line from Pyongyang that all had been approved. Captain Bucher was ordered to step out of the lead bus and ready himself. He was first taken to a waiting ambulance, there to identify in an open coffin the mummified body of fireman Hodges, now eleven months dead. Then Bucher stood, bewildered, in the gathering snow, listening patiently as Major General Pak, the chisel-jawed man who had made the speech starting the negotiations six months beforehand, harangued him for twenty minutes about the evils of his life.

Then one of the better known and least liked of the guards who had presided over the prisoners, a man of studied cruelty whom the prisoners had called Odd Job, pointed Bucher toward the bridge entrance and spoke to him one last time: "Now walk across that bridge, Captain. Not stop. Not look back. Not make any bad move. Just walk across sincerely. Go now."

The red-and-white-striped pole across the entrance was then raised. Bucher stepped onto the bridge and walked nervously, but steadily, across the ten cement arches that supported it above the ice-choked stream. At the far end was a delegation of friendly looking men with, as he came ever closer, ever-broadening smiles on their faces. They were Americans.

When he was just feet from safety, one of them took the final picture. It is black and white, of a now much older and much thinner Lloyd Bucher, in the last moments of his coming home. His mouth is set in a rictus, a faint and sardonic smile of exhausted relief. His left hand clutches the Mao cap he had been given, but had opted not to wear. His dark raincoat is tightly belted, its collar raised against the biting wind. His trousers are too short—flood pants, Americans would call them. His shoes are dark with white laces.

Behind him idles the ambulance bearing the body of Hodges. Back on the far side of the creek are the milling men of his ship, to be sent off on their way to freedom, one by one, by their guards, to follow their captain, to keep a strict twenty paces apart from each other. They had been ordered to walk, not run. They had been told not to dare look back at their captors. They had been told they were forbidden, on pain of being shot, to give the final one-fingered gesture of "Hawaiian greeting" so many had so dearly wanted to give.

Seen from above, from one of the American sentry towers, the handover was like watching sand move through an hourglass. On the far side was the dark crowd of the American captives, a mass that was steadily winnowing itself smaller and smaller as a line of dark-clad men poured slowly down through the narrow neck of the bridge and was then re-formed into an ever-enlarging crowd of men on the other side. The two groups may have looked just the same, like the sand in the glass. But the second of the two groups, even though at one halfway moment identical in size, was all of a sudden different from the first group, in that those in it were entirely free, and at last.

American sentries were watching all this from their towers. On the ground on the far side, also gazing impassively at the unfolding drama, were a score or more of green-fatigued men of the Korean People's Army. They were presiding over a repatriation that was no more dignified than it should have been, and from which no one had drawn much of a victory, or a triumph, or a propaganda score. As the last American limped offstage, the Koreans turned away and got back on the buses to their barracks, to kimchi and cheap shochu and endless propaganda. The Americans went for steaks and showers, orange juice and telephone calls, and flights home, to meetings with wives and girlfriends and the usual celebrations and parades and oompahs that greet returning astronauts and Olympians and such other temporary heroes and survivors that Americans have come to admire and revere so much. In time there would be courts of inquiry and investigations and postmortems, and then memoirs and interviews and reminiscences. But not today: ahead was Christmas, and freedom.

There was one final irony for the men, though it passed unrecognized at the time.

As they stepped off the concrete abutments of the bridge and into a hut decorated with Christmas tinsel and lights, they were each to be greeted in person by the commander in chief of the United Nations Command, the senior Allied soldier sent in to protect South Korea. The C-in-C at the time of this event was none other than U.S. Army general Charles Hartwell Bonesteel III, the same man who, as a mere colonel almost a quarter century before, had so casually defaced a National Geographic map on a Pentagon wall, and whose innocent remark that "the Thirty-Eighth Parallel should do it," led to the creation of North Korea in the first place.

Had Bonesteel not drawn his grease pencil line late that summer's night in 1945, the melancholy chain of events surrounding the capture of the USS *Pueblo* probably would never have taken place.

The seizure of the *Pueblo** may have been one of the most painful episodes in America's entanglement with North Korea, but it
was not to be the last. More than sixty years have elapsed since
the signing of the armistice, and almost every year has been
peppered with events that have been, by turns, lethal, curious,
frightening, or all three. Even a cursory summary of the more
serious happenings confirms the idea that North Korea has been
an interminable nuisance. In 1958 an airliner on an internal
flight in South Korea was hijacked and flown to Pyongyang. In
1965 two MiG fighters attacked an American plane fifty miles
off the coast. In 1969 four American soldiers were ambushed
and killed on the southern boundary of the DMZ. In February
1974, North Korean patrol boats sank two South Korean trawlers. In August 1974, South Korean president Park Chung Hee's
wife was shot dead by a North Korean agent in Seoul. In 1978 a
South Korean actress and her film director husband were kid-
napped in Hong Kong and smuggled to Pyongyang. In 1983 three
South Korean government ministers and fourteen staff were
among twenty-one people killed in Rangoon by bombs smuggled
through the DPRK embassy in Burma by North Korean agents.

One of the more notorious events took place inside the Joint
Security Area, on August 18, 1976, when a group of American
soldiers attempted to trim a poplar tree close to the end of the
Bridge of No Return because it obscured their view of the North
Korean watchtower at the far side, the same one from which Cap-
tain Bucher began his lonely walk to freedom. North Korean sol-
diers protested, absurdly, that the tree had been planted by their
Great Leader, Kim Il Sung, and so should be considered inviolable.

---

* The vessel remains in North Korean hands, currently anchored in the Botong River in
central Pyongyang and freshly repainted as a tourist attraction. From time to time there
are vague hints that the ship might be sent back to the United States, in exchange for some
unspecified high-level political or economic concession. Meanwhile, she is on the active
roll of commissioned U.S. Navy vessels, and still sporting the initials USS to indicate her
exalted status.

When the Americans continued to prune its branches, a posse of North Korean soldiers rushed them with axes and crowbars and bludgeoned and hacked two American officers to death. Four other Americans and five South Korean soldiers were also badly injured.

Washington, it turned out, had been pushed just a little too far. Three days later the so-called ax murder incident at Panmunjom triggered an almost unimaginable American response. It was called Operation Paul Bunyan, and it had the full approval of an enraged President Gerald Ford. Its mission, undertaken in the spirit of the legendary lumberjack, was to cut down the entire tree. On the face of it laughably trivial, a simple enough exercise, a robust response, a small opportunity to redress the sad capitulation of the *Pueblo*.

Except it was very much more than this. The deliberate American plan was to use this single small arboreal incident to demonstrate to the North Koreans that, as Washington liked to put it, "you don't mess around with Uncle Sam." The White House let it be known that the disproportionately immense team assembled to cut down the tree—with twenty-three vehicles, two engineering companies, sixty American security men, a sixty-four-man South Korean special forces team, and a howitzer big enough to blow the Bridge of No Return to smithereens—was to be supported by as much power as the Americans needed in the event any North Koreans dared to retaliate. "Take *that*, President Kim," the Americans seemed to be saying.

So a backup assemblage of armor and weaponry was put together on a scale seldom seen in peacetime, and only occasionally seen in war. Lurking just outside the DMZ, ready to move at an instant's notice, was an entire U.S. infantry company and nearly thirty helicopters to ferry them into battle. B-52 bombers had been scrambled from Guam, F-4 Phantom jets were swooping in from bases all around the region, F-111 fighters had come in from their faraway base in Idaho. There were Hawk missiles at the ready, the carrier group of the USS *Midway* had been moved

close to the coast, nuclear-capable bombers were flying overhead, and twelve thousand additional troops and eighteen hundred U.S. Marines had been put on standby to fly to Korea.

In the end, the tree-cutting party took just forty-three peaceful minutes to bring the poplar down and to trim its remains
to serve as a memorial to the officers who had died three days
before. By the end of that hot summer day, the forces had been
stood down, the bombers sent back to routine patrolling, the
fighters sent back to their revetments in Japan and the Philippines and faraway Idaho. America, for once, felt it had managed
to shock and awe the North Koreans into some semblance of
common sense and good order.

Yet it would not last. The regime constructed by North Korea's
doctrinaire soldier founder, Kim Il Sung, evolved into a dynasty
of unparalleled and ever-increasing cruelty and hostility, behaving toward its own people and the world in ways quite detached from the norms of even the most bizarre. Perhaps only
Pol Pot's Cambodia can offer a valid comparison. Or maybe Al-
bania, during the most extreme excesses of Enver Hoxha's time.
Or perhaps Idi Amin's Uganda, at its worst.

The original national philosophy of *juche*, a form of self-
reliance coupled with extreme nationalism, and which, according to state propaganda, was dreamed up by Kim Il Sung
at an improbably early age (ten years old, according to some),
steadily transmuted itself, as Kim's son and then his grandson
took over the reins of governance. There was a time when some
felt that Korean self-reliance was not a bad thing—after all,
India, with its own rigid application of a similar idea, *swadeshi*,
which forbade the importing of virtually anything from outside, worked well for a time. Moreover, the early ideas of Kim
Il Sung allowed this half of Korea to retain some sense of a
uniquely Korean identity, even as it was being demonstrably

Hypnotic organized performances by thousands of impeccably drilled youngsters—known as "mass games," with any misstep harshly punished—are one of the few signature achievements of North Korea.

lost in the very Westernized and commercialized atmosphere of South Korea.

But *juche* went slowly insane. Lenin, Marx, Engels, and Trotsky, in common with such Communist sponsors who still pay lip service to the old Soviet regime, would be hard-pressed to recognize the autarkic impoverishment that guides North Korea today. In the republic's earliest days, it might have enjoyed some kind of reputation as a failing form of socialism, an experiment that didn't quite work, but that had a kind of nobility of purpose. Instead, the country has evolved into a snarling, spitting, and ceaselessly hostile monster of a nation, an alien life-form that lurks menacingly in the folds of the East.

There is widespread hunger, grinding poverty of a depth and kind unknown elsewhere in the world for decades past. There is almost no personal freedom, and punishment for the slightest of crimes against the state or the dignity of the guiding ideas or their practitioners is swift and pitiless. One senior soldier was executed for seeming to nod off during a cabinet meeting—pounded into oblivion by antiaircraft cannon, some reports said. Another, reported by the lurid South Korean press to have been torn to shreds inside a cage of wild dogs, was the leader's uncle—the fact of his kinship bringing him no mercy. Gigantic prison camps are scattered everywhere across the bleak and famine-racked landscape, and inmates are corralled within them in conditions unimaginable by even the darkest, most Orwellian, most Kafkaesque minds. One arrest can lead to the instant imprisonment of three generations of a family, the deliberate extinguishing of any potential for dissent by the preventive detention of the innocent.

I can never forget a visit I made to North Korea in the mid-1990s. If one consequence of my venture was dreadful, it was all my fault. One afternoon I was driven down to see the Panmunjom truce village from the northern side. I had a government minder with me, as always. He was a friendly man who spoke with carefully calibrated candor about the pleasures of living in the North, but liked also to voice at length his fantastic vision of the decadence and corruption of the South.

Our driver on that afternoon spoke not a word of English and smoked incessantly. He listened to his car radio, which like all Korean receivers had had its tuner welded to pick up just a single government station, one that pumped out a torrent of loud political exhortations, or else saccharine musical numbers of great patriotism and fervor. So when the minder stepped away from the car for a moment, I made what I thought was a kind gesture: I showed the driver the tiny Sony transistor radio I had in my pocket, and tuned it for him to pick up an American forces radio station. We were less than a mile from the border, and the transmitter aerials were visible.

It was a revelation. He had never heard anything like it. For the next ten minutes or so the snatches of music, and the announcer's voice, seemed to give the driver the greatest pleasure, and he drummed his fingers on the dashboard, beamed broadly, and offered me cigarettes. His personality had completely changed. He seemed genuinely *happy*. Then, without warning, the door was wrenched open and the minder got into the car while my radio was still switched on. He barked angry questions at the driver, who had stopped smiling and who returned answers monosyllabically, sheepishly. The minder glowered as we were driven back to Pyongyang that evening in total silence. I never saw the driver again. No one later claimed to have any idea of his fate. In fact, in spite of much questioning, no one claimed even to know of his existence.

The savagery of this most ruthless police state is all but undeniable.* Within the country, though, there is a continuing and skillful attempt to mitigate its horrors by the endless presentation of the Kim family's geniality and benevolence. The Kims are everywhere. The regime's three founders, the Supreme Leaders, the eternal bestowers of guidance, have become so godlike that their given names, Il Sung, Jong Il, and Jong Un, have been banned from use by any other Koreans, and those already using the names have been ordered to change them. Photographs of the three men must now hang in every dwelling, there are immense marble statues of them in public places, and members of the Korean Workers' Party must wear tiny brooches sporting the leaders' enameled faces.

North Korea counts its calendar years from the date of Kim Il Sung's birth, in April 1912, which was Juche Year 1. The year 2016 is Juche 104. On the Great Leader's birthday each spring there

---

* Nonetheless, the authority of the devastating 2014 report of the UN Commission of Inquiry into the alleged barbarities of the regime was somewhat challenged a year later when a key witness, Shin Dong Hyuk, admitted to some fabrications in his testimony. The commission chairman stood by the central message of the report, however, and insisted that the regime be held criminally accountable for its excesses.

are wild demonstrations of impeccably choreographed ecstasy, the so-called mass games, involving thousands of identically drilled children. The games are designed to have a hypnotic effect on a public glued to the state-controlled television. Similar demonstrations occur on dozens of other state holidays and anniversaries; and on more momentous occasions, the armed forces take part, with goose-stepping infantrymen and five-mile-long parades of missiles, tanks, and armored cars—all of them turning a show that is merely chilling into something truly alarming.

The North Korean army is immense—at almost a million soldiers, it is probably the biggest or (after China) the second biggest in the world. The country's 2009 constitution gave the army primacy over all other institutions of state; the new notion of *songun*, as the army's unchallengeable authority is now called, has ever since stood alongside *juche* as the state's guiding philosophy. At the same time, the term *communism*, which in comparison seems almost quaint and harmless, was quietly dropped from the description of the DPRK's central ideological principle.

That North Korea, already a fanatically militarized state, is by now formally ranking its enormous army as the leading instrument of policy worries everyone—this is a matter of ever-growing concern. The country's attainment of a small number of crude but working nuclear weapons, along with sufficient rocketry to propel these weapons beyond its coasts, combined with its declared intention to punish anyone who has ever disrespected its leadership or its aims, presents a threat of real danger. In global terms, it may still be only a nuisance, but it is a serious, grave, and potentially bloody nuisance, which none in the outside world seems to have the power or ability to check.

The Korean DMZ, the central focus of all these nightmares, is a strange and dangerous place for humans. It is a place of searchlights and fortifications, watchtowers and minefields, howitzers and tanks, and the massing on both sides of countless stone-faced soldiers, heavily armed and ready to deploy at an instant's notice. It is a place of strange and dangerous happenings—of shootings and stabbings, of the floating of balloons containing propaganda or poison, of the building of giant illuminations (usually of Christian crosses) that are designed to advertise god to the godless. Loudspeakers of incalculable decibellage blare the sayings of one or another of the Kims to listeners in the South.

Once, when I had completed a three-month walk up the entire length of South Korea and had arrived at the end of the Bridge of No Return, the loudspeakers were screeching out in English, welcoming me, inviting me to walk farther, to cross the bridge and savor the delights of the Democratic People's Republic. It seemed a fine idea, but the U.S. Marines who had escorted me through the Joint Security Area were having none of it. Could I just walk across the bridge? I asked. Definitely not, they replied, and added "sir," for emphasis. And what if I just set off and walked? They drew themselves up to their full, imposing height. We'd break your fucking legs, they said. Again, they added "sir," for emphasis.

But there is more to the DMZ than mere menace. Though it may be a strange and hostile place for humans, it is, for example, anything but for flotillas of Siberian cranes and hordes of brown bears, musk deer, and the goatlike Amur gorals, who flourish in what is for them a sanctuary, a gun-free, human-free four-kilometer-wide swath between the two great fences. The creatures can hardly be petted, or visited, or even accurately counted. But they are there, munching and fluttering and preening under the gun sights of thousands, oblivious to all the anger and ideology swirling around them.

And there are some moments in and around the DMZ that have a certain charm to them. One such took place for me late in the 1990s, when an American magazine of some flamboyance wondered if it might be possible to stage a lunch party in Korea—"somewhere interesting" as the publisher put it, "like the middle of the Korean DMZ."

It seemed at first a quite ludicrously impossible idea. Only wild animals (the aforesaid cranes, bears, and gorals) inhabited the DMZ—or so I thought. On a trip to Seoul to investigate other potential sites—the abbot of one of the loveliest Buddhist monasteries in the world, Haeinsa, said he might know of a nearby hall we could possibly use—I mentioned to a diplomat friend the impossibility of lunching in the DMZ. "Not so fast," he said. "Have you tried the Swiss?"

I had quite forgotten about the Swiss. At the time of the signing of the 1953 armistice, a group of four supposedly neutral countries agreed to monitor the cease-fire. The North had nominated as its two countries Poland and Czechoslovakia; the South had selected Sweden and Switzerland. I telephoned the Swiss embassy for details, and was given the number of the only major general in the Swiss army, a civilian diplomat deputed for five years at a time to take charge of his Neutral Nations Supervisory Commission camp, up by the JSA at Panmunjom. He was delighted to have someone stop by; he saw few outsiders. "Just American soldiers," he said. "You know how that can be."

His camp was in a small spinney right inside the DMZ and just outside the JSA. To get there, I had to drive to the American base camp outside the DMZ and wait for a Swiss guard to come collect me, which he did in a white-painted Mercedes G-wagon. He took me through well-guarded double gates in the fences, along a gravel driveway, and up to a comfortable little headquarters house, a cottage with a mess room and bedrooms for the ten or so Swiss soldiers who had been sent out from Bern to help keep the local peace.

The general was an affable middle-aged officer, clearly weary of his tour between the Koreas, and now readying himself to leave and take up a job as Swiss consul general in San Francisco. He was up for anything, he said; yes, he had a chef, who was, in truth, rather bored cooking for Swiss soldiers; and no, he hadn't given a party up on the DMZ for many months past. But he'd very much like to.

It would be a great relief for him, considering what he called the "comical absurdity" of his situation. Absurd mainly because of what had happened since the fall of the Soviet Union in 1991, and the abandonment of communism by the North's two chosen neutral nations, Poland and Czechoslovakia (and the division of the latter into two brand-new and entirely capitalist countries). North Korea had responded to the apostates by kicking their observers out of the country, leaving only the Swiss and the Swedes to maintain the monitoring. Except that the Poles kept on trying to send at least one delegate to maintain the fiction that the commission still existed, or three-quarters of it anyway. North Korea, which has made denunciation into a cottage industry, still denounces this near-beer body as something "forgotten in history" that is now no more than "a servant of America."

Nonetheless, each Tuesday the countries' representatives meet in formal session—about thirty-five hundred meetings have taken place since the cease-fire in 1953—and discuss and take notes of all the various alleged breaches of the cease-fire and other such matters (tunnel diggings heard, snips in the barbed wire noticed), and write a report. They place these written reports in a mailbox marked KPA, for Korean People's Army. But since 1995 no North Korean has ever picked up the mail, and so every six months an official from the commission empties the overflowing mailbox and puts all the reports into a file cabinet, just in case Pyongyang ever demands to see them.

As it happens, the door of the commission's hut opens directly into North Korean territory, and for a while the Swiss general would unlock it and wave the latest report at the soldiers a few yards away. They turned their backs and ignored him, never came to collect the document, and later complained that the waving constituted an offensive gesture. So with a sigh of frustration the general stopped opening the door, not wanting to provoke or seem impolite, and maybe see for his pains the business end of a Korean-made AK-47.

So, yes, he'd very much like to entertain our party. How many guests?

A date was set, and at an appointed hour, forty distinguished men and women from the advertising world—the magazine I was writing for was eager to impress and show gratitude to those who placed advertisements on its pages each month—duly arrived at a U.S. Army base in central Seoul. They were told what to expect, told how to behave—no pointing at North Korean soldiers, no sudden movements, no loud remarks—and were kitted out with flak jackets and steel helmets "just in case." They were then herded onto two large Chinook helicopters, which rose and then chugged for a half hour up over the mountains and paddy fields north of the city to a river and a ruined bridge, and then settled down on a grassy sports fields at the American forward operating base. A long line of Swiss vehicles was waiting for them, the general in the lead car with pennants, insignia, and the paraphernalia thought likely to appeal to his visitors. The entourage set off through the great fence gates and up to the peaceful-looking wooded grove set down in the very middle of it all.

For the next three hours we sat at a dining table eating rösti and raclette and chocolate fondue. Out one set of windows we could glimpse in the distance American soldiers drilling, cleaning their weapons, changing duties, jogging. Through the other windows, those looking out over the cold mountain to the north, we could see North Korean artillery pieces and armored vehicles and the barracks of scores of soldiers performing precisely the same tasks as the Americans, from whom they were separated by two mighty fences and two enormous minefields and four kilometers of grassland, with their populations of wild deer and bears and goatlike Amur gorals.

What we all remembered most, I suspect, was the constant querulous voice on the North Korean loudspeakers, repeating in a soaring monotone the words and wisdom of Kim Il Sung, who had chosen first to rule this benighted country and whose generations of offspring have ruled it, dangerously and wickedly, ever since.

Charles Hartwell Bonesteel III, who back in the summer of 1945 first drew the pencil line that marks the true epicenter of all this, the line that ran down the center of the Swiss general's Panmunjom dining table, came to know something of the DMZ's dangers when he welcomed Captain Bucher and the *Pueblo*'s crew back across the Bridge of No Return in 1968. This third-generation military man, educated at West Point and Oxford, is now long dead, interred at Arlington National Cemetery. He would without doubt be astonished to see how North Korea has endured, and has so case-hardened and strengthened itself in the years since. Quite unwittingly, the good colonel left the world a powerful legacy—one that to this day, seven decades on, remains memorable, malevolent, unpredictable, dangerous, and a terrible, terrible nuisance.

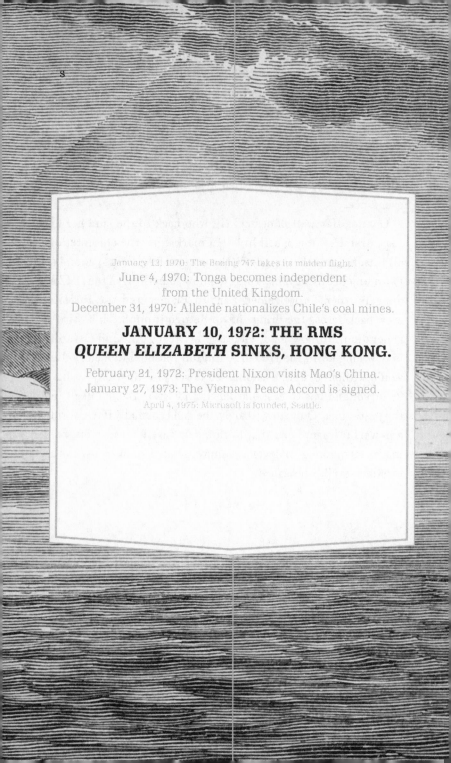

January 13, 1970: The Boeing 747 takes its maiden flight.
June 4, 1970: Tonga becomes independent
from the United Kingdom.
December 31, 1970: Allende nationalizes Chile's coal mines.

JANUARY 10, 1972: THE RMS *QUEEN ELIZABETH* SINKS, HONG KONG.

February 21, 1972: President Nixon visits Mao's China.
January 27, 1973: The Vietnam Peace Accord is signed.
April 4, 1975: Microsoft is founded, Seattle.

Chapter 5

FAREWELL, ALL MY
FRIENDS AND FOES

The enemy has overrun us. We are blowing up everything.
Vive la France!
—French radio operator's final words, May 7, 1954,
Battle of Dien Bien Phu, Indochina

I have relinquished the administration of this government.
God Save the Queen.
—Hong Kong governor Chris Patten,
final telegram to London, midnight, June 30, 1997

She was the loveliest ocean liner the world had ever seen—
eighty-three thousand tons of Clyde-built elegance and pride
and craftsmanship, a ship of longing and allure and fine cui-

sine, of passions promised over her moonlit taffrail, of romances hatched in the sway of her grand saloons. But just after noon on a winter's Monday in Hong Kong, half the world away from her Scottish birthplace, the burning and twisted wreck of this mighty vessel capsized onto her starboard side and slumped down heavily into the shallow, greasy waters of the harbor, never to rise again. The Royal Mail ship *Queen Elizabeth*, the younger of the pair of great and graceful sister ships that had for decades dominated the *grand luxe* transatlantic run, had come to her most wretched and unseemly end.

Unseemly endings are everywhere in this part of the modern Pacific's story. So far as the great old ship is concerned, the British firm that built her, John Brown of Clydebank, vanished in financial ignominy. In 1986 the British firm that operated her, Cunard, was reduced to a mere subsidiary of another giant, and now it possesses only three ships, compared with its fleet of sixteen back when the *Queen*s were sailing. And Hong Kong, the British colony where the liner sank, was wrested from London's hands in 1997 and is now an increasingly Chinese part of China.

Endings of varying degrees of seemliness extend through the Pacific far beyond the waters of Hong Kong Harbor. Since the sixteenth century and the first crossings of the ocean by exploring outsiders, it has provided foreigners (Europeans, mainly) with an immense imperial playground, with territories for all sides to take, either for wholesale exploitation or for the simple amassment of regional power. The Portuguese came first; and then, in quick succession, the Pacific's great coastal states and its long drifts of islands were snatched up by the Dutch, the Spanish, the British, the Russians, the Germans, the French, the Japanese, the Americans, the New Zealanders, and even the Norwegians, all in three centuries of uncontrolled imperial greed.

Yet so many glorious beginnings were inevitably followed by as many inglorious endings—with the result that all these various powers have retreated from the ocean, leaving the Pa-

The graceful Cunarder RMS *Queen Elizabeth*, shown in her glamorous heyday and at her sad sabotaged demise in Hong Kong, had a thirty-three-year life, which marked the beginning of the decline and fall of the British Empire.

cific now almost entirely to its own devices, to be run by its own people.

The foreigners' first withdrawals from the ocean began, effectively, in the mid-1950s, when France came to accept the reality that its once great Southeast Asian peninsular landholding "Indochine" was no more, and had to be returned. For the next forty years, farewell ceremonies seemed to be held almost monthly, with alien flags lowered and swansdown plumes, helmets, and swords being loaded into cabin trunks and sent home to London, Lisbon, Paris, The Hague, and Washington, from islands and outposts dotted in and around the gigantic blue space of sea. Hong Kong was the last of the outsiders' grand territories to be handed back—it was *retroceded*—and this was done with appropriately grand ceremony, half a century after the French pulled down their tricolor in Hanoi.

The sinking of the *Queen Elizabeth*, which took place twenty-five years earlier, more or less halfway between the return of Hanoi and that of Hong Kong, might well serve as a symbol for the frailty of empire. It was a reminder of the temporary nature, the sinkability, of all the foreign majesty wielded in this great expanse of sea. That the sinking occurred to a Western-made sea machine, and under circumstances that were peculiarly and bewilderingly Pacific in their nature, helped make the symbolism of the event all the more potent.

The Cunard Line had first sold off its two *Queen* liners in 1967. The great ships had operated across the North Atlantic ever since the end of World War II, and both had at first been filled to the gunwales with more than two thousand paying passengers each week. But it was a creature of the Pacific that proved their nemesis: the made-in-Seattle four-engine passenger jetliner the Boeing 707.

Starting in the late 1950s, when three airlines began using

these jets to run daily ferry service between London's Heathrow Airport and Idlewild in New York City, all of a sudden crossing by ship seemed quaint and inefficient. Despite Cunard's slogan suggesting that "getting there is half the fun," the paying public decided that getting there and back in half the time was much more sensible. So, in their thousands, they abandoned the ships, leaving them embarrassingly empty, light in the water.

"Space is usually available on all departures," read a dismayed internal note to Cunard directors in 1965. The combination of vacant staterooms and low-budget passengers steadily reduced Cunard's profits to the thinness of the cucumbers in the Britannia lounge's afternoon tea sandwiches. And though the firm added to the *Queen Elizabeth* a three-million-dollar outdoor swimming pool, a Lido Deck, and other touches that it thought holidaymakers might like, and then packed the liner off to try wintertime cruises in the Bahamas,* the figures continued to slide. In the end, the unsentimental clicking of the back office abacuses sounded the death knell. The *Queen Mary* was sold first, in August 1967, and on Halloween she sailed off by way of Cape Horn to the Pacific to be, as she remains today, a cemented-in-place hotel and museum on the Long Beach seafront, a successful and well-liked international seamark now remade as a successful and well-liked local landmark.

The *Queen Elizabeth*'s fate was to be very much more complicated, very much less dignified, and ultimately, a terribly sorry one. The moment that Cunard put the liner on the block, there came a buzz of intense interest—but a buzz that quickly faded. The firm heard about vaguely suitable offers from companies in Brazil and Japan, but these never materialized, and as many as a

---

* The firm gave no thought to installing air-conditioning, however. *Elizabeth* did have a "cool air" system, but Cunard acknowledged, in a classic of British circumlocution, that both vessels were "not entirely comfortable" when berthed in a tropical climate. Moreover, both ships were too broad of beam to pass through the Panama Canal, so their possible destinations were severely limited.

hundred others were haughtily dismissed. Eventually a group of American investors in Philadelphia made a real offer; their plan was to moor her in a swamp on the Delaware River and run her as a hotel, similar to the *Mary* on the far side of the country. But they never checked to see if she would fit in the river (she wouldn't), or how customers would get to her (a brand-new highway would have to be built).

Yet, for inexplicable reasons, the Cunard chiefs stuck with these investors, and a deal was made for the liner to become a hotel not in a Chesapeake Bay swamp, but by a nicely tropical beach in Florida. After a series of formal farewells—one a banquet attended by the Queen Mother, the very Elizabeth who had launched the ship thirty years before—the *Queen Elizabeth* left Southampton at the end of November 1968 for what many hoped was some kind of dignified future.

It was not to be. She scraped into the port with inches to spare beneath her keel and promptly had the word *Queen* erased from her name with welding torches. Soon after, she was declared a fire hazard and then began decaying in the damp Florida heat. Meanwhile, among her buyers and their friends, people involved were shot, whacked Mafia style; others went to prison on racketeering charges; some declared bankruptcy; others appeared in long-drawn-out American court battles in which hapless and bewildered Cunard executives from London were brought across to testify.

When two of the original buyers were jailed, Cunard put the forlorn ship up for sale again—and in September 1970 a Shanghai-born shipowner, Tung Chao Yung, bid three million dollars for her at auction. He would take her to Hong Kong, he said, where he would restore her to her former glory, establish her as a floating center of learning and intellectual discourse, and rename her *Seawise University*. Eight hundred first-class passengers and eight hundred students would sail with her. "She will be more beautiful than ever," Tung promised.

It first took almost one million dollars to make the ship safe enough to take to the seas, and even then the journey to Hong Kong was something of an ordeal. Her boilers broke down. She spent hours adrift, powerless, off Cuba. For days, there was no running water aboard, and welders had to affix three-seater toilets to the outside of the ship's boat deck, where passengers could go to perform their natural functions, using gravity and the sea instead of the flush and the pipe. She then had to be towed to Aruba, where she spent three months having her engines repaired. She eventually arrived in Singapore, and Royal Air Force jets flew over to salute her as she was tugged to her berth. Two weeks later, in July 1971, she finally arrived in Hong Kong—though she arrived a day early, and had to sail back and forth, south of the colony, like an actress with stage fright waiting for her curtain call. It had taken her five expensive months to perform a journey that a cargo vessel would have done in six weeks. C. Y. Tung was twelve million dollars in the hole already.

Upon her entering British colonial waters, a fireboat performed a water storm of welcome—a prescient gesture, considering what would then unfold. For, as she anchored off Tsing Yi Island, and as her first refitting got under way (new paint, new cabins, new boilers), there came ominous warnings, particularly of that most feared enemy of all deep ocean ships: an outbreak of fire.

Ship fires are terrifically dangerous—with all that fuel aboard, with hundreds of tons of combustible materials, with scores of passengers and crew. They are also very hard to fight—foul weather and great distance can hinder any firefighting efforts from outside, and if water is pumped onto the blaze, it may well endanger the vessel's buoyancy. After "the indiscriminate use of large quantities of water," reads one of the standard manuals on the subject, "the ship may be lost as a result of instability, and not because of the fire."

Hong Kong government officials had been openmouthed with

dismay at the lack of fire precautions they discovered when they had toured the great ship. They found sprinklers out of order, an electrical system as frayed as the carpets, fire hoses not working, main pipes cracked and blocked, watertight doors left open, and no fire crews with any idea of how to fight a blaze. Twenty-one recommendations for improvements were swiftly made: if the *Queen* was left as she was, said the government, she "presents an extremely dangerous fire and life risk." Mr. Tung said he would do all that was asked.

But his problem turned out to be political, a classic Pacific collision between ideological systems that were both Asian and American, even between races. For, many years before, Tung had committed what to some appeared a heresy: he had turned his back on Communist China. Like so many Hong Kongers of his generation, he had fled by way of Taiwan to the British colony, and there had run an empire of profit, indulgence, and entre-preneurialism. He had lived a life that was entirely counter to the attitudes and aspiration of the Communists and, more impor-tant, because of his unique situation, of the Communist agents who were operating among the working millions in Hong Kong.

For, in the early 1970s, the colony was an ideological tinderbox. In China the Cultural Revolution may have been starting to wane, but the Red Guards were still ferocious and active, and the turmoil that had convulsed the mainland since 1966 frequently seeped southward, into the British colony, and erupted in demonstrations and riots that had sorely tested the local police and militia.

Strikes and slowdowns were a constant threat, and though by the early 1970s the main sources of disruption had been largely tamped down, there were agitators and troublemakers aplenty still active. Many were active in the labor unions, and most espe-cially of all, there were many among the workers who came each day to hammer and chip, weld and paint, on Mr. Tung's great white whale of—as they liked to see her—a onetime British impe-rialist, white man's ship.

Moreover, the ship now sported a new symbol sure to envenom the more radical of these workers. Tung's company symbol was a plum blossom, and he demanded it be etched onto the old *Queen's* two funnels. The flower was not the Tung logo alone; it was also a symbol of Taiwan, the island republic that had broken away from Communist China, and was particularly loathed and despised for having done so.

But if the workers on the ship felt they had cause to abominate their bosses, the Tung managers had their own reasons to be irritated with the workers. They grumbled, for instance, when the laborers left the ship each lunchtime to eat ashore, declining a bizarre offer by Tung's managers to stage Cantonese operas in the liner's ballroom in an effort to persuade them to stay. Many of the painters and cleaners demanded time to play poker and mah-jongg, ran onboard gambling rings, and demanded to smoke whenever and wherever they liked.

Moreover, it was also believed that many of the dockyard workers were members of the Triads, the Mafia-like secret societies that practiced big- and small-scale villainy in a colony that was riddled with corruption and whose police force did little to clamp down on the gangs' activities. Anyone who objected to the workers' right to smoke, gamble, or eat where they liked could easily be hacked to pieces with a meat cleaver, the weapon of choice, both now and back then, for the Triads' pitiless hit men.

Yet despite all this poisonous stew, C. Y. Tung remained confident that things were going well and that his Seawise University project would work out in the end—that is, until people spotted smoke shortly after 11:00 a.m. on Sunday, January 9. Wintertime weather in Hong Kong is more often than not calm, clear, and, to most non-Asians, comfortably warm. That morning, barbecues were getting under way on a score of rooftops at the western end of Hong Kong Island and up on the mid-levels of Victoria Peak. As the first gin and tonics were poured, those gathered together couldn't help focusing on the great newly painted white liner gleaming brightly in the southern sunshine.

Then, at once, they began to notice that all along the ranks of portholes, from the ship's stem to her very stern, and on three of her decks, black, oily smoke started streaming out into the clear winter air. This joined into a cloud, which the morning breeze blew in their direction. Within minutes, the lunching hundreds could smell an acrid, chemical, greasy industrial smoke, heavy and sinister.

The greatest old ship of the British Empire was on fire.

But no fireboats came, not right away. A local accountant was giving his English fiancée's parents a Sunday boat ride around the harbor, and the four of them stayed, entranced, for three hours. For the first hour no rescue craft came, and they watched with amazement as the blazes consolidated, as explosions began to rock the ship, and as curtains of fire began to race uncontrollably along the vast superstructure. It became swiftly clear that the great liner was doomed to burn, and to sink—and that there were people aboard who needed to be rescued.

It was a full hour before the first fireboats arrived, including the *Sir Alexander Grantham*, the powerful red-painted vessel that had welcomed the liner six months earlier with vertical jets and curtains of colored water. Instead, today, tens of thousands of gallons of hastily inhaled seawater would be thrust onto the burning ship to drench and protect the police and ambulance crews who were trying to get everyone off her. Which they did, and successfully, even rescuing a worker's child (who of course should not have been on the ship), who was tied to a rope and lowered over the stern to a waiting tugboat. They also found and took to safety C. H. Tung, C.Y.'s son, who was making a Sunday visit and who would in time inherit the family firm.

No one died that Sunday; nor was anyone badly hurt, except for one man who broke his leg jumping from a porthole and onto a waiting police launch. Nor was anyone hurt the next morning, after all the hoses had been turned off and when the ship was charred black and smoldering and listing alarmingly to starboard, readying herself to founder.

This she did almost precisely at noon, slumping down on her starboard side and into the mud. She died with more of a whimper than a bang, though, with her port-side hull red hot, with fires still burning and setting off dull thumps of explosions that could be heard from deep in her bunkers.

Tung was in Paris at the time, and wept upon hearing the news. In time, insurance paid up eight million dollars, more than twice Tung's purchase price for the ship, though less than he had spent in total since the ship left her berth in Florida. That insurance payment raised eyebrows; the probable presence of the Triads aboard raised eyebrows; the savage distemper of the Communists in the labor unions working on the ship prompted suspicions; the delay in the fireboats' arrival led to still more puzzlement; and Hong Kong's reputation as a sink of corruption made few confident that the obvious two questions—who had done it, and why?—would ever be either properly asked or fully answered.

And this remains the case. All that is certain today, nearly forty years on, is that fires broke out simultaneously in nine different places and that they were deliberately set. But the official reports do not offer a sophisticated conclusion as to who might have set them.[*] The courts have never decided why or who, and the confidential police file on the matter remains open to this day. Most Cunard officials believe the ship was the victim of sab-

---

[*] During colonial times the Hong Kong authorities came up with some interesting rulings—one of the best remembered involved the violent death in 1980 of a twenty-nine-year-old gay policeman at a time when homosexuality was illegal in the territory. (It was legalized only in 1991.) His death was ruled a suicide, even though the officer had five bullet wounds in his chest. It was widely regarded as improbable that anyone could have shot himself there more than once, let alone five times. The court noted, however, that only one of the five shots was lethal, and that the officer had not shot himself in either the head or the heart. Death by his own hand would have been painful but not impossible.

otage, most probably politically inspired. The Tung family still regards any political motive as wholly improbable.

The ship was scrapped where she lay. About three-quarters of the steel from the wreck was salvaged. Divers employed by a South Korean company worked with acetylene torches to cut her into bite-size chunks, which were then taken away to be smelted into girders for use in Hong Kong's many new housing projects. Her brassware, screws included, was sold to the Parker Pen Company, and five thousand "QE75" pens were made with brass inlays and nibs, and with a solemnly worded certificate of authenticity from the Tung's Island Navigation Corporation.

The salvage work became increasingly difficult, as the divers had to venture deeper and deeper—one of them blowing himself up with an accidentally placed gelignite charge—until work was halted in March 1978. Twenty thousand tons of the liner remained buried in the mud south of Tsing Yi Island—including her keel, John Brown's Keel Number 552, which had formed the enduring base of the largest riveted ship ever made.

For years, her resting place was marked on navigation charts with the green notation that signified a wreck. There was a buoy, and passing mariners were cautioned not to get too close. Then the island near where she sank was expanded with landfill and pilings and cement, and most of the ruined vessel now lies beneath the wharves and walkways and crane tracks of the territory's main container port.

The saga has an interesting coda, which began to unfold soon after C. Y. Tung's death in 1982. His son C. H. Tung took over the family's new-formed shipping company, Orient Overseas Container Line (OOCL), the firm having realized, rightly, that containerization was the wave of the Pacific's maritime future. But it was not a future the company principals were apparently geared to meet, and the firm soon ran into heavy weather, and needed cash. But—and here is the irony—it was not Taiwan that eventually bailed Tung out, but Communist China. China's banks loaned him $120 million, and from that moment on, according to one sardonic Hong Kong civil servant, Tung was effectively owned by the mainland party, and henceforward did Beijing's bidding as his masters saw fit.

They did not wait long to call in his obligations. Britain's century and a half of sovereignty over Hong Kong would end at midnight on June 30, 1997, and the Chinese would take over. Beijing decided that it would be Tung Chee Hwa (C. H. Tung, newly styled) who would become the first Chinese-appointed chief executive of what would now be called the special administrative region of Hong Kong. Mr. Tung was a shipowner, a man with no knowledge of running a country, or even part of one. But that was perhaps not the point. For, ever since China's banks bailed him out, he was in China's debt, and he would be unfailingly loyal.

He had shown this already, in a speech given just one month before the handover of sovereignty. It was a speech that chilled some spines: "Freedom is not unimportant," he said. "But the West just doesn't understand Chinese culture. It is time to re-affirm who we are. *Individual rights are not as important as order in our society* [my emphasis]. That is how we are."

The Pacific was slowly shifting gears. Those Europeans who had for so long pulled the levers of power in the region were gradually leaving, saying their farewells. A new order was coming to the fore. The United Nations had been eagerly promoting the benefits of decolonization ever since the end of the Second World War, when seven hundred fifty million of the world's peoples were governed by outsiders and aliens. By the 1970s the old imperial possessions were beginning to shrink like ice cubes on a stove top. New commanders were on the bridge, giving their directions, mouthing their orders, dictating the region's future with either caution or swiftness or relief, according to their own devices. On great ships, on tiny islands, and along exotic coastlines, all around.

This new order produced many farewells, some poignant, others lethally violent. The sabotaging of the great British ocean liner, and the elevation to power of a man who was so intimately involved with her fate, serve as a potent symbol—but as a symbol only. Other farewells were very much more savage, and some had global repercussions. The most notoriously dramatic and costliest of these had to be the enforced departure of the Americans from Vietnam in the spring of 1975. For it was only then, and after more than a century, that the various states* of Indochina, bastions of the western Pacific hinterland, were at last able to rule themselves again. Foreign domination had utterly defined the recent history of the Southeast Asian peninsula. But the process of restoring governance to the various Indochinese peoples (the Vietnamese, the Lao, and the Khmer) was a far more protracted business than the infamous nine years of America's own ill-judged involvement there, which cost the lives of fifty-eight thousand of its young men and women, and two million or more of those who claimed these countries as their own.

The French had ruled in Indochina—had *owned* Indochina, as colonists like to claim—ever since their capture of Saigon in 1859. Though the French were as imperially oppressive as any, they are generally seen today as having been more benign and cultured than such philistine ruffians as the Dutch and the British, and the legacy of their sovereignty—a local fondness for wine; the number of surviving *boulangeries*; the pidgin French still heard there in the cities from Hanoi to Luang Prabang, from Kompong Som to Hue—is still well thought of, and offers to yesterday's "Indochine" a veneer of exotic and erotic Eastern chic. Even so, all empires, benign or brutal, inevitably fade, and the drawing down of French influence in South East Asia would get under way swiftly, soon after the Second World War.

---

* Thailand, alone in the region, managed, by the skillful diplomacy of a succession of strong leaders, to retain its independence through all the years of foreign domination, never bowing to the demands of either the French to its east or the British to its west.

It is conventional to see France's humiliating defeat at the Battle of Dien Bien Phu in far northern Vietnam in 1954 as the beginning of the end of France's land tenure in the western Pacific. But one other episode, half-forgotten now, marks the ultimate cause of the whole unlovely mess, of which Dien Bien Phu was but one part. It occurred in 1945, and it concerns a much-decorated British Indian Army officer named Douglas David Gracey. The strange events that briefly enfolded him offer invaluable context for what would occur in the years following.

For Major General Gracey had been handed the unusual, unprecedented appointment, at the Pacific War's end, of commander in chief, Allied Land Forces French Indochina. He was dispatched to Saigon—a senior British army officer from India ordered to preside over a French colony that was at the time occupied by defeated Japanese invaders must surely be one of the more curious pieces of political flotsam to wash up in the wake of war. Gracey was aware of the sensitivities of the situation: for six heady months, and from a hastily built British military headquarters in southern Vietnam, he directed his twenty thousand soldiers through one of the most bizarre periods in modern Indochinese history—with the specific avowed aim (since restoring the imperial status quo was Winston Churchill's stated policy) of returning the territory to the imperial rule of the French. His expedition's name was Operation Masterdom.

Gracey and his troops had been ordered into Saigon because those Allied politicians who in 1945 were planning Indochina's fate—a peripheral issue in the Potsdam Conference, held after victory was ensured in Europe—had been blindsided by Japan's unexpectedly swift surrender. Vietnam had long been occupied by Japan, and now the Japanese soldiers involved in the mechanics of that occupation had all to be disarmed and packed off home, much as their brother soldiers were to be sent home from various other places in the Pacific.

But what Gracey had not expected was the impassioned oppo-

sition to his mission by the Viet Minh nationalists in Saigon.[*]
No matter that he was there to turf out the Japanese occupiers:
as soon as he arrived, in September 1945, he noted that the road
from the Saigon airport was lined with people waving Viet Minh
flags and holding posters that supposedly welcomed him and his
forces, but that demanded also that the French colonists leave.
Since British policy (Churchill's policy) was precisely the opposite, Gracey smelled trouble.

He first refused point-blank to cooperate with the Viet Minh.
His job, as he saw it, was simply to free all Allied prisoners of
war, to ease the French back into running the country, and to
get the Japanese garrisons out of the country. The Viet Minh did
not take kindly to the general's insouciantly dismissive attitude.
They staged strikes and closed down the Saigon market. Gracey
retaliated by shutting down the newspapers; declaring what was
effectively martial law; and freeing a particularly violent group
of former French soldiers, who promptly armed themselves,
initiated a citywide version of a coup d'état, and embarked on
acts of vengeance against everyone who stood in their way—Viet
Minh nationalists most especially.

Fighting erupted, and quickly spread everywhere. Gracey and
his infantrymen and his kukri-wielding Gurkha battalions from
Nepal, tore into the fight with gusto. His superiors back in Singapore told him to stop, saying the battles were none of his business, were nothing to do with Britain. But such was the ferocity of
some of the attacks (one of them with rifles, spears, and poisoned
arrows) on British positions that he was eventually given carte
blanche and told his new duty was to "pacify" the region.

---

[*] The Viet Minh's leader, Ho Chi Minh, had issued his Vietnamese Declaration of
Independence just two weeks before. It begins, famously, and familiarly to Americans:
" 'All men are created equal. They are endowed by their Creator with certain inalienable
rights, among them are Life, Liberty, and the pursuit of Happiness.' . . . Those are
undeniable truths. . . . Nevertheless, for more than eighty years, the French imperialists,
abusing the standard of Liberty, Equality, and Fraternity, have violated our Fatherland
and oppressed our fellow-citizens."

That's when Gracey made one of the most curious of all post-war decisions, aware that because he had not had time to ship them out, he had thousands of disarmed Japanese troops still in the area. Since they knew the city and knew how to fight, Gracey gave them guns and demanded that they stand alongside his British soldiers against the Viet Minh nationalists.

Of all the many bizarreries of the time, this was among the most extreme. The notion that Japanese troops would be armed by those who had recently vanquished them and that they would then be compelled to fight under a British flag alongside Nepalese soldiers for a French colonial ideal against a Vietnamese force that was demanding its own people's independence is well-nigh incomprehensible.

But it worked. With the help of the Japanese—"they did their job with characteristic efficiency," said one Gurkha officer, noting, in addition, that they may have reduced "casualties among our own troops"—the British did in the end succeed in pacifying the region.

By October, matters had quieted down enough that the French could indeed return, as Churchill had ordered. The magnificently aristocratic French general Philippe Leclerc de Hauteclocque assumed the reins in Saigon, and early in 1946, Major General Gracey was able to conclude his tenure in Vietnam, to take his soldiers back with him to India, and to resume something of a quiet life.*

Unwittingly, though, Gracey had so angered the Viet Minh by his disdain, his arrogance, and his brutal battles against them that some insist it was he who case-hardened Ho Chi Minh's opposition to the West, and indeed to all continued Western interest in the region. Ho's opposition to all future outside in-

---

* Gracey was later appointed commander in chief of Pakistan's new army, following independence in 1947. He was made a full general, and with his knighthood and assorted honors, he ended his life in 1964 as General Sir Douglas Gracey, KCB, KCIE, CBE, with two Military Crosses for gallantry in the First World War.

terference in Indochina became, from that time onward, unyielding, implacable. Critics of Gracey's "ruthless" and "overtly
political" pacification campaign have long blamed him for
standing in the way of what could have been peaceful progress
to self-government.

The Vietnamese path to independence, of shedding their sub-
mission to a European power, was long and bloody. The Vietnamese today speak of the First and the Second Indochina Wars—the
first, pitting the Viet Minh against the French; the second, pitting Vietnam's North against its South, with the Americans in
this case heavily and vainly trying to keep the two young countries from becoming dominated by Communists. The fighting
involved in both wars lasted for thirty years. The first war cost
500,000 Vietnamese lives and 90,000 French. The second resulted in more than 1 million North Vietnamese dead, 200,000
South Vietnamese, more than 58,000 Americans, and an assort-
ment (Australian, Koreans, Thais) of more than 5,000 others.
All told, well over 1.5 million Vietnamese died, and well over
160,000 Europeans and Americans—and in the end, the Indochinese were fully back in control of their own affairs.

The way stations of the two conflicts are fading fast along history's conveyor belt. Some of the names and events once so famed
have receded with pitiless speed. Who now recalls Bao Dai, the
perfumed and bejeweled final emperor of Annam, the last ruler
of the Nguyen dynasty, who gave Vietnam its name and who was
a puppet of the French, more famous in Monte Carlo than he
ever was in the valleys of the Mekong or the Red River? What is
now known of the Navarre Plan of 1953, a scheme by which the
French had hoped to win back their influence from the guerrilla armies of the Viet Minh, and which was formally approved
by the American headquarters in Hawaii? Do any now recall
General Francis Brink, the Cornell-educated American infan-

tryman who headed the first-ever U.S. Army headquarters in Saigon, established back in August 1950—and who shot himself an improbable three times in the chest in his office at the Pentagon because, the army said later, he was depressed? Questions about what drove him to his death—or of how anyone could shoot himself three times, anywhere—have surfaced now and then; but General Brink's medical records were accidentally burned, and the suggestion that he stumbled on some misappropriation of funds or the smuggling of drugs, and was silenced, has been discounted and, like so much else from Vietnam, forgotten.

Dien Bien Phu lingers somewhat in the memory, though. In November 1953, three battalions of French paratroopers dropped from squadrons of aircraft, and established their new fortress in a long valley on the border with Laos. Within weeks this swath of low-lying territory had been transformed into a formidable-looking base. There were two long airstrips, scores of gun emplacements, and subsidiary hilltop forts with winsome female names such as Beatrice, Huguette, Gabrielle, Claudine, and Eliane—which were supposed to help win the support of the war-weary *citoyens* back home, but which actually did the opposite.

Ho Chi Minh had the measure of the giant base almost from the start. It was said that he once took his topee from his head and turned it upside down. He thrust his fist into the concavity: "[T]he French are here." Then, with a sly grin, he traced his fingernail around the rim: "[A]nd we are here."

His equally sly military commander, Vo Nguyen Giap, had been preparing for weeks, bringing in artillery pieces bolt by bolt, trunnion by trunnion, barrel by barrel, along the maze of jungle paths. In total silence his soldiers dug rabbit warrens of trenches to within feet of the French lines. And once the howitzers up on the hillsides had been trained and the trench mortars readied, on March 13, 1954, one of the greatest, saddest, most heroic, most Orientally impudent and Asiatically triumphant battles of recent times got under way.

At a signal from Giap, a thunderous artillery barrage was unleashed from up on the surrounding hilltops, announcing an assault that would go on, uninterrupted, for a horrendously lethal fifty-four days. Day by day the French were pushed into what must have seemed like the unforgiving jaws of a meat grinder. Discipline held, and there were heroic displays that have never been forgotten in France to this day. But it was hopeless. The position was entirely surrounded; the odds, overwhelming.

The French artillery commander committed suicide, killing himself with a grenade for bringing dishonor (as a Frenchman naturally would put it) to his country. The most senior French general was captured in his bunker, well-nigh unimaginably. So desperate was the situation that brief consideration was given to asking the Americans to use tactical atomic weapons, to help drive the unstoppable Viet Minh away.

But that never happened; and for all its ardor, the fighting by the French turned out to be, essentially, all for nothing. They finally surrendered when their last central redoubt was overrun by Viet Minh troops at 5:30 p.m. on Friday, May 7, 1954. The very next day, the topic of Indochina was formally added to the agenda of the Korean War peace conference that was under way in Geneva; and there, in full view of the international community, the French announced their formal withdrawal from all of Indochina. They were done. They were out.

More ominously, the conferees also then divided Vietnam into two. A demilitarized zone was established around the Seventeenth Parallel. To its north were Ho Chi Minh and Vo Nguyen Giap and their Democratic Republic of Vietnam, a brand-new state that was now backed by Moscow and Beijing. To its south was the State of Vietnam, nominally ruled by Emperor Bao Dai, but in effect run by the United States and its chosen surrogates. The mutual hostility between the two states simmered through the remainder of the fifties; there were insurrections and outbreaks of dissidence and monkish demonstrations and assassinations, met with so swift a gathering of support from Washington that President John F. Kennedy was formally warned he was doing no more than replacing the French, and that he was likely to bleed just as badly as they had bled.

But both he and his successor ignored the advice; and on the flimsiest of pretexts, President Lyndon Johnson won congressional approval in the summer of 1964 to send troops to Vietnam without any formal declaration of war. American involvement in the heartache of the Second Indochinese War accelerated mightily from that moment onward. Yet the arc of progress for this conflict ended just as it had for the French, and just as Kennedy had been warned: with defeat, withdrawal, collapse, and humiliation.

The way stations and the dramatis personae of this second war were once familiar icons written in universally known shorthand—there was the Ho Chi Minh Trail, there was Operation Rolling Thunder, there was the My Lai Massacre, Agent Orange, Khe Sanh, the Siege of Hue, the Tet Offensive, Hamburger Hill, Da Trang, William Westmoreland, Hanoi Jane, Le Duc Tho, Operation Linebacker, the Cambodian evacuation operation known as Eagle Pull, and its Saigon equivalent Operation Frequent Wind. Many of these people, places, or events have now to be looked up in indexes; and as one generation is succeeded by the next, those who struggle to remember are fast being overturned by those who never knew. Sixty percent of today's Americans were unborn when the war came to its end. It was not entirely incredible when some late-night comedian remarked that a sizable number of modern high school students fully believe that the Vietnam War was fought against the Germans.

Nine million American men and women served in the military in Vietnam. At the fighting's bitterest, in 1968, more than half a million American troops were in the country. Thereafter, as domestic distaste for the conflict grew, the numbers began to drop, by many tens of thousands every year—until, at the very end, at the close of 1974, just fifty soldiers and marines remained. And then Operation Frequent Wind was staged in the final days of April 1975, to get those final fifty out, and to collect all available others and their friends, and to bring America's formal role in Indochina to its sorry conclusion.

The last American hours of Saigon in 1975 were both wretched and poignant. Even the most Panglossian in the capital knew that their city, by then quite surrounded by Viet Cong army units, was about to fall. Elaborate plans had been laid for a helicopter evacuation of all remaining Americans and those friends and helpers and foreign journalists who were known as "at-risk aliens." Booklets were published with instructions, to be kept as secret as possible: if a radio broadcast began with the phrase "The temperature in Saigon is 112 degrees and rising," and was followed by thirty seconds of Bing Crosby singing "I'm Dreaming of a White Christmas,"* then it was time to go, with all deliberate speed, to one of a dozen designated spots where U.S. Marine Corps and Air America helicopters would be waiting.

But over the din of shellfire and shouting, few in Saigon heard the broadcast; fewer still understood it. Rumors spread wildly. Thousands crowded the landing zones, panic-stricken, frantic, desperate. Previously chosen landing sites came under fire; new sites had to be found in double-quick time—marines were to be seen cutting down tamarind trees to clear the way for the giant flying machines, which came stuttering to the city every few minutes, from a mighty armada of ships hurriedly assembled twenty miles out in the Pacific. The American ambassador was one of the last to go, shortly before 5:00 a.m. on Wednesday, April 30. "Tiger is out" was the coded signal, meaning that the American official presence was officially at an end.

Except that it wasn't. Someone had forgotten to collect the ten

---

* Japanese journalists unfamiliar with the song had to have it sung to them by American colleagues, and committed it to memory.

remaining marine bodyguards, and a final helicopter had to be sent for them. At 7:53 a.m. this final machine took off; and at 8:30 a.m. it landed aboard the amphibious assault ship USS *Okinawa*.

Helicopters employed in the final frantic hours of the evacuation of Saigon were tipped into the Pacific, useless symbols of American power squandered in the hopeless quagmire of an end-of-empire war that should never have been fought.

Three hours later, precisely, North Vietnamese tanks smashed through the gates of the Presidential Palace and raised the Viet Cong flag.

Western occupation of the continental coastline of the far western Pacific was now, after 175 years, at an end. Those foreigners who would later come to this part of the world would do so only by invitation, and with the permission of those whose land it was now, at last.

The Pacific's British colonists, by contrast, departed with rather less fuss. Instead of being run out on a rail as the Americans had been in Southeast Asia (and as had the French, Germans, and Japanese), they went like the country house guests who, on a muffled cough from the butler, suddenly realize they have over-stayed their welcome. So they leave in a state of mild confusion, dropping things, tripping over their shoelaces, shutting their fingers in doors, and saying to their beaming hosts all too many farewells, in their embarrassed and befuddled haste to get away.

Until the 1970s, maps of the Pacific, like those of the rest of the world, were still awash with British imperial pink. The leg-atees of Captain Cook and Stamford Raffles still reigned over tiny but critical morsels of land in the ocean west of the dateline. This meant that British officials governed untold numbers of dark-skinned native peoples—Kipling's "lesser breeds without the law." Britain's islands there were never to be abandoned, never to be forsaken. They were a reminder of what John Milton had once called England's "precedence of teaching nations how to live." This notion recalls the old joke of many a 1950s mother, who would tell her irregular child to eat prunes because the fruit had much in common with British missionaries in these parts: since prunes, like missionaries, "go into dark interiors and do good works."

Malaya, Singapore, Papua New Guinea,* Brunei, Sarawak, and North Borneo—they all form the fortress of the western flank. Then, farther eastward and out at sea, there are the poster chil-dren of Britain's blue-water Pacific empire: the Gilbert Islands, the Ellice Islands, the Solomons, the New Hebrides, and Ocean Island. More distant still are the lonelier imperial holdings: Fiji, Tonga, and the Pitcairn Group. Finally, up north, alone and pre-siding magisterially over all else in the Pacific, is Hong Kong, an

---

* As perhaps befits a country with 848 national languages, Papua New Guinea's history is deliciously complicated. The southern half of the island's east (the west belonged to the Netherlands) had first been annexed by a policeman from the Australian state of Queensland; the northern half, which had been German, was then taken also by Australia, during the Great War, and the annexation was ruled legal under the 1919 Treaty of Versailles. Both halves were then placed within the British Empire, but were run by Australia, and won their independence in 1975—with Britain's Prince Charles presiding over the ceremonies.

outpost of Britishness then little more than a century old, quite alone on the underbelly of China, looking especially vulnerable and impermanent. Few in the 1970s, though, could imagine just how little time was left.

So far as Britain's long invigilation in the region was concerned, the clock was now ticking, insistently. There were three essential reasons. London's treasuries had been emptied by the Second World War, and it had become too expensive to maintain all the imperial flummery and panjandrumry around the world. In the new, postwar world, there was a growing feeling that empires were unfair, immoral, and unfashionable, and this led to stirrings of revolt and disaffection among more than a few of the ruled peoples. All of a sudden London was facing the gnawing realization that this empire really had to be brought to an end.

A program for the empire's divestment was devised, and the whole carelessly assembled confection of countries, enclaves, islands, lighthouses, and reefs and redoubts, on which no sun had set for scores of years, was let go, and Britain started to ease herself out.

Malaya was the first to go. The Duke of Gloucester came and read a message from his niece, Queen Elizabeth, wishing all the Malays deserved good fortune. The British rulers of North Borneo then packed for home six years later. Then, out at sea, the various colonial entities that occupied the 2.5 million square miles of ocean (and about ten thousand square miles of land within it) that were ruled as one by the British High Commissioner, as his full title had it, "in, over and for the Western Pacific Islands," were given their independences in stages.

One of them, the Kingdom of Tonga, a centuries-old monarchy that had been under British protection since 1900, wrested itself free of this most cumbersome arrangement in 1970. So, too, did Fiji: Prince Charles did the honors there, though he was four hours late because his royal plane broke down in Bahrain. He was given twelve live Fijian turtles as a gift, and a set of gold cuff links. The prince said he was glad they hadn't given him a silver spoon, since he and his siblings had been born with those already. As it happened, the violence for which the Fijian islands had long been known—the fork used to eat the Reverend Mr. Baker is still in a display case in the Fiji Museum—was perpetuated in postcolonial times, with coups d'état and constitutional crises on a heroic scale.

The imperial crown jewels of the Western Pacific, the Gilbert and Ellice Islands, achieved their self-rule in 1976. They had been an immense Pacific possession of two million square miles, first made famous abroad by their former governor Arthur Grimble and his 1932 book, *A Pattern of Islands*, about running them—every British schoolboy raised in the 1950s knew passages from this classic book by heart and *by jingo*! The island groups had been stitched together for reasons of London's administrative convenience—notwithstanding the fact that the Gilberts were populated by Micronesians and the Ellices by Polynesians, peoples not always on the friendliest of terms. So once the islanders scented the merest whiff of freedom, they voted to break away, not just from Britain but from each other. Today the former are known as Kiribati and the latter as Tuvalu, separate entities in the wide and decidedly non-British Pacific.

The British Solomon Islands Protectorate—best known in America either for the savage fighting on Guadalcanal or for being, during their wartime occupation by the Japanese, the place where Lieutenant John F. Kennedy was shipwrecked and where he carved the famous message on a coconut that led to his rescue—then became independent in 1978. (The American astronaut John Glenn, who had served as a marine in the South Pacific, was part of the celebrations.)

The New Hebrides, nearby, went two years after. Since 1906 these islands had been run for complicated reasons by a condominium of two uninvited European powers, the British and French. Two bureaucracies had been set up, exactly mirroring each other—the French official in charge of drains in Port-Vila, for example, had a British counterpart who was charged with exactly the same task.

The language of New Hebridean administration had to be translated twice, Canadian style, and sometime thrice, since the doubly colonized citizens actually spoke a third, Creole tongue called Bislama, and many of the territory's more important legal documents had to be rendered into its mysteries as well. The colony's laws were in consequence a magnificent mess: any New Hebridean could choose to be tried or to sue under either Napoleonic or Magna Cartan legal principles, or else by Melanesian local law in a native court—the chief justice of which was appointed by the nominally neutral king of Spain.

Two police forces, their officers wearing different uniforms, did their best to keep civil order in turns, performing their respective duties on every other day. (History does not record which police officers were the more lenient.) Finance was exceptionally cumbersome—French francs and British pounds sterling being interchangeable, Australian money accepted, and banknotes issued by the Paris-based Bank of Indochina in stores everywhere. National holidays were so numerous and so keenly celebrated in the perpetually torrid climate that little work was performed anyway—and in time, the whole unholy and intractable mess of governance exhausted everyone, collapsed internally, and was finally called to a halt in 1980, with the new and present entity being called Vanuatu. The French, who had wanted to retain a foothold in this corner of the ocean, sulked mightily when independence was eventually declared, and their officials stormed away from the islands with all their telephones, radios, and air-conditioning units, in a display of official petulance seldom rivaled.*

---

* Though the French left their colonies in Indochina and the New Hebrides, they still retain ultimate control over two mid-ocean Overseas Collectivities, fully represented in the Paris parliament: French Polynesia and Wallis and Futuna. There is also the Sui

In the wider Pacific, foreign empires were nearly all done with by the 1990s, after this two-decade cascade of self-determination. Only one pair of imperially run places of significance then remained. By happenstance they were places that in all senses were the polar opposites of each other—and both of them were British.

One was the hugely populous, hugely rich, and world-renowned colony of Hong Kong. Four hundred square miles in extent, it was a crowded mountain home to six million people, most of them Cantonese-speaking Chinese, administered as they'd been since the middle of the nineteenth century by a tiny and privileged corps of very British diplomats, civil servants, and politicians.

The other was the tiny (just two square miles) Pitcairn Island group, which was hardly populated at all, a starveling child of empire (since 1838) with fewer than sixty people. It remains a persistent and unwanted colony today; and though its origins as the refuge of the nine men who, under Fletcher Christian's leadership, mutinied in 1789 against Captain Bligh and his captaincy of HMS *Bounty* are seen by some as the stuff of swashbuckling romance, its present situation has been clouded by a deeply unedifying scandal.

I first visited Pitcairn in 1992, by courtesy of the New Zealand captain of the HMNZS *Canterbury*, who gave me a ride on his frigate as she was making her way from Auckland to Liverpool to take part in a naval review to mark the queen's fortieth jubilee.

We took ten days to reach the island—the colony in fact comprises four: Pitcairn itself, together with nearby Henderson, and then the further atolls of Ducie and Oeno, each about a hundred miles from the others. We saw not a single ship on our passage,

---

Generis Collectivity of New Caledonia, which has proved politically intractable because of the large number of French settlers, but is expected to become fully independent by 2018; and there is an uninhabited mystery island, Clipperton, off the Mexican coast.

sliding into an ever-lonelier sea each day that we pressed farther eastward.

Finally we spotted Pitcairn, a tiny speck of green volcanic hills that rose abruptly out of an otherwise empty tropical ocean. We were met by a flotilla of longboats. Islanders are always on the watch for any sign on the horizon that might suggest an approaching ship; when they saw us, dozens of miles away, they launched every craft available to make sure as many of them as possible got to see a modern ship of war—which, after all, was what the *Bounty*, burned to the waterline but still vaguely visible, had been.

We surfed in, dangerously, to the tiny concrete pier. I walked, painfully, up the viciously steep Hill of Difficulty, the one approach to the colony's small shanty settlement of Adamstown. There was not the vaguest hint of anything untoward. The island was warm, sleepy, friendly. The walking—up and down red laterite roads onto green peaks from which all else was blue, an empty, cloudless pale blue sky and an empty, shipless deep blue ocean—was wearying, hot, lonely.

But there were always stories to be found. At the top of one hill, for instance, I met a Japanese man with a pup tent and a large radio transmitter: he was a ham operator who had come to Pitcairn to broadcast messages and persuade those who heard him to write the so-called QSL "I have heard your transmission" cards, asking for a dollar or so each time in return. People had sent him ten thousand dollars thus far, he said, and felt it was time to go home. Could we take him on to Panama? The captain said no. The Japanese man, only a little crestfallen, crawled back into his tent with a book. He said he was content to wait for another passing ship, maybe in a month or so.

There was a pineapple plantation in a nearby meadow, and a couple of Pitcairners with a Cryovac food packaging machine told how they once had had plans for exporting air-dried pine-

apple to the outside world. But then the French resumed testing nuclear weapons on Mururoa Atoll, six hundred miles to
the west; and even though the prevailing winds were blowing
away from Pitcairn, such of the world as might perhaps have
been interested in buying Pitcairn pineapples decided, in short
order, that Ecuadorian and Philippine pineapples were safer
bets, less likely to be radioactive. Later experiments with Pitcairn honey—a New Zealand beekeeper was brought in to offer
training—proved more successful, though; and *Bounty* products
can be seen today in high-end grocery stores in London. Otherwise, only the sale of postage stamps and of carvings made from
the rock-hard miro wood found on Henderson Island* provide
some islanders with a modest income. The island government,
such as it is, canvasses for outsiders to come and settle. Few have
taken the bait.

The recent scandal hasn't helped. It started to unfold in 1999,
when a young female police officer from England was sent out to
Pitcairn (which had never had a regular police force) for a six-
month training exercise and discovered a widespread culture of
sexual abuse. It seems that Pitcairn men regularly had sex with
girls as young as ten, and it was not uncommon for girls to have
their first pregnancies when they were as young as twelve. The
islanders said they saw nothing unusual or improper about the
practice, and claimed they were following established Polynesian custom—the British mutineers having brought Tahitian

---

* In the 1980s a West Virginia coal mining magnate named Smiley Ratliff tried to lease
Henderson Island to establish a mid-ocean colony. He promised to build an airstrip
and to provide a ferry plying to and from Pitcairn, which would connect the island with
the outside world. The British government gave his plan serious consideration, but it
was pointed out that the rare and flightless Henderson rail as well as the fruit-eating
dove, an endemic warbler, and a lorikeet lived on the island, and Ratliff was asked to
look elsewhere. Some Pitcairners remain chagrined, and the British government now
subsidizes an occasional ferry service out of the French-owned Gambier Islands, where
there is an airport.

wives with them in 1789, and the island stock ever since being an admixture of Anglo-Pacific genes and cultures.

When the police officer reported her findings, the British courts were not so understanding. Detectives promptly descended on the Pitcairn community; then lawyers—some to be involved in historical challenges over exactly who had sovereignty and legal jurisdiction over an island that had only infrequently paid more than lip service to any legal system at all—began what would amount to a five-year field day.

Victims were found. Witnesses were identified. Charges were brought. Seven men, including the island mayor and longboat coxswain Steve Christian, a direct descendant of Fletcher Christian, were arrested. It was first argued that the case should be heard in New Zealand, but the courts decided it should be heard on Pitcairn—with the result that more judges, lawyers, witnesses, and reporters suddenly arrived in Adamstown to take part in the trial than lived in Adamstown in the first place. A satellite system was set up so that witnesses could testify remotely—it turned out that every one of the women involved now lived overseas. All the islanders' guns were confiscated, to prevent any possibility of violence on an island that was now bitterly divided over the issue.

The trial took forty days and cost some twelve million dollars. There was then a series of appeals, which wound their steady and convoluted way through the maze of the British legal system, right up to the then-supreme authority of the Privy Council in Buckingham Palace, which summarily rejected them. Six of the seven men charged were then sentenced to prison—except there was no prison on Pitcairn, and one had to be specially built.

Fears were promptly expressed by many in Adamstown that the whole affair was a devilish plot that would enable Britain to get rid of the costly annoyance that was Pitcairn. For the six able-bodied men put in jail would now not be able to man the longboats that brought in vital cargo from the island supply ships—with the result that the settlement would wither and die, and the remaining islanders would head west to join the refugees from an earlier crisis (a famine, and overcrowding in 1856) on the former prison colony of Norfolk Island, off the Queensland coast.

But wiser counsels prevailed. The men were locked into their cells in a jail they built themselves, from a prefabricated prison kit sent down from Britain. But every time a ship hove to off the island to unload cargo, the men were briefly freed (supervised by corrections officers) and paid to get out the island longboats and transfer goods to the quayside. After which they were marched back up the Hill of Difficulty and into their cells, and locked in again until the next smudge of smoke was spotted on the horizon.

All six men eventually served out their sentences and were released back into the population. Since only one woman on the island is currently of childbearing age, and since these six presumably still fertile men are bound over to be exceptionally well behaved, it is now gloomily assumed that the colony of Pitcairn will wither and die anyway.

Until it does—most demographers believe it can last only until about 2030—some small number of tourists will come, once in a while, to see the relics of the famous mutiny: HMS *Bounty*'s anchor, her Bible, and her cannon. They will sample the local honey. They will buy tchotchkes made of miro wood and postcards with postage stamps. And they will stay in the newly made but quite empty jail—which, since it is one of the few structures on Pitcairn to have proper plumbing, is thought best used as a guesthouse. And those who visit this most remote outpost of Britain's former empire in the Pacific may choose to reflect, and wistfully, on the droll reality: that a place with such a lawless and inglorious beginning is now suffering, as a consequence of lawlessness, an inelegant and inglorious end.

An end rather less celebrated and memorialized than that of

its polar opposite in the Pacific, eight thousand miles farther northwest: Hong Kong.

It was only toward the very end of London's rule in the southern Pacific that those in the one colony in the northern Pacific started to question their own status. Hong Kong's bankers in particular began to fret—especially those who held or gave or thought themselves likely to extend mortgages, fifteen-year mortgages most commonly. These anxious moneymen started to ask a simple but obvious question: would those mortgages still be valid after fifteen years if the government of Hong Kong were no longer capitalist and British, but Chinese and doctrinally Communist?

They began asking this question in the spring of 1979, because fifteen years from that time very roughly spelled the date, in small print, on the half-forgotten last treaty that the British and Chinese had signed together, a treaty that, most important, had said that a major part of Hong Kong's territory was to be ruled by the British *as a lease*. The document had been signed in June 1898. The lease was for ninety-nine years. The bankers were wise enough to know what many ordinary colonial citizens had forgotten, had never known, or had chosen to ignore: that this lease would expire at midnight on June 30, 1997.

So the question the bankers wanted answered was: were the Chinese going to extend the deadline and allow Hong Kong to remain under British rule for some long and indefinable period far into the future?

The bankers asked the then-governor of Hong Kong, Murray MacLehose, to put this question to his Chinese counterparts in the Chinese capital, in what was then generally known as Peking. His Excellency duly flew there and was given banquets and taken to the Great Wall, the Winter Palace, and the Forbidden City and was accorded appropriate respect. And he got the answer to the bankers' question, from Deng Xiaoping, the Chinese leader at the time. It was not what he, the bankers, or anyone else in the colony wanted to hear.

There was absolutely no question, the diminutive Deng declared, of extending any lease. Hong Kong was most assuredly being shepherded back to its motherland. The Chinese wanted all of their territory returned. They wanted the New Territories. They wanted Kowloon. They wanted Hong Kong Island. They wanted, in short, everything. There was no point in any clever British lawyers spluttering that the three treaties signed during Victorian times gave some of the territory to Britain *in perpetuity*. As far as Deng Xiaoping was concerned, all three treaties were unequal and unfair, had no standing in law or modern reality, and could be torn up and turned into confetti at will. Deng insisted that everyone understand that Hong Kong's existence as a British overseas territory was coming to an end. June 30, 1997, was the fixed date when the bills came due. The timer was running down. Talks had better get under way to sort out the details.

The only thing that gave MacLehose any kind of reassurance was a phrase from Deng, uttered just before the pair made their farewells, and that the British governor then went on to repeat endlessly to the nervous, skittish, unhappy millions he greeted upon his return home, after flying over the dark plains of China to the brilliantly lit jewel that was his colony: "Set your hearts at ease." He said it a score of times. It was what the Chinese leader had asked him to tell Chinese citizenry down in Hong Kong. "Set your hearts at ease."

Yet this mantra was not at all helpful. The next years for Hong Kong would be an agony of a thousand cuts. Talks did indeed get under way, but they were dispiriting affairs. I flew to the Chinese capital for one of the earliest sessions, a mid-level British ministerial mission designed to prepare the ground for a visit the following year by Prime Minister Margaret Thatcher. Ordinarily, if Britain has influence in such talks, they have a sense of style, brio, confidence, fun, purpose. But what I watched in

Beijing was more like a solemn requiem in the key of gray, played out in the bitter cold of a northern Chinese winter week.

The city was a place of dreary monochrome—and was gritty, cold, smoky, unnaturally quiet, as if everyone were half-hidden and spoke in whispers. This, in 1981, was the neon-less, barely prosperous, cold-comfort Communist China of quite another era: there were armies of bicycles everywhere; clusters of soldiers huddling for warmth around street corner coke burners, seething masses of laborers in *hutongs* of amiable slums; with hugger-mugger gatherings of workers clad in gray or dark-blue Mao suits and with exhortative Communist slogans everywhere.

The officials who came in from Britain, men and women likewise swathed in gray (though some wore Savile Row pinstripes) would meet each day with lantern-jawed officials from the Chinese ministries, and then retire to their miserable gray-painted rooms in a huge gray-walled hotel on Chang'an Avenue, just along from Tiananmen Square. Each evening they seemed listless, unimpressive, puny—and to the extent that they talked out of school, it was to express their disappointment that the Chinese would not budge, would not agree to any kind of compromise or concession.

When Mrs. Thatcher came to China, she looked puny and unimpressive, too, and she proved no match for her Chinese counterpart. She winced visibly when Deng told her in undiplomatic terms that he could march into Hong Kong anytime he wanted, and retake it in an afternoon if he was so minded—the implication being that he was doing her a considerable favor by restraining himself from doing so. Whatever the tenor of her retort, the spirit gods were not with her that day, for on her way out of the meeting, she tripped and fell onto the steps of the Great Hall of the People in a mess of gabardine and mussed hair, looking vulnerable and weak as she was hoisted upright again by her team of men in gray.

Back at home, she may still have cut a heroic figure, fresh

from having had her soldiers win the return of the Falkland Islands from their Argentine invaders. But this did not interest Mr. Deng. Not one whit. This Pacific colony was not one she was going to win back from history or from him, now or ever.

This somber fact was eventually formalized by a joint declaration of the British and Chinese governments, made public shortly before Christmas 1984. All knew such a statement was coming, but a number of Hong Kong government officials wept openly at hearing the words. The only way of life they had ever known was coming to an end. Soothing nostrums were offered, suggesting that *nothing would change*, but everyone shrewd enough in matters Chinese knew that *everything would*, slowly but surely.

A mood of deep pessimism then settled on the territory—especially now that the Tiananmen Square demonstrations in 1989 that ended with the killings of so very many protesters had reminded residents of the terrors of totalitarianism. Over the next several years, thousands of Hong Kong residents were prompted to leave; to settle in Canada, Australia, the United States, and, to a lesser extent than seemed proper, the United Kingdom, which did not exactly open wide its doors to its nonwhite colonial subjects.*

Embassies and consulates in Hong Kong were thronged with visa applications. The lines outside the U.S. consulate snaked uphill for hundreds of yards. Sixty thousand people left in 1992. Parts of Vancouver changed their appearance almost overnight: sedate suburban houses on the airport road, which for decades had looked as though transplanted from Tunbridge Wells, were

---

* For many years only Falkland Islanders, Bermudians, and Gibraltarians enjoyed near-automatic right to British citizenship. Colonials from such territories as Anguilla, the British Virgin Islands, Saint Helena, the Cayman Islands, and Hong Kong were, so far as immigration to Britain was concerned, treated almost as Congolese or Brazilians. A more liberal policy was brought into force once Hong Kong had passed back into Chinese hands and the risk of London's airports being mobbed by Cantonese multitudes had abated.

sold to fleeing Hong Kongers for many times their asking price. They were then torn down and replaced by enormous mansions, all marble and onyx, without the fringing flower gardens for which the newcomers had no use.

As 1997 approached, an enormous digital clock was set up in Tiananmen Square, counting down the days and hours until the completion of what most people called "the handover." (Official London called it "the retrocession.") Fireworks were prepared, and local Beijing residents were urged to celebrate the territory's return to the motherland after its century and a half away. Not a single Chinese could be found who wished the territory to remain British—and small wonder. Hong Kong, after all, was one of the spoils of the Opium Wars, a legatee of the time when cocksure Britain had tempted the impoverished Chinese with cheap drugs from India and then reacted violently when the emperor tried to ban the commerce, and had used warships and overwhelming force to press home the British right to trade.

In Hong Kong the approach of its tryst with destiny—to use Nehru's phrase in independent India, won from the British exactly a half century before—was regarded very differently. Few in the territory could be found who regarded with equanimity the ticking down of the clock, and as hot April became sweaty May and then sweltering June, there was widespread apprehension that something, unspecified but ominous, would occur when the deadline passed. Television journalists flooded in from all over the world, and there were scaremongers among them, who spoke of seeing Chinese heavy armor on the far side of the border, wild-eyed men of the kind who had perpetrated the Tiananmen tragedy eight years before, massing beyond the wire and ready to storm in and make dreadful mischief once the territory was restored.

The British took months to pack and get ready to leave. I watched many of the details, most especially those involving the military, always a great component of maintaining any empire.

Ammunition, thousands of tons of naval artillery shells and torpedoes, was taken from the tunnels on Stonecutters Island and put on ships to go back to armories in England. The Sikh security men who had acted as its guards for generations (chosen to preside over flammable explosives because of their absolute religious prohibition on smoking) were offered jobs elsewhere. (Local hotels liked to employ Sikh doormen because of their magnificent, guest-impressing headgear.)

Likewise, thousands of Gurkha soldiers were stood down, and most of those who chose not to go home to Nepal were employed as security guards by one of the British firms that planned to remain in the territory post-handover. The doughty little motor yacht that used to take the official with the job of "District Officer, Islands," on his enviable inspection tours of the scores of little rocks and their fishing communities in his imperial charge, had its Union Jack replaced by a burgee sporting the insignia of the bauhinia, a sterile, orchid-like flower.

The last keeper of the Waglan Island Lighthouse was retired and replaced by a machine: a sorry fate, though it meant that the "Light at the End of the Empire," as we all knew the summit tower on the colony's eastern tip, would continue to shine out long after its owners had departed, winking its reassurance to ships gliding into and out of one of the world's busiest and richest ports.

The British Forces Broadcasting Service transmitters were unbolted from their barrack block studios, where they had been transmitting to British soldiers and sailors since 1945, and moved onto a waiting destroyer. The plan was to continue to broadcast until midnight, when the ship would leave Hong Kong waters and the station's final refrain (the national anthem, or maybe Vera Lynn's "White Cliffs of Dover") would fade into the ether as the craft moved away to sea.

The royal yacht *Britannia* arrived in port, in the very last few days of British rule, tasked to take away Prince Charles and the governor and the various diplomats and officials who would participate in the final ceremonials. The weather on the afternoon and evening of June 30 was ferocious, and driving tropical rain and wind turned Britain's final sunset military parade into a soggy maelstrom of misery, with the symbolism not lost on either side—either the skies were weeping for the departure of the British or the winds were raging to drive them from the scene.

The ceremony of the transfer of sovereignty, choreographed to the microsecond around the midnight hour, was sturdily impressive. The Chinese leadership had been flown in from Beijing by 747. Some five hundred truckloads of Chinese soldiers came across the border three hours before the deadline. The tallest and best dressed of them took part in a performance of crisp goose-stepping discipline that was as chilling as it was majestic. And when they raised their national flag, the pictures, transmitted

The June 1997 night of Hong Kong's long-awaited "retrocession" from the British Empire was cold and rainswept, drenching the ceremonial and those who attended, and making for an unseemly end to Britain's presence in the North Pacific.

live onto giant screens a thousand miles to the north, prompted Tiananmen Square to erupt in paroxysms (whether enforced is still not known) of fireworks and wild enthusiasm. The British forces who then took part in the flag lowering, moments before midnight, looked by contrast worn, weary, and unkempt, their uniforms still damp from the rainstorm, their performance to be seen as either shabbily charming or unhappily threadbare.

Once the flag was down, Hong Kong was no longer a British colony; and the governor's formal telegram was transmitted to the queen, a relinquishment done, the retrocession achieved, the ills of the Opium Wars overturned and finished with.

No one said sorry, though—the messages were all merely of farewell, and of Godspeed, and good fortune for the future. Prince Charles had declined to bow to the Chinese and promptly left for his safe haven on *Britannia*, and as soon as was decent, and with Chinese warships watching every move (though with the frigate HMS *Chatham* escorting), he sailed west out of Hong Kong Harbor, bound for the Philippines, and home.

In the small hours of that postimperial night, Prince Charles wrote a diary entry, which was somehow leaked and which made it clear that he had hated every minute of the experience. "After my speech," he wrote, "the president [of China] detached himself from the group of appalling old waxworks who accompanied him and took his place at the lectern. He then gave a kind of 'propaganda' speech which was loudly cheered by the bussed-in party faithful at the suitable moment in the text." The goose stepping, said the heir to the British throne, was unnecessary and ridiculous.

The two British ships moved silently through the night and into the wind—with Hong Kong Island glittering on the port side, Kowloon and the blue remembered hills of the New Territories on the starboard. In the distance ahead were the lights of Macau, still as Portuguese a territory as it had been since the sixteenth century, but now due for its own peaceable return to China in 1999.

There comes a point, fifteen minutes into the journey, where there is a slightly awkward maneuver for any vessel outbound from Hong Kong. It occurs shortly before the steersman must begin the long and lazy turn to port that brings his ship down into the fairway and out onto the powerful Pacific swells of the South China Sea. It occurred that final June night—or was it actually July now, the pitch-dark start of a brand-new day?—when the two captains, one of the *Britannia*, the other of the frigate, had to make the well-known and very slight course adjustment, to avoid one well-marked obstruction, underwater.

This was the submerged wreck of the RMS *Queen Elizabeth*, thirty feet down and sunk there, by sabotage and fire, a quarter century before. The wreck had significance that night: the foundering of this great old British ship, the finest and grandest vessel of her time, could be said to have marked the beginning of the end for all the imperial powers in the Pacific.

But the yacht and the warship passed her by that night without remark, and then, with the sunken wreck falling fast astern, spooled up their engines to full ahead all and officially left Hong Kong and their remaining British corner of the North Pacific Ocean, forever.

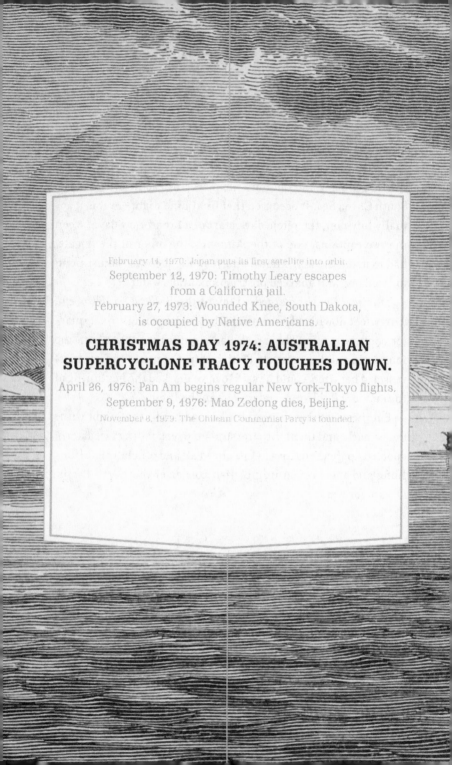

February 14, 1970: Japan puts its first satellite into orbit.
September 12, 1970: Timothy Leary escapes
from a California jail.
February 27, 1973: Wounded Knee, South Dakota,
is occupied by Native Americans.

CHRISTMAS DAY 1974: AUSTRALIAN
SUPERCYCLONE TRACY TOUCHES DOWN.

April 26, 1976: Pan Am begins regular New York–Tokyo flights.
September 9, 1976: Mao Zedong dies, Beijing.
November 8, 1979: The Chilean Communist Party is founded.

Chapter 6

ECHOES OF DISTANT
THUNDER

—⊶⊷—

I am the Rider of the wind
The Stirrer of the storm.
—LORD BYRON, *Manfred*, 1817

At first there seemed little reason to fret. Darwin was a tough little frontier town, a hard-drinking, broken-jaw kind of place— Australia in the raw, ready for most kinds of trouble. Over the years it had suffered gamely through a variety of storms, natural and man-made. During the war, more Japanese bombs had rained down on Darwin than on Pearl Harbor, and the scores of subsequent Japanese raids made for a certain proudly held resilience of spirit, still evident today in this far "top end" of Australia. That, and the weather—perpetually hot and humid; and during the summertime rainy season, which the locals still call "the wet," it is often ripped apart by frequent and spectacular tropical thunderstorms.

When a gathering storm was first noticed out in the Arafura Sea, at the start of Christmas week of 1974, no one thought it was anything special. Cyclones were part of the usual patter of midsummer weather. It was coming up to the centenary of the opening of the Port Darwin cable station, which had first connected Australia to Java and the rest of the world, and over all the years since then, fierce tropical cyclones had regularly barreled in from the seas to the north and hammered the crude government shacks in which the first hundred had lived, and now hammered the tin-roof shanties that were home to most of the forty thousand present-day Darwinians. Old-timers would tell of the unique sounds of a Darwin cyclone—of the screechings of thousands of corrugated-iron roofs being gale-blown along the roadways, of the noise of breaking glass, of the endless howls of wind, of lashing rains, and, as *basso continuo*, of the furious pounding of ocean.

That was not likely to happen this time. Christmas 1974 would surely be hot and steamy, but peaceful enough. The storm out in the sea was just a small one, and it was heading south, well away from the city. It would press on down through the Joseph Bonaparte Gulf and make eventual landfall in the Kimberleys. There would be rain in western suburbs, perhaps, and fine displays of lightning. But that wasn't unusual: Darwin always had rain and lightning at this time of year. Each season had a full dozen cyclone alerts, and each time the ABC sounded its sirens, the announcer went on the air with the usual warnings about tying down loose objects and filling the bath with emergency water. Everyone heard, but few listened. The ABC was crying wolf, people grumbled.

The weather bureau tracked the storm as it passed slowly to the west of Bathurst Island, heading south. Most in town remained complacent—it was Christmas, after all: there was church to attend, presents to wrap, trees to decorate, children to persuade to sleep.

This time, though, a very few suspected that something was up, mainly because the air in town felt somehow different. "Jerky" was how a Chinese shopkeeper described it. The songbirds had fallen strangely quiet; and according to the long-grass dwellers, the Larrakia aboriginal peoples who (then, as today) liked to camp in the tall grasses outside town, all the usual populations of green ants they would normally have seen seemed to have vanished. "Something dreadful is going to happen," a woman named Ida Bishop said to her boss, the manager of a fleet of prawning boats. The gathering quiet was ominous. The clouds were too high, were strangely shaped, and were vivid with purples and greens, and with other bizarre colors that just shouldn't be there. Some reported seeing what they described as a black velvet cloud hanging in the air five miles above the sea, rolling and pulsing, and blotting out the sun.

Then, out at sea, the storm changed direction. Quite unexpectedly, it made a sharp right-angled swerve to the east. At exactly the same time, it contracted, like a tightening sphincter, and to the growing consternation of officials, it began to bear down with withering accuracy toward the dead center of Darwin.

The storm was called Cyclone Tracy, and there has never been a more dreadful and destructive event in recorded Australian history. Once the Meteorological Office forecasters had confirmed the approach, ABC radio turned on its sirens. The chief announcer, a man named Don Sanders, whose voice was known for its confidence-inspiring depth and richness, was brought in to warn of the coming danger. It may have saved lives. It didn't save the town.

When the storm hit shore a little after midnight, it crushed building after building like a giant's hand. Ten thousand houses (80 percent of the city) were totally destroyed. They were near-instantly demolished, reduced almost to matchwood and pulverized concrete. The process was identical, house after Christmas-decorated house: First, the roof was ripped off its stanchions and whirled away into the rain-soaked night. Then the windows shattered, slicing people with slivers of glass. The walls next blew out, one by one—people would speak of running in darkness and panic from room to room, locating by feel the bathroom doors and racing inside in the belief that the smallest room would be the strongest—only to find the outside wall gone and only the darkness beyond, a terrifying frenzy of gales and oceans of pounding hot rain.

Darwin was brought to its knees. Everything failed: The telephones were out. Electricity was down. Antennas were blown flat. Aircraft had been tossed about like chaff, smashed beyond recognition. Ships broke loose in the harbor and either sank or drifted far from their moorings, useless. Scores of people who might have been useful were away for the Christmas holiday. The broadcast stations had only skeleton crews, but no light or water—though one of them, the local ABC affiliate, did have a generator, and managed to get a message out to a remote sister station in the Queensland outback. This tenuous link provided the only communication Darwin had with the outside world for the first three days after the catastrophe.

Word got out late on Christmas afternoon. It was then that the rest of Australia came to realize that its most northerly capital city had been laid waste by a terrible storm. Ministers in Canberra, and others in Sydney and Melbourne and Brisbane, were roused from turkey- and mince pie–induced lunchtime slumbers to be told of the ruin and devastation thousands of miles away.

And when the first rescuers got there, they all made the same comparison: Hiroshima, or Nagasaki. The comparison is invidious, of course; the casualty tolls were incomparable—seventy-one were killed by Tracy, not even one-tenth of a percent of those who died in Japan—but the physical destruction of Darwin was

The near-total destruction of the city of Darwin by the unfor-
gotten Cyclone Tracy, which struck on Christmas Day. Storms of
this size and power are increasing in frequency and ferocity in
and around the Pacific.

total, and the images more than amply resembled the familiar
photographs of the two postnuclear cities. Roads were no more
than pathways through scores of square miles of broken rubble
and splintered timber. People were wandering around glassy-
eyed, bewildered. Hundreds of dogs, frightened and unfed,
emerged from the ruins to forage, and the first rescuers to arrive
were struck by their snarling presence, adding to the menace in
the air. There was a real threat of typhoid and cholera. Police had
to find guns (shotguns, mainly, from the nearby sheep stations)
to deal with looters.

In the end, almost the entire city had to be evacuated. Forty-
one thousand of the forty-seven thousand were without home,
shelter, water, food, medicine, or communication. The government arranged shuttles of aircraft—slowly at first, because the
ruined Darwin airport could accommodate only one flight every
ninety minutes. Over the next five days, a total of more than
thirty-five thousand people were flown or driven out of the city—

and by the time the year ended, it had been all but emptied. More than half of those who left never came back.

Darwin is a wholly rebuilt city now, slick and modern, with most of its high-rise buildings apartment blocks rather than the banks and insurance companies that usually dominate downtowns. And everything now is claimed to be cyclone-proof—because, if nothing else, those who live at the top end of Australia learned from the disaster of the 1970s that the Pacific can be a place of extraordinarily violent weather.

There is more. The storms created in this ocean are coming to be recognized as the harbingers of the weather in the rest of the planet—the very first indicators of, maybe even the generators of, the swirls of wind, pressure, and humidity that sweep from west to east as the world turns beneath them. Since these cyclonic storms* are said, both anecdotally and statistically, to be becoming ever more furious as the earth and its seas warm up, as the climate changes, as the oceans rise, the consequences for the world are at the very least potentially troubling, maybe even dire.

Time and history have their understandable ways of turning tragedy into statistics, and in matters such as storms, earthquakes, and eruptions, statistics often boast superlatives, turning contemporary misery into historical pride. The people in Darwin who were forced to hide for hours under their bathtubs in their utterly ruined homes and to fend off starving dogs as they tried to keep themselves safe and alive may not quite see it that way, but the fact remains that theirs was the most lethal miniature storm ever recorded.

For, in areal span, it was very small: it extended over only 24 miles, side to side and through its eye. It was quite Lilliputian

---

* Clockwise-spinning cyclones in the Southern Hemisphere, counterclockwise typhoons (Chinese for "big winds") in the Northern, and similarly configured hurricanes in the Americas comprise this Aeolian family, together with the plagues of tornadoes and waterspouts in the cadet branch.

when compared with monsters such as Hurricane Katrina, which was 400 miles across when it struck New Orleans in 2005. It was minuscule when set against Typhoon Tip, which roared across the tropical Pacific in 1979 with a record-breaking diameter of 1,380 miles and an eye not much smaller than the entirety of Cyclone Tracy, which could have been pushed inside it quite handily.

Pacific storms have clearly been getting ever more menacing in recent years. Cyclone Tracy in 1974 was close to the beginning of this development: Typhoon Haiyan, which struck the Philippines in November 2013, suggests how truly ferocious things can get. The four decades between these two storms saw two developing trends: bigger and increasingly troublesome storms and an ever-greater accuracy in pinpointing where they might make landfall. More lives were at risk from the storms' gathering power; more lives were saved by the gathering boon of science.

Typhoon Haiyan makes the point well. It was first spotted far away, by observers in Hawaii. The four duty officers who arrived for their night shift on the early evening of Friday, November 1, 2013, in the drab Pearl Harbor building that houses the offices of the Joint Typhoon Warning Center were the first to notice something unusual. The satellite images for their routine sweep of the far western Pacific (where it was already the afternoon of the next day, Saturday) were just scrolling onto their monitors. And while most of the ocean was quiet—just a single disorganized cluster of squalls, a small and nameless tropical storm, that was wandering aimlessly westward toward Mindanao—a new pattern of clouds had just sprung into view in the central ocean, and this one had an ominous look about it.

The proto-cyclone, if that is what this was, must have been developing rapidly during the day, for the earlier shift had reported nothing from the satellite when it last transmitted pictures six hours previously. But now, quite clearly, wisps of cloud had arranged themselves in a manner suggesting that a definite and organized pattern was forming, about two hundred fifty miles to the southeast of the tiny island of Pohnpei, in Micronesia, and that it was changing its appearance fast. Real-time imagery then showed that it was assuming the all too familiar, vaguely swirling cyclonic appearance that betokens danger. The suddenness of its appearance and the fast lowering of pressure beneath the cover of clouds all struck the weather analysts as noteworthy, at the very least.

They promptly sent a message across the road to the operations room of Pacific Fleet headquarters: U.S. Navy ships in the area might want to know that wind and rain could well affect any vessels heading for that quarter of the sea. A routine message. No alarums and excursions. Not yet.

By November 3, the Japan Meteorological Agency outside Tokyo had assigned the now swirling wisps of cloud a number, Tropical Storm 31W. By the next day, the swirls had grown very much more powerful, and the storm had been upgraded to full typhoon status. It was given the preassigned name Haiyan, the Chinese word for "petrel," a bird that in mariners' lore is often associated with foul weather. The fast-gathering beast by now appeared to be moving in a westerly direction, traveling directly toward the barrier wall of the Philippine Islands, where the local weather agency, following its own naming rules, had confusingly decided not to follow the international rules, but to call the storm Yolanda.

The situation was becoming alarming. The American and Japanese weather forecasters, and later those watching the big weather radars in China and Hong Kong, knew this was going to be a monster storm. They began to issue warnings to the civil defense agencies in the southern Philippines—the accuracy of their forecast allowing them to offer some days of preparation for the onslaught of what was now clearly going to be a storm of a power seldom seen before at sea, and perhaps never before experienced on land. Evacuations were ordered, and people began to stream away from the country's southeastern coasts, where the storm was predicted to land.

The forecasts were right, nearly to the minute. Typhoon Haiyan struck head-on into the eastern Philippines, hitting the islands of Samar and Leyte almost simultaneously, at about 9:00 a.m. on Friday, November 8. By the time it reached land, it had become the fiercest typhoon to have done so in the world's recorded history. When the northern eye wall of the storm struck the village of Guiuan, such anemometers as hadn't already whirled off scale recorded wind gusts of 196 miles per hour—greater by far than anything previously known.

The physical and human damage was terrifying in its extent and consequence, although the warnings and the precision of the forecasts certainly helped keep down the total of human casualties. Sixty-five hundred people were killed, twenty-seven thousand were injured, and more than a thousand were missing. Just as in Darwin forty years before, whole cities were flattened, every building reduced to mere debris as if by an earthquake or an atom bomb. The city of Tacloban, the biggest in the region, was almost unrecognizable after first being hit by the full force of the storm, and then being swamped by the corrosive seawaters of the thirteen-foot storm surge that followed.

There was an uncanny coincidence in the Philippines that did not pass unnoticed. Landfall in 2013 of Haiyan, the most savage of all the world's storms, was made along the coast of Leyte Gulf, the site in 1944 of the most savage of all the world's naval battles. Two of the nearby villages have since been named for Douglas MacArthur, the general being a great hero in these parts. His famous "I Shall Return" promise is commemorated by a bronze statue depicting him and his staff striding through the ankle-deep waters to resume control of the Philippines; it stands by the beach where the event took place, in the tiny town of Palo. All three of these places, Palo and the two villages called MacArthur, were savagely damaged by the violence of the typhoon.

Typhoon Haiyan, which devastated much of the Leyte Gulf region of the southeastern Philippines in late 2013, brought a massive international response, most notably by the U.S. military—a classic demonstration of Washington's use of "soft power."

American military forces were quite as heavily involved in dealing with Haiyan's violence now, seventy years after the Battle of Leyte Gulf. Thanks once again to the accuracy of the forecasts, U.S. Marines and U.S. Navy ships were already on standby in Japan and Okinawa, or they were out at sea. Once the signal came that the State Department had agreed to respond to Manila's official request for help, the well-oiled machinery of a full-blown American-led rescue operation got under way.

Operation Damayan, the $21 million quick in-and-out rescue operation, formally began the next morning, Saturday, November 9. The night before, however, when a stunned Tacloban was still crawling out from under the storm's wreckage, a small flotilla of helicopters quietly brought in members of a U.S. Special Forces team that was already in-country, secretly helping to deal with a long-running insurgency.* They set up radios and began talking to the incoming armada of ships and the waves of marines who would soon fill Leyte Gulf with as many vessels as had been there seventy years earlier, during the legendary naval battle.

More than eight hundred U.S. Marines from the Third Expeditionary Brigade in Okinawa were on the ground in the Philippines by the Saturday afternoon. A survey ship already working in the gulf was on station the next day, and then a submarine tender, filled with emergency supplies and drinking water. The high point, at least cinematically, was the arrival on the Thursday of the huge nuclear-powered carrier USS *George Washington*, with her attendant strike group of destroyers and frigates. She anchored in the bay and, for the next eight days, served as a floating headquarters for a relief operation that ultimately involved twenty-two hundred U.S. military personnel, thirteen warships, twenty-one helicopters, and the distribution of two thousand tons of American food, blankets, tents, generators, water purifiers, and myriad other kinds of aid invariably needed in the aftermath of such a calamity.

The sight of this vastly impressive, hulking, broad-shouldered behemoth of a warship, and of the squadrons of lesser vessels anchored around her (these later included two British ships, a destroyer and a carrier) was at once powerful, comforting, reassuring. It served as a reminder, important in the propaganda wars, that American military influence in the world is predicated not only on war and the projection of hard power. This was so-called soft power projection at its most effective—and once the immediate storm crisis was over and the American carrier

* The indigenous Muslim Moro people, living largely in Mindanao, in the southern Philippines, have been engaged in independence wars with various Manila regimes for many years. For many reasons—not least because the Moro leadership has openly allied itself with China—the United States has been offering the Philippine government military aid, with some U.S. personnel on the ground on the Mindanao battlefields.

had weighed anchor and slipped off back into patrolling the China seas, Washington propagandists publicly pointed out how little the Chinese had done to help. Beijing had initially offered a laughable $100,000. Only when stung by the world's response to their seeming niggardliness did they increase the aid to $1.6 million, and send down from Shanghai a new hospital ship, on what turned out to be her maiden voyage. She arrived too late to be of much use.

Beyond the melancholy dramas of their immediate impact, storms like these have proved of great use in recent studies of the world's climate. They and their kin offer in particular many clues for understanding one thing and for realizing another. They have allowed for an ever-greater understanding of the dramatic recent changes in the planet's atmospheric environment. And they and their like have served to offer a confirmation of something long suspected but never firmly proved: that whatever the changes in the earth's climate may be, it is and always has been the Pacific Ocean that is the generator, the originator, of much of the world's weather.

Tracy and Haiyan were more than simple events, however individually tragic and dramatic. They were bookends to a catalogue of atmospheric occurrences in the Pacific Ocean that have been getting steadily more ferocious in recent years, and to some they tell a much greater and more significant story—a story that many are now beginning to relate with ever-greater urgency.

In March 2013, for example, six full months before Haiyan hit the Philippines, Samuel Locklear III, the American four-star admiral who was then in charge of all U.S. forces in and around the Pacific (three hundred thousand navy, army, marine, and air force personnel, ranged around more than half the world), detected a pattern in the frequency and violence of recent Pacific typhoons, and then made what seemed to many an eccentric prediction. His initial observation was factually unremarkable: "Weather patterns are more severe than they have been in the past," he declared at a meeting in Boston. "We are already on super-typhoon 27 or 28 here in the Western Pacific. The annual average is about 17."

The conclusion he drew from the trend, though, was quite unanticipated. In spite of the tensions between China and Japan, between North and South Korea, between Beijing and Washington, the admiral declared his belief that it was actually changes to the climate—changes that were powerfully suggested by typhoon clusterings that he and his weather analysts had observed—that posed the greatest of all security threats in the region.

"Significant upheaval related to the warming planet is probably the thing most likely to happen . . . and that will cripple the security environment. Probably that will be more likely than the other scenarios we often talk about." A ripple of amazement coursed through Washington. Significantly, no one in the White House or the Pentagon, however, chose to challenge the admiral's view. It was clear that he spoke with gazetted authority.*

Not that the admiral's stated concern over the number of storms that gathered within his AOR (navyspeak for "area of responsibility") was meant to imply that storms in the other, lesser oceans were any less daunting. Notorious monsters such as Katrina, Camille, Andrew, Ike, Sandy, Hugo, Wilma, Rita, the Labor Day hurricane of 1935, the Okeechobee hurricane of 1928—all these were great Atlantic storms of truly historic proportions, and all were hugely destructive and frightening.

"Destructive" and "frightening" are not true measurements, however. Nor is the most commonly used metric of a storm's financial cost. In America, Atlantic hurricanes tend to be popularly described by their eventual price—the quoted losses for the

---

* It was left to Senator James Inhofe, a Republican from Oklahoma, to take immediate issue with the admiral, by famously declaring that God was the only one with the authority to alter the world's climate.

insurance companies of $108 billion in and around New Orleans in 2005 have made Katrina come to be seen as the absolute worst storm in American history. But cost can hardly be a neutral descriptor: Storms that strike American cities are expensive because they wreck expensive things. Storms that strike isolated cities in the eastern Philippines may cause just as much devastation, but in dollar terms are much less costly. Human damage, of course, is different—still, that is not neutral, either, since a typhoon hitting a crowded slum will kill far more than one that sinks ships and swamps atolls in the middle of the ocean.

We do have scales to measure storm intensity, but they are not perfect. Most use wind speed as a categorizing device because wind is what does the greatest damage. It also suggests the overall energy (the kinetic energy of a fast-rotating body of air) of the storm as a whole. Critics reasonably complain also that it is imprudent to ignore the amount of rainfall dropped by a storm, or the speed with which the storm develops, or the surge it creates in the sea. They insist that a wind-only classification is of somewhat limited use—at least, beyond the television news.

Arguably the most ideal and neutral way to describe a storm is much simpler, if hardly telegenic, and that is according to the lowest pressure in the storm's eye. The lower the pressure, the more intense the storm. The more numerous the isobars, and the more tightly these imaginary lines of equal atmospheric pressure are wound together, the more vicious is the weather below.

This measurement of a storm's *minimum central pressure*— something that was not always easy to acquire in the days before satellites and storm-hunting aircraft, and even today often requires that a dropsonde be trickily inserted into the storm's eye— makes comparison much easier. It makes it possible to measure one ocean's violence against that of another, for instance. It makes it simpler to compare the storms of one year with those of previous years, to range decades against earlier decades. In short, and therefore most usefully, it allows climatologists to spot and identify real climatic trends.

By employing this one measure, scientists are able to determine that the apparently biggest and costliest storms do in fact, and much as expected, tend to be the deeper and more isobarically intense ones. By this same measure, oceans can now be compared with other oceans. And recent data show that most Atlantic hurricanes, when measured according to the lowest pressure in their eyes, yield overwhelmingly in their strength, power, and destructiveness to those gigantic storms (such as Tracy and Haiyan) that now regularly cannonade across the broad reaches of the Pacific.

The key number that the World Meteorological Organization has chosen as a baseline for assessing a storm's strength is 925 hectopascals, or what used to be called millibars (mbar). Any storm eye with a pressure measured as less than 925 mbar is one for the books, intense enough to be worthy of record.* And when one looks at the Pacific Ocean using that measure alone, it becomes swiftly clear that this body of water is beyond any other when it comes to playing host to a number of the world's truly intense tropical storms.

The figures are telling. In the Atlantic since 1924, just nineteen hurricanes made it into the list of storms with eye pressures of less than 925 mbar. Just one out of five of those (the hurricanes known as Labor Day 1935, Allen, Gilbert, Rita, and Wilma) were superintense, with eye pressures of less than 900 mbar. Neither Camille nor Katrina managed to figure below the 900 mbar number. Hurricane Sandy, infamous in recent New York and New Jersey history, did not even make the World Meteorological

---

* Normal sea-level atmosphere pressure is 1013.25 mbar—or, in the old style, sea-level atmosphere would typically support a column of mercury that rose 29.92 inches in a vacuum tube. The highest pressure is generally to be found in parts of Siberia, up to 1050 mbar. The lowest pressures at sea level are invariably to be found in the eyes of tropical storms.

Organization cut, registering a comparatively benign 940 mbar in its nonspinning center.

In the western North Pacific, however, atmospheric violence as measured by intensely low eye pressure is much more common, almost routine. Since 1950, there have been fifty-nine fully formed typhoons north of the equator, and in the western South Pacific and off Australia there have been twenty-five similarly rated cyclones since 1975. In the Atlantic, the rate of occurrences of the sub-925 mbar storms runs at about one every five years. In the western Pacific they are much more numerous, with about one every single year.

Moreover, the large, sprawling, ultra-low-pressure storms occur five times more often in the Pacific than elsewhere in the world. They are generally much more intense, with thirty-seven of the northwest Pacific's fifty-nine having pressures lower than 900 mbar. Typhoon Tip, the deepest of them all, recorded an eye-watering low pressure of just 870 mbar—and enjoyed the unique distinction of being both the deepest and the widest of all tropical storms on record, with an edge-to-edge spread of 1,380 miles—meaning that if superimposed on the United States, it would have extended from the Mexican border to the Canadian border, and from Yosemite to the Mississippi River, with its eye directly above Denver.*

The best explanation for why the Pacific storms are now more numerous and violent has much to do with the ocean's vast size and, most crucially, with the near-unimaginable amount of heat that its waters collect from the sun. And this, it seems, is key to everything else: if the Pacific Ocean is the principal generator of the world's weather, then the ultimate source of all the Pacif-

---

* Most of Tip's two-week progress was through the open ocean, so casualties were lighter. But there were a number of collateral incidents—most notably at a U.S. Marine base near Tokyo, where the intense winds collapsed a wall that in turn dislodged fuel pipes from a farm of gasoline bladders. A river of fuel coursed down a hill and into a barrack block, and was ignited by a space heater. The resulting fire killed thirteen marines.

ic's extreme meteorological behavior is the initial presence of its massive aggregation of solar-generated heat. This changes the long-term phenomenon we know as the climate. The climate in turn brings about the short-term phenomena we know as the weather.

The Pacific Ocean is broiled by the sun, whatever the season. Given the tilt of the planet, the 23.5 degrees offset from vertical of the axis around which the world spins, the ocean's northern parts are broiled in the northern summer, and the southern parts in the southern summer. The immense region of sea that lies between the Tropics of Cancer and Capricorn are being broiled all the time.

The heat, the thermal energy, that blasts endlessly down on the planet is dealt with differently depending on whether it strikes solid or liquid below. When intense sunshine radiates down onto the solid earth, the rocks become very hot very quickly—but then, because of the immutable physics of solids, they release this heat equally fast, return it to the atmosphere, and retain little. To a wanderer in the desert, a rock at nighttime can be blessedly cool.

It is different when the same intense heat is radiated down onto the ocean. Initially the water warms slowly, but then, and crucially, it retains the heat it has absorbed for some long while. Because it is a liquid, mobile entity, it then shifts this captured heat about, three-dimensionally. Under the influence of its currents and its surface winds, it drives the captured thermal energy either laterally, from east to west, or from north to south. Or else, by way of a pattern known as thermohaline circulation, it shifts the heat deep downward into its depths. Since the Pacific is by far the deepest ocean as well as the broadest and longest, the amount of heat it can incorporate within is almost beyond imagination.

Heat, in immeasurable quantities, is stored in the world's oceans generally. The Pacific, which occupies one-third of the planet's entire surface area, is responsible, then, for storing a very great deal of it. Much of this stored heat then warms the atmosphere. It does so most especially where the sea is subject to the most intense solar heating, along that wide band of ocean between the tropics and along the equator, a band that shifts to the north and the south as the seasons change.

Within this well-defined area, the intense heat causes the seawater to evaporate and the warm air above it to rise—so gigantic banks of cloud form and billow skyward. As they do so, they lower the air pressure in the void they leave behind them. Cooler and heavier air then pours into the low-pressure zone, from the north and the south. Thanks to the west-to-east spinning of the earth, this air tracks in a more or less westerly direction as it cascades inward: the air from the north heading toward the southwest, the air from the south tracking to the northwest. Since it is the custom to name winds for the direction from which they are coming (whereas currents, confusingly, are named according to where they are traveling *to*), these inrushings of cool air become the famous trade winds: the northeast trades in the Northern Hemisphere, the southeast trades below the equator.

A great deal goes on within this specific area, where the world's climatic business begins. It is where the trade winds blow. It is the site of the so-called Intertropical Convergence Zone, where the hot, humid, and (infuriatingly, for those traders who used sailing vessels) windless doldrums lie. It is where all the cyclones, hurricanes, and typhoons are born. It is where the monsoons begin their lives.

In the Pacific portion of this tropical area (which is by far the world's largest portion, since the Pacific is the sea of greatest extent), a series of curious atmospheric and oceanic events occurs, and which now seems to be most crucial of all to both the marking and the making of cycles of the world's weather. These events, long known, were once reasonably predictable, and were gathered under the general rubric name of El Niño. Now, though, such occurrences seem to be becoming both more frequent and increasingly irregular. Their timing these days is perhaps somehow linked to the undeniable warming of the ocean, as the global climate (for whatever reason, man-made or not) continues to alter.

These oceanic events have long been marked initially by sudden strange changes in the business of fishing. They have been long recorded, and scrupulously so. As far back as the late sixteenth century, Peruvian fishermen working out of ports in the north of the country (from Tumbes, on the Ecuadorian border, to Chimbote, close to Lima) would make careful note of the changes in the local fish population, since their livelihoods depended on what was taking place.

Chimbote was once known as the World Anchovy Capital, because of the small, silvery, and memorably pungent anchoveta fish that were to be found in staggering abundance in the cold waters just twenty miles offshore. Few fish have ever known such a boom as the Peruvian anchoveta: from the early settlement of the country, fisheries would spring up in every possible harbor along the coast, and thousands of men would work the waters, eventually making the anchoveta the most exploited wild fish in the world. Thirteen million tons of it were hauled into nets in 1971. Most of it was ground into fishmeal and sent off to fertilize fields and feed livestock all across the world.

The abundance of the fish was a fitful thing, though, as the Chimbote fishermen came to know all too well. Once every five or six years, with some ragged regularity, and most usually in November or December, the anchoveta would all but vanish. One day there would be darts of quicksilver shoals all around; the next, nothing but the blue silence of the deep. And there was another thing: the cold waters offshore, which would bring in the evening fogs so welcome in seaside desert towns such as Chimbote, would at the same time become mysteriously warmer. The fogs would vanish, too, and the skies would magically clear.

The want of catch would frustrate the fishermen, to be sure, and they would curse their empty nets. The absence of fish had an effect that then spread insidiously all the way up the food chain. The gannets, cormorants, and pelicans that fed on the anchovies died, or else they flew long distances in search of food, abandoning their nests and leaving their waiting chicks to die in their stead. Squid, turtles, even small sea mammals would pass away also, either because of their intolerance of the warmer water or because of the sudden strange voids in the food chain, spawned by the lack of anchovies. Then, compounding the misery, large numbers of these dead creatures would float to the surface and create small islands of decay, the foul gases they emitted so acidic as to blister the hulls of the fishermen's boats.

The loss of anchovies was an economic disaster; the smorgasbord of other deaths and absences made the event curious and more sinister. Because the happenings invariably arrived around the anniversary of Christ's birth, the fishermen would name it, with a bitter and sardonic humor, El Niño de Navidad, the "Christmas Child."

The term *El Niño* first appeared in the English language at the end of the nineteenth century—and not so much because of the fishermen's melancholia, but as a name for the change in the current in the waters below. What happened was that the cold waters of the Humboldt Current, part of the normal pattern of Pacific circulation that powerfully sweeps Antarctic waters northward up along the South American coast before the waters head west along the equator, become on occasion mysteriously disrupted. Instead they are replaced, or nudged farther out to sea, by an irruption of warm water that bullies its way down from the equator—and in the case of the anchoveta, this warm water smothered the upwelling of nutrients on which the little fish fed. The fish then went elsewhere, well beyond the ken of the Peruvian fishermen. They simply vanished from Peru.

Early on it was nothing more than the change of current, this unusual warming of the sea, that came to be named El Niño—and it remained so until oceanographers and climatologists in the mid-twentieth century realized that the change of currents off Peru was just one of many features of a much larger and more important phenomenon, one that had its impact across the entire breadth of the ocean.

Many names are associated with the research that confirmed this. One in particular belongs to a civil servant in British India, Gilbert Walker, who had a meteorological epiphany in 1924 that helped secure what would become the Pacific's reputation as weather maker for the world.

Sir Gilbert Walker, described in the closing paragraph of his 1958 obituary as "modest, kindly, liberal-minded, wide of interest and a very perfect gentleman," was a classic of his breed, a polymath of the old school. He was first and foremost a Cambridge mathematician—no less than the 1889 Senior Wrangler, meaning that he had achieved his country's greatest intellectual achievement of the year. But he was many other things besides: a designer of flutes, a keen student of the boomerang and of the flight paths of ancient Celtic spears, an authority on the aerodynamics of birds' wings, a passionate advocate of the sports of skating and gliding, a wizard in the more arcane uses of statistics, and a recognized expert on the formation of clouds.

He loved India and the Himalayas, and when he was appointed director-general of observatories in India, he spent twenty years trying (in vain, as it happened) to figure out a mathematical means of predicting monsoons. He had been led to this obsession by a monsoon failure in 1890 that caused a terrible famine. His frustrated quest might well have caused him to leave India somewhat deflated—except that, as it happened, Walker's work on the monsoon prompted him, during his retirement years, to come up with quite another and rather more globally significant discovery.

He was a habitual collector of what turned out to be colossal tonnages of meteorological statistics. His exhaustive analysis of these, of decades of weather records from all across the British Empire, allowed him to demonstrate incontrovertibly that the El Niño events occurring off the Peruvian coast—the fishermen's phenomenon was by now well known to scientists around the world—were part of an enormous and all-encompassing transpacific set of weather patterns. These patterns turned out to be mirror-image combinations, in which precisely opposite meteorological manifestations were occurring on one side of the ocean or the other, in one season or another, for one extended period or another.

Periods of warming here led to episodes of cooling there. The Peruvian sea starvation during a locally warm-sea El Niño event

Sir Gilbert Walker, a British meteorologist based for decades in India, is memorialized by one of his discoveries, the Walker Circulation, a main driving atmospheric force behind the making and unmaking of the cyclical El Niño phenomenon.

would in time be followed by a local sea cooling and return to abundance, and that would be called (keeping to the Christmas-themed naming practice) a La Niña time. Floods on one side of the ocean led to droughts on the other. There were periodic swings in weather and in the human response to it. There were times when there were more cyclones and times when there were fewer. Some years when the Indian monsoon never happened, the fields were parched and crops failed. Other years were marked by luxuriant summertime drenchings. There were years of famine and years of abundance, of dust bowl summers and harvest-rich autumns, years of concomitant prosperity and ruin, periods of consequent peace and turmoil—within the Pacific, around the Pacific's coasts, and even, perhaps, beyond them and around the globe.

And all of it—Sir Gilbert Thomas "Boomerang" Walker realized through all his Renaissance man amusements of flutes and bird flight and skating techniques—was due to a hitherto unseen natural phenomenon. Walker declared that what drove the regular and dramatic changes in the Pacific weather must be some kind of repeating mechanism high up in the atmosphere. Whatever it was, this pattern of invisible winds and movements seemed to him to operate as a kind of unseen atmospheric seesaw, a beam engine balance around a central pivot lying somewhere smack dab in the center of the ocean.

The axis seemed to hover where the International Date Line crossed the equator, in the middle of that sprawl of limestone specks then known as the Gilbert Islands and the Phoenix Islands, now the Republic of Kiribati. Up on one side of this fulcrum meant down on the other; high pressure here meant low pressure there; hot here, cool across the other side; cruelly wet in this place, bone dry in that. It had a beautiful logic to it; and measurements taken over the years that followed have proved that Walker was exactly right.

This transpacific atmospheric wind pattern he discovered was in time to be named, and in his honor, the Walker circulation. This was the engine, the mechanism, that then produced what Sir Gilbert himself went on to name, for the back-and-forth, hot-and-cold, wet-and-dry, stormy-and-serene periods that appeared to dominate the tropical Pacific's weather: the Southern Oscillation.

ENSO—the acronym is formed from the combined initials of the El Niño and the Southern Oscillation—denotes what today is recognized as undeniably the planet's most important climatic phenomenon. If the Pacific is truly the generator of the world's weather, then ENSO represents the turbines that give it the power to do so. And the Walker circulation is the force that sets the turbines spinning in the first place.

The Walker circulation's basic structure is made up of long-lasting cells of pressure in certain places around the ocean. The eastern Pacific generally has high atmospheric pressure. Correspondingly, the western Pacific generally has a large low-pressure area, most notably around the sea-spattered islands of Indonesia and the Philippines, the area that oceanographers and meteorologists like to call, if oxymoronically, the Maritime Continent. The air above the ocean then moves, as physics demands, from the high-pressure area to the low—in other words, from east to west. The trade winds at the surface, which blow nearly constantly in this direction, are of course this movement's very visible and familiar manifestation.

As the winds blow in this manner, they help push the warm waters of the tropical seas below them in the same direction. Incredible though this may sound, the sea then piles up very slowly and deliberately as a huge wave of water passes steadily across and into the western reaches of the ocean. The western Pacific can sometimes be a full two feet higher than the waters in the east. Some of this warm water evaporates as the huge cyclonic storms and typhoons, such as Tracy and Haiyan, form over the western seas. Some of it dives deep back into the ocean, cools, and is returned to the east by the work of deep ocean currents. In a normal series of years, this pattern is repeated again and again: the Walker circulation of the air above, the migration of the seawater below, the explosive growth of storms in the far western Pacific, the return of the cool and dry air and upwelling cool water (and with them the anchovies) to the Pacific east. As a result, calm and stability reign.

But sometimes, and for some still unexplained reason, the Walker circulation changes. The trade winds weaken or falter or even reverse their direction, and then an El Niño period occurs, and the system changes with it, and dramatically. It can sometimes strengthen in the opposite manner and with equal drama—and then the reverse, the phenomenon of La Niña, dominates the weather picture instead. Tracking and measuring the arrival of the two phenomena, El Niño and La Niña, have lately become major elements in worldwide weather forecasting and climate modeling. It is safe and reasonable to say that in the computation of the planet's weather, all eyes are on the happenings in the Pacific and the behavior of the Southern Oscillations. For as the Pacific oscillates, so oscillates the world.

These days the Southern Oscillation that Gilbert Walker defined is measured by the careful tracking of the atmospheric pressures and the sea temperatures in the region. Pressure is measured at two key points: one in Tahiti, the other down in Darwin. If the pressure in Tahiti falls significantly below what is normal, and at the same time the pressure in Darwin rises above what is normal, then an El Niño period is declared to be under way. The American and the British weather services also like to measure sea temperature along the narrow equatorial zone that is (in more ways than one) central to the development of an El Niño. If the water temperature in the eastern part of this region (that closest to the South American coastline) rises by half a degree Celsius, and if, in accordance with the British weather researchers' rules, it then holds that tem-

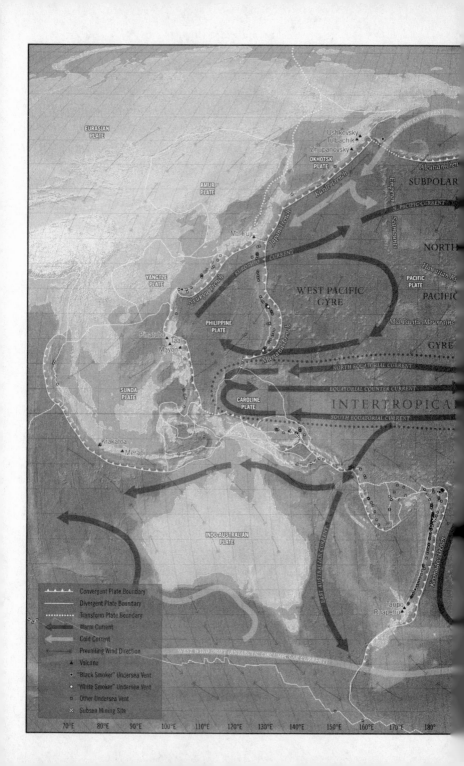

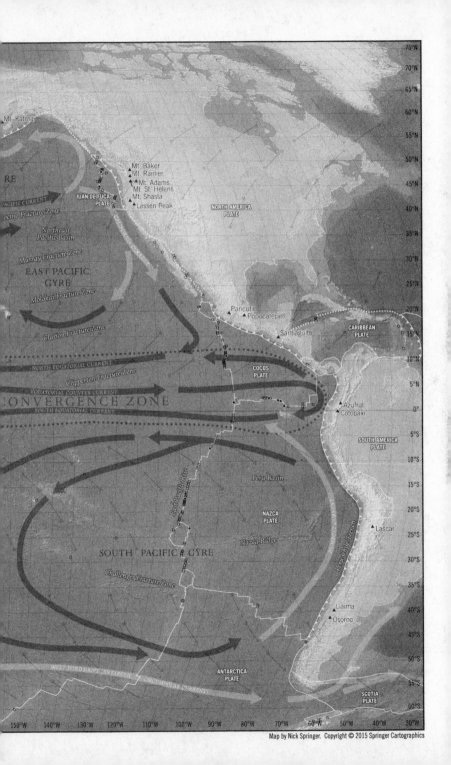

Map by Nick Springer. Copyright © 2015 Springer Cartographics

perature for nine full months, then an El Niño is declared to be under way.

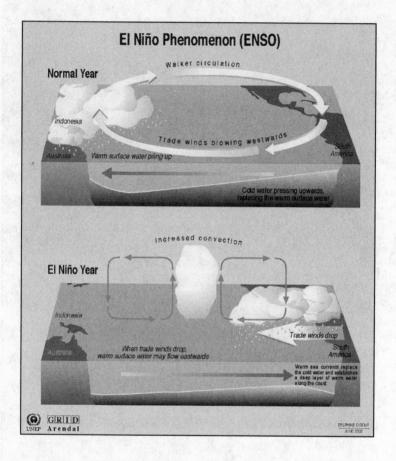

The El Niño Southern Oscillation, ENSO, is a still unpredictable and irregularly cyclical movement of waters and winds in the Pacific Ocean south of the equator, and a factor that determines much of the world's major weather patterns.

The Japanese government in particular is investing millions in its own studies of El Niño, and for good reason. Japan has historically been a magnet for highly destructive Pacific typhoons, storms that, along with the earthquakes, tsunamis, and volcanic eruptions that bring regular ruin, have helped forge the national character traits of stoicism and mutual philanthropy. Forecasting such traumatic occurrences would of course be a fine thing, for the national economy, for the nation's morale. The recent accelerating ability to forecast the eruptions of volcanoes may still not have been matched by an ability to predict earthquakes. But to balance that, a major effort is now being made in Japan to fine-tune global long-term weather forecasting, and in particular to investigate the possibilities of predicting when an El Niño—with its clustering of typhoons—is most likely to occur.

This task is being handled by what has been claimed variously to be one of the world's largest, most powerful, fastest, and most efficient supercomputers. It is grandly known as the Earth Simulator 2, and it is sited in a suburb of Yokohama, west of Tokyo, in the offices of JAMSTEC, the Japan Agency for Marine-Earth Science and Technology.

As a record breaker, the Simulator has recently been knocked off its perch by a new superfast machine in China, but it is still a remarkable and home-built device, which keeps being improved and upgraded: currently it can calculate at the rate of 122 teraFLOPS (122 trillion floating-point operations per second). A recent live test—which was conducted in its building with no more than the usual electrical hummings and air-conditioning whirs and the flashing of thousands of light-emitting diodes, and with only the operators squinting rather tensely at their terminals hinting at any anxiety—showed that the contraption can produce global weather analyses of witheringly complex detail: several times a day it can produce a three-dimensional map of the world's atmosphere, showing the climatic details every three miles horizontally and through more than one hundred slices of the atmosphere vertically.

It is so costly a creation, and Japan is so seismically unstable an island chain, that its guardians protect it as if it were the

Mona Lisa or the Hope Diamond. It has its own building mounted
on gimbals and rubber feet; metal-mesh ceiling nets to diffuse
lightning strikes; and special metal shields to keep stray magnetic fields at bay.

Swaddled in care, Japan's Earth Simulator quietly crunches
away at the numbers of what it sees. Its operators continue to try
to divine, as do many others in similar laboratories around the
world, whether the onset of an El Niño or a La Niña can be declared not merely under way but about to happen. Can it be predicted, in other words, just like any other weather forecast?

Most recently the Japanese team working on El Niño has been
able to show that the onset of an ENSO warm phase is often
preceded by a machine-gun-like series of small and intense
storms north of Australia, in the waters off Papua New Guinea.
The storms are small enough to be known as westerly wind
bursts; and though for a while they were dismissed as random
events, unconnected to the happenings on the far side of the
ocean, nowadays scientists believe they may be linked. But as to
whether they indicate the onset of an El Niño, or whether they
are *the result of* the onset of an El Niño, is a matter of much debate
in the meteorological community.

No one yet has come up with a way to forecast the onset of
ENSO—which is a major concern at a time when the worldwide
weather has become such a worldwide obsession. For El Niños
wreak havoc everywhere.

Locally, their effects have been noticed for centuries. The first
is familiar, but it bears repeating: the inflow of warm water in the
eastern Pacific halts the cold upwelling rush of nutrients; all the
anchovetas vanish from the waters off Peru; other sea creatures
die; noxious gases from rotting marine carrion bubble up from
the sea; boats have their paint blistered by the scum of acidity.

Globally, a host of other phenomena can develop in tandem,
all probable knock-on effects of these particular changes in
this, the world's biggest expanse of seawater. The expanse, it is

worth restating also, of heat-catching seawater—for the catching of solar heat is ultimately what this meteorological drama is all about.

So during an El Niño* there can be, among other things, major flooding on the South American west coast (with the current-warmed humid breezes rising above the Andes, the humidity condensed and then precipitated out as rain or snow). During this phase, there can be droughts in northern Brazil, but severe rainstorms near Rio. Cyclones and typhoons tend to form in the Pacific more centrally than usual during an ENSO warm phase, and since the storms spend longer times tracking their way westward over larger expanses of warmer seas, they can grow and accelerate and deepen, and thus can be much more violently destructive when they finally reach land.

The 1982–83 El Niño, one of the strongest ever known, was memorable for its cascade of events. The trade winds weakened. Sea levels in the eastern Pacific began to rise (up to a foot higher along the coast of Ecuador). Eastern sea temperatures shot up. Fur seals and sea lions began to die off the coast of Peru. Deserts in eastern South America were drenched with rain, grasshoppers swarmed, toad populations went through the roof, mosquitoes came in clouds, and malaria cases skyrocketed. There were droughts and forest fires in Java, terrible storms wracked the coastline of California, there was flooding in the American Deep South, ski resorts in the American Northeast reported warmer weather and lackluster business. All told, the economic cost of the 1983 El Niño was estimated by the U.S. government at eight billion dollars—and of course for every malaria-related death in Ecuador or for every village burning in Sulawesi, boundless misery.

That was an extreme event. Even during a modest El Niño

---

* Some meteorologists now refer to this as the ENSO warm phase and to the reverse, what is still traditionally called La Niña, as the ENSO cool phase. The relative lack of ambiguity and confusion will be likely to widen this practice in coming years.

the effects can be widespread and unexpected. Drought can affect Hawaii, drastically lowering the sugar crop (and before 2009, when Dole pulled its business out, the pineapple crop, too). Forest fires can and do sweep across Borneo; monsoon-dependent crops can wither and fail in India. Sea lions and elephant seals die in the waters off California; unanticipated fish and squid appear in the waters off Oregon and British Columbia. A supposedly moderate El Niño in 1877 triggered a two-year drought in China and the deaths of nine million people from starvation. The polar jet stream can be nudged farther southward during an El Niño event, making winters in Canada more acutely cold, forcing more rain to fall in the southern states, cooling everything down—and shortening the growing season for Florida oranges. Northern Europe is colder and drier, Kenya wetter, Botswana drier. The effects are legion, the lists endless (and at times, seemingly contradictory), the concerns global.

And the matter of global warming is a constant concern, underpinning or overlaying everything. The differing scales of the events involved are irksome to those employing statistics to help spot trends and links. The wildly complex mechanics of an El Niño oscillation occur over fairly short intervals, for instance. The best-known parts of that equation, the Walker circulation and the ENSO, operate at fairly short intervals, in three- or four- or five-year cycles.

It can become very much more complicated than this. Yet other, more arcane atmospheric and oceanic phenomena—such as the Kelvin waves and Rossby waves that move quantities of ocean hither and yon around the subsea boundary between warm and chill waters known as the thermocline—are similarly swift in their operation. As are the so-called Hadley cells and their more northerly cousins, the Ferrel cells, which operate in the atmosphere rather than the sea, and which bring much rain, and swirl about under the majestic impress of that westward and world-dominating force which was discovered in Victorian times by the Frenchman Gustave-Gaspard Coriolis, and which bears his name today.

All these named phenomena operate in relatively rapid, time-lapse motions. As does a final main, named component of the entire Pacific process, and perhaps the fastest of them all: it is known as the Madden-Julian Oscillation. This is best described as a traveling wave of unusual atmospheric behavior. When functioning as normal, it brings periods of hot storms and blustery rains to the tropical western Pacific, and it does so every thirty to sixty days.

Global warming, though, operates with very much more languor than this. Most mathematical models suggest that the central tropical Pacific will not rise in temperature by three degrees Celsius* until the end of the twenty-first century. At the same time, the level of the sea will have risen between one and three feet, according to the Intergovernmental Panel on Climate Change. How will those two long-term changes affect, or else be affected by, the El Niño warm phases expected during the remaining decades of the century? It would be at least convenient to know this, because the world's weather is entirely born of this phenomenon, and substantial changes like these either are caused by it or are the cause of it.

Little is certain, though the computers hum. One recent observation that has produced consistent enough results to be called a discovery is that the Walker circulation, Sir Gilbert's lasting legacy, has steadily weakened over the last sixty years. Moreover, it has done so, with weakening trade winds its most obvious demonstration, at a rate that is entirely consistent with the rising temperature of the Pacific's surface. And as will be re-

---

* The warming rate slowed down dramatically in 2008, to the glee of those who believe that climate change alarmism is all piffle. Climate statisticians, who acknowledge the existence of a global warming hiatus, insist that this is merely a cyclical event, and that the upward trend will resume and continue so long as fossil fuels continue to be burned with such careless abandon.

called, a weaker Walker circulation is linked with the start of an El Niño warm phase—which suggests, to put it most crudely, that if this trend persists, the world could find itself in a state of more or less permanent El Niño conditions.

And that, with its corollary of ever more extreme weather events in the western Pacific and over the North American continent (to say nothing of a total permanent collapse of the Peruvian fishmeal industry), could cause long-term changes to human behavior, to the siting of cities, to the planting of crops. But little is certain. Thanks to the new computers, and to the fascination with Pacific weather, global forecasting is less of a crapshoot than once it was. But out in the Pacific it remains a mystery of daunting complexity.

Yet a consensus of a sort appears to be building. It is all to do with heat, with the radiation from the sun, and with the manner in which the planet deals with it. Not a few climatologists are coming now to believe that because of its immense appetite for absorbing the solar heat, the Pacific could in time actually be seen as the savior of the world's living creatures. It will be so by taking in all that destructive heat from the sun and from the excesses of carbon emissions and, rather than allow it to scorch dead the inhabited earth, employ it to warm itself up, slowly and sedately, as befits the dominant entity on the planet, and thereby enable itself to carry the world's heat burden on its own.

The effects of all that absorption of warmth will be locally dramatic, for sure. As the American admiral fretted, there will be bigger and more destructive typhoons; there will be more super Tracys, more stupendous Haiyans. There will perhaps be a more urgent need to evacuate islands that will be inundated more swiftly than was thought. Maybe there will be bigger snowfalls in the Cascades and the Sierra. Maybe no anchovies will ever be caught again off Peru. Maybe the forests of Sarawak will be consumed by fire.

Locally, there will be mayhem. But globally, less so. The planet, perhaps, will manage to heal itself. The world and its creatures will survive, and all will eventually allow itself to come back into balance, just as the geologic record shows that it survived and returned to balance after any number of previous cycles of excess and danger. And once that happens, the Pacific Ocean will be seen, uniquely, for what many climatologists are coming to believe it to be: a gigantic safety valve, essential to the future of the planet

The ocean's monstrous size puts it in a position to let the planet go thermally wild for a time, to wobble dangerously. But then, like a formidable gyroscope, the Pacific will dampen the excess, will help bring sanity back, and will restore calm, serenity, and normality.

The Pacific Ocean as the world's pacifier—the thought is maybe born of all too little science. But it is a thought endowed with poetry, and is now held by many. And in the gloom-dimmed world of today, even such a thought is surely a most welcome one.

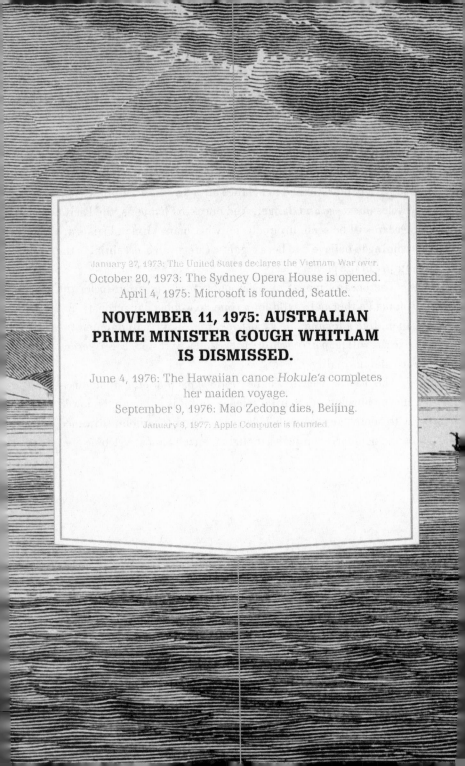

January 27, 1973: The United States declares the Vietnam War over.
October 20, 1973: The Sydney Opera House is opened.
April 4, 1975: Microsoft is founded, Seattle.

NOVEMBER 11, 1975: AUSTRALIAN PRIME MINISTER GOUGH WHITLAM IS DISMISSED.

June 4, 1976: The Hawaiian canoe *Hokule'a* completes
her maiden voyage.
September 9, 1976: Mao Zedong dies, Beijing.
January 3, 1977: Apple Computer is founded.

Chapter 7

HOW GOES THE
LUCKY COUNTRY?

───◄▓▓▌⟋▐▓▓►───

*. . . Australia! You are a rising child, and doubtless
some day will reign a great princess in the South.*
—CHARLES DARWIN, *Voyage of the Beagle*, 1836

*Australia is a lucky country run mainly by
second rate people who share its luck.*
—DONALD HORNE, *The Lucky Country*, 1964

The moment when, on the perfect, warm spring afternoon of
Armistice Day 1975, a serving Australian prime minister was
suddenly sacked by a representative of the British queen, ten
thousand miles away, is still known and remembered, from Perth
to Sydney, from Hobart to Darwin. It is recalled simply and starkly

as the Dismissal. As with Watergate, the Blitz, or the Tsunami, the economy of the description belies the enormity of the event.

It was a quite unprecedented happening, unforgettable in its staging and its consequences. It was the highest of dramas in a country long burdened by the lowest of politics. Its leading characters were petulant, pretentious, and power-hungry martinets. No one came out of it well: when the dust from the fight had cleared, there were, and deservedly so, no identifiable winners.

But it marked a turning point for Australia, by some accounts a belated coming-of-age for the only country in the world, and a very new one, that so massively occupies an entire continent. If this country of twenty-two million is now starting to play a major role in the life of the new Pacific—and it is by no means entirely certain that it has the will to do so—then this one November moment, this rather ludicrous demonstration of faraway Britain's dwindling power over its long-ago colony, was when it all began.

The man sent packing that day was Gough Whitlam, a sleek, imposing, silver-haired and silver-tongued barrister who was seductively charming but with the temper of a honey badger. He had been Labour prime minister for three years, and his rule, which started in 1972 after a quarter century of indecorous political drift,* had sent shock waves through the Australian establishment. In his first ten days in office, he and a colleague took charge of all the government ministries (ruling as what Whitlam called his "duumvirate"), pulled the last Aus-

---

* Few of his predecessor premiers are today well remembered outside Australia, save perhaps for Robert Menzies, who held office twice, for a total of eighteen years. His successor in 1966, Harold Holt, is recalled for a curious reason: he vanished while swimming in the sea near Melbourne, his body never found. His governance of Australia ended once he was declared dead, a tragic end to a long political career—and to his reign, as it happens, as one of Australia's best-dressed men.

tralian troops out of Vietnam, ended the conscription that had put them there, and freed all those imprisoned for evading the draft. He supported equal pay for men and women, increased funding for schools, and ensured land rights for the country's aboriginals.

He gave independence to Papua New Guinea and formally opened diplomatic relations with Mao's China, and then continued on his merry reforming way to change, and drastically, the inner workings of his country: by introducing universal health care, free university schooling, no-fault divorce laws, votes at the age of eighteen, and a set of swingeing tax reforms.

He ended the British honors system, which had long allowed the monarch in London to bestow knighthoods and medals on the citizens of a country that now delighted in its classlessness, and wanted little or nothing to do with the fiddle-faddle of nobiliary enrollment. (The system was revived in 2014.) Under Whitlam, the country also began its abandonment of the British anthem "God Save the Queen," and eventually replaced it with, not the jaunty and traditional "Waltzing Matilda," but the anodyne and fantastically dull "Advance Australia Fair."

The nation, "God's Own" as its happier residents have long thought of it, had never seen anything like it: a politician doing in government exactly what he had pledged to do during his campaign, and doing it fast—"crash through or crash" was Whitlam's mantra—and overturning Australia's social status quo in a matter of weeks. His popularity among the no-nonsense armies of Australia's working "blokes," and many of their spouses, soared. He was for a short while widely regarded as the best prime minister the country had ever had,

But only for a short while. All his achievements, so many of them won at enormous cost to the taxpayer, helped to concoct a formula that was ready-made for political disaster. And given that so many popular politicians unwittingly flirt with hubris,

the disaster was not long in coming. It had much to do, as ultimately the whole scandal had, with money.

It was one specific spending scheme that triggered his government's spectacular fall.

Late in 1974, in the aftermath of the worldwide oil shock and
its associated economic turmoil, Whitlam launched an attempt
to insulate Australia from any such energy-related problems
in the future by boosting the country's immense and untapped
supplies of energy. Specifically, it needed to create a number of
large new mines to extract coal and other of the many minerals
with which Australia had been blessed, to build a giant new gas
pipeline, and to electrify a long series of freight railways in the
country's southeast. Constructing all these would cost the then-
staggering sum of four billion dollars, an amount of proposed
spending very much in keeping with the Whitlam government's
reputation for profligacy on a heroic scale.

The energy minister, a financially unsophisticated former
used-car salesman, heard rumors at a late-night cocktail party
that a Pakistani trader in London could lend such a sum, denom-
inated in then freely available petrodollars. Moreover, the trader
could lend the money at knockdown interest rates, an arrangement that could be secured simply in exchange for paying the
trader a $100 million commission.

Such a plan seemed to the minister not only convenient, but
also *just*—in that by doing the deal, the impoverished West (as
Australia regarded itself) could now get back some of those dollars it had given to OPEC and assorted other oil-trading villains
who had conspired in the decade's savage oil price increases, and
which had wreaked so much damage worldwide. The minister
didn't smell a rat; nor did he ever imagine that the trader might
be a less-than-honorable man, though he had been given advice
from London that the trader was, if not necessarily a crook, then
at least deeply unreliable. So secret negotiations got under way.

Inevitably, with the Australian press being tireless in pursuit of malfeasance, news of these secret talks broke open, and Whitlam's political opposition scented blood. A Melbourne newspaper produced telexes showing that the car salesman—minister had lied to Parliament about the talks, and though he was promptly fired for doing so, the opposition (a coalition of anti-Whitlam parties led by a smooth, Oxford-educated up-and-coming politician named Malcolm Fraser, and which narrowly controlled the Australian Senate) forced Whitlam into a corner. They pressed home their advantage with a simple threat: Whitlam must call a general election, which by now he would be likely to lose; and if he didn't, the Senate would refuse to authorize any further money for the running of the government.

Whitlam, stubborn and proud, refused to give way. Accordingly, Fraser did as threatened, called his parliamentary colleagues to order, and cut off the government's money. It was a simple, if devastating, crisis. And because of the uniquely British manner in which the Australian government was run, its sole possible solution was a uniquely British one. The queen had to become involved. No matter that she sat on a throne halfway around the planet. She now had a constitutional role to play, and given Whitlam's intransigence, she was obliged to do so.

Not the queen herself, however. Her Majesty had a local representative. He was a former boilermaker's son from Sydney named John Kerr, a cherubic but leonine figure who had worked his way up through the Australian education system to become one of the country's prominent labor lawyers. To add irony to this particular situation, he had been appointed to be the queen's man in Australia, the governor-general, by none other than the politician who was now at the center of this imbroglio, Gough Whitlam.

The office of governor-general—a figure who in Australia is splendidly uniformed and copiously bemedaled, and has a flag of his own, an emblem of his own, a fleet of large cars without license plates, a flotilla of boats, the use of a private aircraft, and servants staffing a pair of fine and opulently furnished houses—is one peculiar to the collegial and world-girdling body known as the British Commonwealth. All the countries with a governor-general today are former British territories that have chosen to retain the British monarch as head of state. In other words, they have opted not to become republics, with presidents of their own, heads of state who are homegrown and homemade. And the Pacific, as an earlier chapter has demonstrated, is amply supplied with such former British possessions: notably Australia, Canada, Fiji, Hong Kong, Kiribati, Nauru, New Zealand, the Solomon Islands, Tonga, Tuvalu, and Vanuatu. Only four of these, however (Australia, Canada, New Zealand, and the Solomon Islands), still recognize the queen of England as head of state. So only these have governors-general as Her Distant Majesty's local representatives.

And only one of these, the eighteenth governor-general of Australia—who in full was in 1975 titled "His Excellency the Right Honourable Sir John Robert Kerr, AK, GCMG,* GCVO, QC, of Government House, Canberra and Admiralty House, Sydney"—ever had the temerity to wield the ultimate reserved power of his office. And he shocked the world by doing so.

In normal circumstances an Australian governor-general—usually a white man, until 1965 wellborn and British, only once a woman (named Quentin), and thus far not even once an aboriginal—does little more than put on fancy clothing and wander about opening things. Technically he is the country's head of state, so he receives foreign ambassadors, represents the country on overseas excursions, and is the commander in chief of the armed forces.

Yet, technically, while he is the representative of the British

---

* The GCMG, the Grand Cross of the Order of St. Michael and St. George, is a British order of chivalry traditionally given to British and Commonwealth diplomats. There are three orders, in ascending seniority, CMG, KCMG, and GCMG; they are popularly, if a little unkindly, said to signify "Call Me God," "Kindly Call Me God," and, as in the case of Sir John Kerr, "God Calls Me God."

monarch, and is bound to perform the monarch's wishes, he also wields some power of his own, as an Australian. And this power can be considerable. In particular he has what are known as reserve powers, one of which is the ability, under certain constitutionally defined circumstances, to sack a serving prime minister.

Which is what Sir John Kerr did—and to the very man who had appointed him. He put the boot in shortly after lunch on Tuesday, November 11, 1975.

This would have been a busy day in ordinary circumstances: it was the day when Angola, on the far side of the world, became independent from Portugal; the day when Australia marked the anniversary of the 1880 hanging of its most notoriously romantic criminal, the armor-wearing Ned Kelly; and the day when millions would stand in silence to remember the dead of various world wars, the eleventh day of the eleventh month having been chosen to mark the memorial of all.

But for this singular Australian crisis, the date remains seared into the soul of every Australian living at the time. For no one imagined this would or could ever happen.

Whitlam and Fraser had been sparring in complicated and deviously political ways for several days. On that Tuesday morning, Whitlam insisted that if no money was forthcoming, he would, in fact and at last, call an election. He telephoned Kerr to make a formal appointment to come and tell him so. Kerr, however, decided to act preemptively on his own—being only too aware that Whitlam might well telephone Buckingham Palace and have the queen remove him, Kerr, from his job, as Whitlam had every right to do. So, in his view, he had no alternative but to remove Whitlam from his job first, before Whitlam could possibly move against him. He accordingly telephoned Fraser in his capacity as leader of the opposition, and told him to report to Government House (the governor-general's official house), clandestinely but immediately. It was all memorably Machiavellian.

So Fraser was already there, carefully hidden away in an outer room, when Gough Whitlam, fifteen minutes late and quite unsuspecting, arrived for his own appointment. He was shown into Kerr's official study, saying he had the documents for signature calling for the long-awaited election. Kerr said that before he would look at those documents—which in any case he could not sign because an election could not be held until the money supply was restarted—would Whitlam kindly read the letter he now handed him. Whitlam sat back and read the four-paragraph document, with mounting astonishment.

"Dear Mr. Whitlam . . . In accordance with Section 64 of the Constitution," it began, and with much highfalutin quasicolonial rigmarole went on to authorize the "dismissal of you and your ministerial colleagues . . . great deal of regret . . ." and ended, with magisterial defiance: "I propose to send for the Leader of the Opposition and commission him to form a new caretaker government, until an election can be held."

Such a thing had never happened before in Australian politics, and in modern times, seldom before anywhere. It was the modern version of the ax and the block: sudden, swift, and dramatic.

So there was much spluttering. Whitlam first tried to telephone Buckingham Palace, to demand that, instead, Kerr be sacked. But it was 2:00 a.m. in London, and no one answered the phone. Whitlam left Government House shaking with rage. Fraser, waiting sheepishly in an anteroom, was then ushered into Kerr's office and told that, provided he untied the cords of his reticule and permitted the flow of money to resume, he would now be Australia's prime minister. He agreed, wrote and signed a letter, and was formally sworn in just after lunchtime that very day.

Word leaked out within moments. Radio stations broke into their regular programs. Crowds began to gather. Fraser, it was widely assumed, had played a dark game, and stones were thrown

The silver-haired lawyer-politician Gough Whitlam at the moment of his dismissal as Australia's prime minister at the hands of the Queen's local representative. The event proved a watershed in the governance of this former British possession.

through the windows of his political party headquarters in Melbourne. Dockworkers immediately went on strike, and all the country's ports were promptly closed to foreign trade. Australia briefly shut itself down. People spoke of a coup d'état. Whitlam put out an inflammatory statement, urging his supporters to "maintain your rage." The capital, Canberra, was in an uproar, and in a flurry of frantic activity, votes were being called, statements were being made; the House of Representatives refused to accept Fraser's elevation, and then refused to adjourn itself; and it was only when the Senate accepted the transfer of power and passed the money supply bills to get the nation under way once more that some sort of legislative calm settled on the nation.

The drama's final act played out on the steps of Parliament, with a bemused Whitlam, jostled by scores of heavily sideburned reporters (this being the mid-seventies; no doubt their trousers were flared), listening to an icy, unsmiling civil servant, David Smith, who had been deputed to read out the governor-general's formal proclamation. Whitlam, with his swept-back mane of silver hair, stood tall, cutting a praetorian figure. When finally Smith ended his peroration with "God save the queen," Whitlam drew himself up still more majestically and rumbled famously into the microphones, "Well may we say, 'God save the queen,' because nothing will save the governor-general."

This was followed by disturbances and marches and protests of one kind and another—"We Want Gough!"—for several days to come. But the deed was done. When the election was held the next month, Fraser won the country by a thumping majority, the largest in the country's history thus far, and he ruled through three further elections, until 1983.

Though Whitlam had once called Malcolm Fraser "Kerr's cur," the pair became firm friends in the aftermath of the affair, and when Whitlam died in 2014 (just a few months before Fraser himself died, in 2015) his successor had the kindest of words to say of him. Whitlam, indeed, was viewed as an elder statesman, a man whose three years of policies had changed Australia for all time. "He had the sense of Australian identity," said Fraser. "He had the vision for an independent Australia. He had a grand idea for the country."

Other former prime ministers agreed. "A true international-ist and regionalist," said one of them, Bob Hawke. "He helped Australia earn the world's respect," said Paul Keating, another. "The country came very close to the same marginalization that South Africa experienced over racial discrimination—our salva-tion from that occurred through and with Gough Whitlam."

And most agreed that the Dismissal eventually achieved the opposite of what John Kerr had intended, which was to remind the country that he, the queen's man, still wielded in extremis, and on her behalf, the ultimate authority over her faraway do-minion. For the office of Australian prime minister has since become hugely enhanced, London's authority has dwindled to little more than a vague notion, Canberra's power as a federal capital has been greatly magnified,* and the endless disputes among the country's various states and territories have been generally reduced to a dull roar.

Coincident with this affair, Australia's presence in the Pacific, and as a regional power, has become steadily more visible. The Australians' vocal determination that no such thing as the Dismissal should ever happen again appears to have triggered a sense of national resolve, an empowerment that has advanced the country and its standing in the world no end.

The luckless John Kerr himself was to be the butt of hostility and acrimony for the rest of his days. At almost every occasion, when he was asked to cut a ribbon or lay a foundation stone, he was mobbed by demonstrators or subjected to a withering barrage of catcalls. He took first to drink, and then he took to London, and he ended his days miserably, staggering and disdained. When he died in 1991, his family buried him quietly and in secret, sparing the government of the day—by now Labour once again, just as Gough Whitlam's government had been in 1975—from having to contemplate a state funeral. It almost certainly would have denied such an honor to one of the most despised men in recent Australian history.

If the crisis of 1975 provides a convenient marker year for Australia and for its standing as a Pacific regional power, then there was something else about the mid-seventies, something less specific, less amenable to definition, but nonetheless wholly recognizable. For the time can perhaps also, if very roughly, be

---

* Even though, near-uniquely, it does not have an international airport. Even Ottawa has one; though Monaco, Lichtenstein, Andorra, and the Vatican do not, and the Pacific archipelago of Tokelau, which is entirely without an airfield, has no capital, either, but rotates its administrative headquarters annually among its three main islands.

said to have ushered in a sense of a quite new Australian style, a phenomenon that promptly (and at last!) lent a touch of swagger to the country's newfound authority.

Even Australia's most ardent admirers would probably admit that style was a commodity not much in evidence in the country's immediate postwar years. This, after all, was a nation that for decades had displayed *cultural cringe* prominently on its sun-bronzed forearm, and seemed to wish to bask in intellectual underachievement and a mulish incapacity for any kind of distinction, except in matters relating to sport. The reputation that Australians had then was one born of a gallimaufry of clichés: of meat pies; of grubby pubs; of kangaroo hunters, perpetual sunburn, the outback, larrikins, a brutish kind of football, poisonous spiders in the dunny, Anzac biscuits, Vegemite and lamingtons, Castlemaine XXXX, the tall poppy syndrome, blackfellas, barbies, "G'day, sport," an enviable competence at cricket, an enviable concept known as mateship, the White Australia Policy . . .

Yet if anyone dared be critical of such things, or cringed at their existence, then all and any of the ill will such as these provoked would be quite trumped by one other singular and undeniable fact: the Aussies' singular courage and determination in war, and the simple sad affection of their countrymen and -women who sent them off to fight. At Gallipoli the no-nonsense nature of the memorials to the eight thousand Australians who fell to the Turkish machine guns make this point with a special eloquence. A passerby will find, instead of grand marble graves tricked out in gold, small cards leaning by more modest tombstones, many a one left there by a mother, or else by friends who came when the mothers couldn't afford the fare. "You did your best, son," the cards say, or something like it. Or else there are notes placed by their mates, who had come from back home: "Good on Yer, Billy-boy," or "You done good, Jack."

A kindhearted country, a visitor would be likely to say—on visiting both the country and, more decidedly, the country's faraway memorials. A goodhearted, kindly people. And if with not too much style about them, then so be it.

But come the mid-1970s, a new affect began, if timidly at first, to edge out the old. Australia started to present itself in a quite different manner. Within just a few months of each other in 1974 there appeared two markers of note: one that served to remind outsiders, if only satirically, of the adherent nature of the old Australian ways; and the other that, in a far more substantial manner and not satirically at all, delightedly and majestically exalted the new.

The first—and it has to be stressed that this is satire writ large—was the unanticipated appearance on the world scene of an antipodean archetype: the entirely memorable Australian diplomat named Sir Leslie Colin Patterson—Les Patterson, as he would genially remind his audiences, for short.

I first encountered Sir Les in Hong Kong in the autumn of 1974, when he made a speech in the swanky comfort of the Mandarin Hotel, formally announcing his appointment as Australia's cultural attaché to the Far East. He had been officially sent out from Canberra "to impugn," as one commentator had it, "the fundamental refinement of the Australian character." He was just thirty-two years old, with a career already hinting at greatness.

Thanks to family connections from his schooldays in Taren Point, in the southern Sydney suburbs, he had been plucked out of a soul-destroying job in the Literature Division of Her Majesty's Customs and Excise Office, and given the portfolio of "shark conservation minister" in the government of Sir Robert Menzies. He adroitly managed to survive the change to Gough Whitlam's Labour government and became "minister of drought." Next came his appointment as a cultural attaché, from which he would go on, two years later, to be posted in this capacity in London, then recalled to be chairman of the Australian Cheese Board, to become founder of a private school of etiquette, and later to be given the title of "adviser on etiquette and protocol" to the Australian federal government.

Sir Les's sense of vision was evident even during his youthful first appearance in Hong Kong. That evening, if I remember correctly, he wore a vividly iridescent blue suit, with a yellow check lining, which had clearly seen many better days. His tie, the wide style of which owed much to cinema noir films of the 1940s, bore encrusted evidence of his many earlier dining experiences. His teeth were long, and they protruded, and they were stained the same coppery color as his fingers, which seemed always to be clutching a cigarette from which dangled an ash of improbable length. His hair, long and amply greased, hung over the none-too-well-laundered collar. One of his shoes was missing a lace.

He appeared that evening to have had a fair amount to drink and needed some support from diplomat colleagues when attempting to stand. He was easily tempted into making remarks about human reproductive activities, and made it clear he was not in favor of men lying with men. However, he had no particular problem with women lying with women, since he could not entirely understand what they did when they lay together, and no one in his various Roman Catholic schools back in Taren Point had ever found the time to explain it to him.

He was by all appearances overfed; was often overcome with what appeared to be libation-triggered tiredness and emotional excess; and to judge from his utterances, was evidently wildly oversexed. His outbursts caused many of his more sensitive listeners to recoil. His favorite nonsexual pastime, which he often indulged in while giving the very speeches that so marked his career, seemed to be either nasal excavation and gastronomy or competitive wind breaking. He was a caricature, in short, of a certain kind of Australian of old, an amalgam of bronzed ocker and working stiff, of corrupt pol* and journeyman sheep stealer. I daresay some, even in the eighties and nineties, when he was most visibly on the world speaking circuit, imagined that he did represent some kind of an Australia that still existed, if only just.

Les Patterson was, of course, entirely fictional—a character created by the writer, comedian, and artist Barry Humphries, whose other alter ego is the somewhat more lovable and acceptable Dame Edna Everage. It would be idle to read too much into what is essentially and only a fictive creation of contemporary comedy. Yet Sir Les Patterson, who performed for more than thirty years after his creation, well into the twenty-first century, is still quite recognizably emblematic of an Australian type—a type that most modern Australians hope is fading in the rearview mirror as the country eases itself steadfastly into the more respectable and re-spected role that it increasingly likes to play today.

A role that owes much to, and is symbolized by, the creation of one quite remarkable building, and one that happened to be completed at almost exactly the same time that Sir Les Patter-son first made it onto the stage. This structure is the Sydney Opera House, and it was formally opened in October 1973 by Queen Elizabeth, in her role as queen of Australia. The Ameri-can architect Frank Gehry once said it was a building that, quite simply, "changed the image of an entire country."

Until that moment, Australia did not possess one national con-struction that the world would see and instantly mouth, "Austra-

---

* A template might well have been Hon. Russ Hinze, Queensland's minister for racing, minister for main roads, minister for police, and supposed "minister for everything" in the 1970s. Fat, uncouth, and incorrigibly corrupt, Hinze is best known for having a state freeway diverted several miles to bring customers to a pub he owned. His genes almost outlived his reputation: his granddaughter Kristy Hinze went on to model for Victoria's Secret.

lia!" There was of course the immense sandstone upwelling of Ayers Rock, Uluru, which spoke of outback, of remoteness, of the aboriginal peoples and the continent's serene inner space. But other nations that had natural wonders on a similar scale had man-made marvels, too. America had its Grand Canyon and its Empire State Building. Britain had both its White Cliffs of Dover and Stonehenge. Egypt had the Nile and its Pyramids. Australia did not.

The best that Australia could muster, and which might be seen to complement its own natural wonders, was the great "Coat Hanger Bridge" that spanned the narrow entrance to Sydney Harbour. But it wasn't really Australian: the Sydney Harbour Bridge was more properly a monument to empire, since it was constructed by a firm in Middlesbrough, and made mostly of steel that had been shipped down from England. To regard that as an Australian monument might be akin to a Delhi resident showing off the Viceroy's House by Edward Lutyens, instead of the Taj Mahal.

But to see the Sydney Opera House—and yes, framed by the now venerable bridge, for together the two offer a quite incomparable spectacle—and to see its jumble of graceful white sails, its soaring peaked seashells all apparently floating beside the blue ship-busy harbor waters, is to experience the sublime lyricism of one of architecture's greatest moments. You see it when you fly in to Sydney, you see it when you drive beside the harbor, you glimpse it peeking between the skyscrapers of the business district. And each time you see it, you notice it. It is a building impossible to take for granted. It is a masterpiece.

One recent sighting rekindled a bittersweet memory. Just after dawn one summer's day in Sydney, I was telephoned from a hospital in New York. Would I speak, the caller asked, some last words to a dear friend of mine, an elderly editor beloved around the world, and who now was dying in a room surrounded by friends. He was unconscious; his breath was labored. But they put the telephone to his ear, and I described to him as best I could what I could see from my hotel window: the sun gilding the top of the great bridge, the little insect-like Manly ferries skittering along on white-waked water, bringing commuters across the bay, and then, rising from between the office towers like a clutch of lilies, the soft peaks and angles in a sun-washed pink of the Opera House.

When they took the phone from his ear, they told me his breathing had paused. They were sure he was listening to every word, for he loved Sydney as I loved it, and my sending him a farewell from there brought him serenity, if only for a moment. He died an hour later, calm and at peace.

But there was little peace in the making of the Opera House—and the saga of its construction amply displays many of the same contradictions evident in the makings of today's very new Australia. The dramas of the years of its building (together with one terrible, coincident tragedy and another subsequent sex scandal) were legion: there were vicious tugs-of-war between forces old and new, between provincialism and globalism, between old-school philistine Australia and twentieth-century visionary Australia. The battles spoke volumes about the process; yet the result is memorable and wonderful, and could not have been made, it often seems, anywhere else in the world

All began peaceably enough. It was an Englishman who first conceived the need for a dedicated opera house in the country's cultural capital: the conductor Eugene Goossens, the London-born scion of a distinguished Belgian musical family, who had been invited to Australia in 1947 after a highly successful two decades in America, to conduct the Sydney Symphony. But not long after he arrived he was heard grumbling that his new orchestra's home, the ornate Victorian town hall, even though it had one of the world's largest pipe organs, was far too small. Sydney, he insisted, deserved better. It could be a world-class city; it should have a world-class opera house.

His six years of energetic lobbying eventually bore fruit in

1954, when the then–New South Wales premier, a former railway worker and insurance salesman named Joe Cahill, threw his weight behind Goossens's idea. He agreed to clear space for an opera house by demolishing a city-owned tram depot on a delightful little peninsula, Bennelong Point, just north of the city's beautiful Botanic Gardens. Cahill staged an international contest to find the best architect: more than 230 men and women from more than thirty countries submitted drawings. It seemed as though the entire architectural world wanted Sydney, so spectacular a city in so visually blessed a country, to have something special.

The winner was an almost unknown architect from Denmark, a forty-year-old would-be sailor and admirer of Mayan temples, a man with not a single memorable building yet constructed, named Jørn Utzon.

The story goes that Utzon's vaguely realized sketch, all elliptical shells and curves that seemed to burst organically upward and outward into the harbor like spinnaker sails, or like a huge billowing flower, was initially rejected—but that Eero Saarinen, the Finnish jury member already known for his futuristic designs in the American Midwest, pulled Utzon's sketch from the reject pile, declared it a work of total genius, and said he would support no other competitor. "So many opera houses look like boots," he said, a little oddly. "Utzon has solved the problem." The Sydney city assessors, who had the final vote, were equally enthusiastic: "We are convinced that they present a concept of an Opera House which is capable of becoming one of the great buildings of the world."

Utzon was sent a telegram informing him of his success. His ten-year-old daughter, Lin, intercepted it and brought him the news, pedaling her bicycle furiously across the flat Danish countryside to his studio, and then demanding, "Now, can I have my horse?" He could well afford it: Sydney wired him the prize money of five thousand pounds* and told to come on down and get weaving.

The weaving, though, proved to be something of a trial. Like so many of architecture's stars—Gehry, Calatrava, and Frank Lloyd Wright come to mind—Utzon was big on vision, short on details. For example, he was never quite certain that it was possible to make the ogival shells for the roofs—especially because of their different sizes, with varying arcs and angles. The cost of all the eccentrically shaped wooden forms that would be needed to support the drying concrete for each one was going to blow a massive hole in the budget. Moreover, the total weight of these extraordinary roofs exceeded the strength of the concrete pillars that were to support them—a potentially lethal problem for audiences, a death knell for the building.

As the costs mounted, the project slipped behind, and the state government organized an emergency lottery—first prize, one hundred thousand pounds—to help raise further funds. Morale was helped in 1960 by the impromptu appearance on-site of the American singer Paul Robeson, who gave an unexpected concert among the cranes and scaffolding towers. He sang "Ol' Man River" to a throng of Italian and Greek construction men, his dark face and their olive complexions prefiguring Australia's future of multiculturalism.

Then, in 1961, the design teams came up with the technical breakthrough, the aha! moment, that finally made Utzon's dream possible. All the shells, the technicians declared, could be thought of as parts of a single enormous sphere, like segments of an orange. Since they would all now have a common radius, they all could be cast from a common mold, and then cut down to smaller sizes in those places where Utzon wanted them. It seems

---

* Australia, still at the time clinging to the mother country's apron strings, used British-style pounds, shillings, and pence until 1966.

such an obvious solution now, but in 1961 it took hundreds of computer hours (when computers were rarely used to solve architectural problems) to come up with the final answer to a taxing technical conundrum. There has been some controversy over whether it was Utzon himself or some other mathematically inspired architect who enjoyed the necessary epiphany. But in either case, the project was then promptly freed to race toward completion.

Jørn Utzon, the unassuming Danish architect plucked from obscurity to design Australia's best-known structure, the Sydney Opera House, never saw it opened, but was honored posthumously for the creation of one of the world's greatest public buildings.

Or it should have been—but then the old Australia briefly reared its head. In 1965 a new state government took office, with the Opera House still a long way from being finished. Two of its most senior figures happened to be politicians who, in terms of their deep disdain for high art and culture, could give Les Patterson a run for his money. They were the new leader, Bob Askin; and, more notoriously, his public works minister, Davis Hughes, a figure described by the Australian critic Elizabeth Farrelly, in

Utzon's obituary, as "a fraud and a philistine," a man who falsely claimed to have a university degree and who had "no interest in art, architecture or aesthetics."

Hughes wanted Utzon out, denouncing the architect as a foreigner, a prima donna, and in the very worst sense of the word, an *artist*. Using the want of taxpayer money as the excuse, the minister gradually pared down the budget, slicing away at the architect's ability to pay his bills or, indeed, his staff. "How can you alter everything against my advice?" Utzon bleated pathetically during one meeting. "Here in Australia," Hughes responded tartly, "you do what your client says"—the voting people of New South Wales, of course, being collectively *the client*.

Day by day through the southern summer of 1966, Utzon's situation worsened; and when, in February, he totted up the figures to show that the government owed him some one hundred thousand dollars* in fees, and threatened to resign, Davis Hughes called his bluff and accepted.

Utzon, shattered, left Australia six weeks later, traveling under an assumed name to avoid the press. A thousand protesters marched to the half-completed building, many of them architects. A local sculptor went on hunger strike to demand that the Dane be invited back. And though he expected to be recalled, he never was.

Instead, several Australian architects were hired in his place, and they took seven further years to complete the building's initially drab and uninspiring interior. While the outside sailed itself into architectural history as one of the great creations of the twentieth century, the inside was riddled with imperfections and crabbed spaces—early operatic orchestras had to have their percussion sections cordoned off behind plastic screens so the violins could hear themselves; ballet dancers exiting the stage

---

* By now the country had left the pound and was using the dollar, which was initially to be called the royal, but which ended up as the Australian dollar.

had to have catchers stationed in the wings to stop them from hurling themselves into the walls.

Jørn Utzon never came back to Australia, and he never saw the completed Opera House in person. When the queen opened the building in October 1973, twenty years after it was first conceived, ten years later than scheduled, and 1,400 percent over its original budget, Utzon was not invited; nor was his name mentioned. Publicly, he was an unperson; privately, he remained stoic and unembittered. His most sardonic comment was simply to call his experiences in Australia an example of "Malice in Blunderland."

And he continued to believe that history would eventually judge him more kindly. His faith would be borne out in his later years, when Australia effectively apologized to him. He was given an award, the Companion of the Order of Australia, in 1985. But then, more important, he was given work. The inadequacies of his building's interior were deemed so egregious that, in 2000, Utzon was approached to help undo the botches of the Sydney architects, and to redesign it. He said he was minded to accept the commission—though on hearing him say so, a clutch of guardians of the old Australia briefly stirred themselves to life once more. The ever-querulous Davis Hughes, for instance, swiftly went to the papers: "There's obviously a need to upgrade the place," he allowed, "but why do we need Utzon? Why can't we get a competent Sydney architect?"

Yet, in the end, the Dane did do the work, though from long distance, by airmail and couriered blueprint. The city was so duly delighted with the result that the Utzon Room (light, airy, sparely furnished, and with views of the sparkling harbor below) was named in his honor. The old man, now unwell and living in Mallorca, was thrilled beyond measure to receive the news, and reacted with undeserved magnanimity. "The fact that I'm mentioned in such a marvelous way, it gives me the greatest pleasure and satisfaction. I don't think you can give me more joy as the architect. It supersedes any medal of any kind that I could get and have got."

The queen came back once again, in 2006, to open the refurbished interiors that Utzon had designed. Utzon himself was by now too ill to travel so far, but his son was there, and he made a wistful speech in which he said that his father "lives and breathes the Opera House, and as its creator just has to close his eyes to see it."

Utzon died in Copenhagen two years later. A year after his passing, the city of Sydney staged a concert, as both memorial and official reconciliation. An apology, if you will. But the event paled before the world's recognition of what he had accomplished. Shortly before he died, Utzon heard that UNESCO had declared his creation a World Heritage Site. The proclamation was lengthy, as befits a structure of such complex design and tortured history. Its preamble was eloquent: Jørn Utzon's building, a gift from Europe to the Pacific, and thence from the Pacific to the world, was "a masterpiece . . . its significance is based on its unparalleled design and construction; its exceptional engineering achievements and technological innovation and its position as a world-famous icon of architecture. It is a daring and visionary experiment that has had an enduring influence on the emergent architecture of the late 20th century."

As befits so troubled a passage to completion, there were two codas to the Opera House story, both of them melancholy—one quite tragically so, the other more curious and bizarre.

The first related to the lottery, which had been staged in 1960 to raise additional funds for a project whose costs were at the time beginning to spiral beyond control. The prize offered was one hundred thousand pounds; and on June 1, the winner's name was announced in the newspapers: a Mr. Bazil Thorne, who lived with his family in Bondi, where the surf famously pounds in from the South Pacific Ocean. There were no privacy laws at the time; the family's full address was listed in the paper.

A week later, the Thornes' eight-year-old son, Graeme, was picked up at a street corner near his home, to be taken to school— only, he never arrived. That night a man called the house demanding twenty-five thousand pounds in ransom. A massive police search began, ending a wretched month later when the child was found bludgeoned and suffocated.

The killer was captured three months later, after a triumph of forensic detection that involved pink paint chips, mismatching flower types, stolen cars—and the discovery that the man allegedly involved had just left Australia aboard a London-bound P&O passenger ship, the SS *Himalaya*. Australian federal police were waiting for the vessel when she arrived in Colombo; and after much legal complication (since the Sri Lankans did not at the time have an extradition treaty with Australia), the man, a Hungarian immigrant named Stephen Bradley, was arrested and returned to his adopted country. Aboard the plane, he confessed to the child's murder. He was sentenced to life imprisonment, and died in his cell eight years later.

The other coda is more simply bizarre, and involved a train of events that provide their own commentary on the fifties Australian zeitgeist. For Sir Eugene Goossens, the towering and talented figure of English music* who, while conductor of the Sydney Symphony, had begun the process that led to the building of the Opera House, turned out to be a man of highly exotic sexual tastes. And that, to the Australia of the time, was most decidedly *not on*.

While in Sydney, Goossens became romantically involved with a woman named Rosaleen Norton, who was a pagan, a keen practitioner of the occult, and a lady who had a liking for both

---

* It was Goossens who, in 1942, with the aim of finding stirring music to help with the war effort, wrote to Aaron Copland asking for a composition. The result was Copland's famous *Fanfare for the Common Man*.

flogging and unusual kinds of misbehavior with animals, mostly goats. The popular press in Sydney—then, as now, eager for London-style sensation—liked to call Miss Norton the Witch of King's Cross, and her studio and place of work were very much a cornerstone (a socially unacceptable cornerstone) of this louche and seedy Sydney neighborhood.

Goossens, who himself was an admirer of the British occult-ist (and would-be climber of Kanchenjunga) Aleister Crowley, would occasionally bring Miss Norton gifts from London. In March 1956, after he had returned from being awarded a knight-hood at Buckingham Palace, customs rummagers at Sydney airport found in Goossens's suitcase large numbers of dubious photographs, together with rolls of film and what were described as "ritual masks." He was promptly arrested, and threatened with the serious charge of "scandalous conduct." He was not un-reasonably terrified by the prospect of spending a lengthy time in prison, and eventually agreed that he had violated a section of the Customs Act that banned the import of "blasphemous, in-decent or obscene works" into Australia. His crime carried the lesser penalty of a fine, and he eventually was obliged to pay up the not insubstantial sum of one hundred pounds.

But what he didn't reckon on was the publicity, which was im-mediate and merciless. The event, and its prominence in the more raffish newspapers, brought to an abrupt end what had been a glittering musical career. Goossens, now utterly shamed in public, immediately resigned from both the symphony and the New South Wales State Conservatorium, and fled to London on his sixty-third birthday. Just as Jørn Utzon would do ten years later, Goossens chose to slink out of the country under a pseudo-nym and in disgrace. He was subsequently described by friends back in England as having been "absolutely destroyed" by the affair. He was dead six years later.

Yet, as with Jørn Utzon and the room now dedicated to his memory, there would in time be a more kindly end to Goossens's story, too. The Opera House, the grand realization of his long-ago vision, was opened ten years after his death. And when it opened, the foyer held a commanding sculpture of the conductor, honoring the contribution that he had made to the building of one of the great monuments to music in the twentieth century.

One television journalist in Sydney later wrote that Goossens had surely been a victim of the times, of what she called the wowserish,* churchly, prudish, censorious, hypocritical, and deeply conservative Australia. His offenses were, by today's standards, entirely venial, unworthy of remark. Though the law that Goossens broke remains, technically, on the books, no case has been brought under its strictures for many decades. The Australia of those times, one can be certain, has now been all but submerged, almost forgotten. On the surface at least, Australia is a liberal and tolerant society, multicultural in nature and cosmopolitan in attitude, its politics progressive and forward-looking, and with a reputation and a standing that in consequence have changed in the past half century almost beyond recognition.

Or have they changed? Few would dispute that if Australia ever wants to enjoy a degree of respect in the western Pacific commensurate with its wealth and power, it has to be taken seriously by its neighbors. Most especially by its Asian neighbors, those who inhabit a vast slew of north-running countries from New Guinea up to Siberia, by way of Indonesia and Indochina, the Philippines, Korea, Japan, Taiwan, and mainland China itself. For many years, that had not been the case at all: Australia was regarded principally as a source of minerals, little more than an immense quarry, and was much caricatured as such, as an un-

---

* H. L. Mencken liked the word, though it was an Australian who invented it, as a mild insult for a moral scold. Or, as the ribald twenties poet C. J. Dennis put it, "*Wowser:* an ineffably pious person who mistakes this world for a penitentiary and himself for a warder."

cultured, socially conservative, unsympathetic, misogynistic, and racist outpost of the British Empire.

Geographically and geologically, and with its own distinct and ancient native anthropology, Australia is properly and undeniably a component of Asia. Yet in societal terms, and as reflected in its press and by way of its politicians until as recently as the 1970s, it seemed to see itself differently—not as part of the East at all, having made no serious attempt ever to be so, and with the great majority of its people shuddering at the thought of ever becoming so.

It was the country's infamous White Australia immigration policy that first set the tone. Laws were enacted in 1901, from the country's very beginnings as an independent federation, to protect it from Asians, "from the coloured races which surround us, and which are inclined to invade our shores." The whites-only policy was declared by its supporters to be the nation's Magna Carta, the ultimate shield that would prevent the country from being "engulfed in an Asian tidal wave."

This was motivated by fear, of course. It was much the same fear as was enacted on the American side of the Pacific, and that hastened the passage of the various exclusion acts that kept "Orientals" so firmly at bay in California and beyond. Fear of the Chinese among Australian miners who couldn't dig as fast or as furiously. Fear of the Pacific Islanders who would work in the cane fields of Queensland for much lower pay than their white counterparts. Fear of the Filipinas who might launder the linens and cook the pork stews with more alacrity and eagerness than would the ladies from Ireland and South Wales. Fear of the Indians and Malays who could labor on the stations of the outback with far less complaint about the heat than the wild but pale-skinned Australian boys whose ancestors had emigrated from English cities such as London, Leeds, and Liverpool.

Wars with the Japanese hardly helped. The original act that had been passed after the First World War was to be hailed by successive immigration ministers as "the greatest thing we have ever achieved," and in keeping out nonwhites (Japanese now most especially), it seemed an even more blessed creation at the start of the Second. The prime minister of the day backed the policy wholeheartedly: "This country shall remain forever," he declaimed, "the home of the descendants of those people who came here in peace in order to establish in the South Seas an outpost of the British race."

The Labour Party, purportedly the champion of the working man, turned out to be the most vocal in keeping Australia as pure as pure could be. "Two Wongs don't make a White," said a Labour Party immigration minister in 1947. Under the strictly enforced rules, no madmen could come in, no one afflicted by an illness "of loathsome or dangerous character," no prostitutes, no criminals; nor could any "Asiatics" or any "coloureds" enter, either; and for good measure, no one who failed a written dictation test, an examination that could be given to an unwary applicant at a moment's notice, and in the language (not necessarily English) of the immigration officer's spontaneous choice. Sometimes the officer would, for his own amusement, choose to have his applicant write out the test in Gaelic, to be quite certain of a ban.

This couldn't last, of course, this fantastic notion of keeping Australia an antipodean refuge for snow-white Britons, for simon-pure Englishmen. Soon after the end of the Second World War, when war brides began knocking at the country's doors, they were opened a crack, and somewhat reluctantly, to some of the swarthier-looking Europeans: Greeks and Italians at first. They soon came in waves, found the climate and the scenery and the city life much to their liking, and were publicly welcomed in return to a far greater degree than the politicians had supposed. Melbourne in particular soon became the most populous Greek city outside Greece. And these immigrants were well liked.

"Better a dark-skinned Greek than a Japanese," one historian commented.

Then, in the 1960s, the Japanese and the Chinese started being allowed in, too—"distinguished and highly qualified Asians" only, at first; then, as these bellwether arrivals were found acceptable, the restrictions were eased still further. Before long, members of the Oriental races who were vaguely described as "well-qualified" could apply to enter, too. And then, by 1973, Gough Whitlam, as part of his abruptly instituted reform program, ended all such restrictions. The dictation test had already been scrapped. The degrees of qualification were now dropped. The question of an applicant's race vanished from the forms.

All who now wanted to come, and who met the none-too-strict criteria of entry, were welcome to apply. The seventy-year-old White Australia policy was swept into history. The country now became, and in short order, a great multicultural experiment. A country first manufactured as the colony of interloping white men could now reinvent itself as a brand-new community born of the entire world. It was a locally novel type of national entity, a western Pacific version of two tried-and-tested equivalents— Canada and the United States—on the ocean's faraway east coast. All three experiments were at last joined to an ocean that was now fast turning itself into a test bed, a place where the future of human society would begin to be charted. The U.S. president Bill Clinton seemed to understand this when he came to Sydney in 1996: "I cannot think of a better place in the entire world, a more shining example of how people can come together as one nation and one community."

It was quite an endorsement. Except that inside Australia, there were still legions of highly vocal opponents of any policies like this, policies that might draw the country more closely, as they saw it, into Asia's too foreign maw. Some of the shriller of these have adamantly refused to quiet themselves, to accept the realities of change. They have on occasion managed to tap into an alarming groundswell of very ugly popular opinion, and by doing so have managed to set back somewhat Australia's gathering reputation as a fully functioning member of a new pan-Pacific society.

Pauline Hanson is perhaps the most egregious recent example. This was a lady who came to brief prominence in the autumn of 1996, on the heels of a savage outbreak of race rioting that briefly convulsed the country. She was a twice-divorced mother of four, of very limited education, and the owner of a fish-and-chip shop near Brisbane—who yet managed to win election to the federal parliament in Canberra on a platform of undiluted racism and xenophobia. Her views were primitive, direct, and aimed at readily identifiable targets, both at home and overseas.

The aboriginals who lived among her own people were bad enough—they were a lazy, ill-disciplined, and grubby population of hard drinkers who, according to a book to which she gladly put her name soon after she won her seat in Parliament, ate their own babies and regularly cannibalized one another. Yet they were handed privileges in abundance, and they vacuumed up public money that should by rights have been spent on "mainstream Australians."

Her views were no less sparing of peoples living beyond Australia's coasts: she was especially contemptuous toward the Asians to her north:

"I believe we are in danger of being swamped by Asians," she told Parliament in a truly splenetic inaugural address—in which she could say as she wished quite uninterrupted, as one of the courtesies that is customarily accorded to a maiden speech. "Between 1984 and 1995, 40 percent of all migrants coming into this country were of Asian origin. They have their own culture and religion, they form ghettos and do not assimilate. Of course, I will be called racist, but if I can invite whom I want into my home, then I should have the right to have a say in who comes into my country.

"A truly multicultural country can never be strong or united. . . . The world is full of failed and tragic examples, ranging from Ireland to Bosnia to Africa and, closer to home, Papua New Guinea. America and Great Britain are currently paying the price. . . . It is a pity that there are not men of . . . stature sitting on the opposition benches today. . . . Japan, India, Burma, Ceylon and every new African nation are fiercely anti-white and anti one another. Do we want or need any of these people here? I am one red-blooded Australian who says no, and who speaks for 90 percent of Australians."

For a short while her message won a great deal of domestic traction. She wanted Australia out of the United Nations, a total end to Australian foreign aid, and, as her career progressed, ever-harsher limits on nonwhite immigration. The newspapers splashed her over the front pages for weeks. The popular Australian art form of talk-back radio was dominated by her, even though she had a voice that was as penetrating as a dentist's drill. Television interviewers managed to find Mrs. Hanson Sr., who, over sweet tea and sticky buns, voiced her own fear, evidently inculcated in her daughter, that "the yellow races will one day rule the world."

They also found Hanson's senior adviser and speechwriter, and wondered why, as a man named Pascarelli, he should be so vehemently opposed to immigration. His answer was glib: "I was de-wogged." And when Mrs. Hanson was asked if she was xenophobic, her lack of schooling offered up a reply, made after a brief and bewildered silence, of studied artlessness: "Please explain."

But neither did her political opponents do much to advance the standing of Australia to the outside world, which watched bemused, even appalled. During a heated television discussion about why aboriginals had an alcohol problem, a well-meaning political critic demanded of Mrs. Hanson if she knew who the world's greatest drunks happened to be? She didn't. "White Australians," he declared. "The biggest drunks ever known." The notion that twenty-first-century Australia might ever revert to its old idea of keeping Asians at bay, while at the same time taking pride in a permanent national inebriation, caused a wave of shame to engulf the continent.

Which is perhaps why, by the turn of the millennium, the phenomenon of Pauline Hanson had begun to fizzle away. The country seemed swiftly to weary of her. She started to lose elections, then she lost money, and she went briefly to prison on fraud charges, though she was acquitted on appeal and released.

She tried hard to turn such occurrences to her political advantage. During her rise to prominence, she frequently claimed to have survived a childhood of "hard knocks"—and these new stumblings, she claimed, were just more of the same. Her remaining supporters found the suggestion engaging, as endearing evidence of her humanity. So, after only a brief hiatus, she was back in the running, and today she is still present, a slowly dimming star in the country's political firmament, her drill-bit voice little more than background noise. But all the while, her political views have managed to hold the attention of not a few Australian voters, and so long as they do so, they manage, if unwittingly, to dull some of the luster of her country's otherwise brightening public image.

Australia's current harsh treatment of asylum seekers, most particularly those who attempt to come from Asia by boat, has served only further to damage this image of regional congeniality. The country's current stand toward those would-be migrants hoping for safety and protection is that it will detain anyone who comes into its waters in the hope of refuge. No matter how violent the war back home, or how harsh the regime you are escaping, or how severe your risk of persecution, or how desperate your voyage—if you arrive by boat in "the lucky country"* without a valid visa, you will be locked up. Where you will go, under what conditions, and for how long are the only variables—and in recent years, the world's human rights community has declared itself, over and over again, deeply troubled at the manner in which Australia, now essentially alone in the democratic world, has been dealing with what its politicians see as an intractable problem.

Boats laden with hungry, sick, frightened people, fleeing from a variety of conflicts and inhospitable situations in a variety of Asian nations, have been arriving in Australian waters since the mid-1970s, when the Communist takeover at the end of the Vietnam War first caused flotillas of crowded, unseaworthy, near-sinking vessels to set sail into the relative freedom of the South China Sea. For five subsequent years, the exodus of such Vietnamese went on. Most sought asylum in Hong Kong, nearby, or slightly farther away, in the Philippines. But the braver souls, or those in better-equipped boats, managed to navigate their way through the mess of Indonesian islands to the northern coast of Australia. The Canberra government of the day—Malcolm Fraser's post-Dismissal government, as it happens—took a kindly view: more than fifty thousand were admitted. Then the situation in Vietnam eased, and the country's frontiers were more keenly guarded. The boats stopped leaving. The South China Sea stilled. The Hong Kong camps were emptied. The coastal waters off Darwin and Cairns and Broome quieted. The problem, so far as Australia was concerned, seemed to be over.

But not for long. In 1989 it all started up again, this time with Indonesians, fleeing from poverty, dictatorship, and summary justice—or else making a quick run across the Arafura Sea in

---

* Donald Horne's famous book of this title was in fact a harsh critique, and his title suitably sardonic. So far as today's boat-borne migrants are concerned, it is most apt. Few who come up against the harsh realities of today's Australian immigration law experience good luck of any kind.

the hope of a better and more prosperous life. Or else they were Papuans trying to make it across the Torres Strait for much the same reason. Or else Afghans or Pakistanis, Burmese or Cambodians. Australia was to such people so close, so very tempting, so empty, so rich, so clearly in need of those who could, and would be willing to, work.

Invariably by now the fleeing thousands had the help (bought at great cost) of gangs of "snakeheads," the people smugglers eager to cash in on those impoverished Asians desperate to get to the bright lights and big opportunities of Australia. But this time Australia reacted, harshly. It had no room for more, it said; Australian workers were bitterly complaining about the low-cost newcomers who were now plundering their jobs. The detention policies that still exist today—draconian, harsh, criticized—were slowly and steadily brought into force. And since that time, with the policies and their manner of implementation ebbing and flowing and shape-shifting as the various governments in Canberra have changed their views, so the matter of dealing with arriving boat people has proved to Australia a practical and humanitarian challenge without end or answer.

A so-called Pacific Solution was brought into force in 2001. It ebbed and flowed and shape-shifted, too; and though it now has an extra aspect with an extra name, Operation Sovereign Borders, Australia's Pacific Solution remains, in essence, the policy of today.

Under its rubric, three island camps were opened to accommodate the hopeful masses, and all remain busily active today. One is on Christmas Island, an Australian territory in the Indian Ocean. Two others are on foreign-owned islands, and the Australians pay to have them there. One is in Papua New Guinea, on the Admiralty Island of Manus, which was made briefly famous by Margaret Mead, who lived in its rain forests after World War II. The other is on the former phosphate-rich equatorial Pacific island of Nauru, which in the 1960s claimed to be the richest per capita state on the planet, but which is now a devastated wreck—a played-out environmental disaster zone; an independent state associated mainly with flagrant corruption, money laundering, a population of nine thousand, and, in recent years, the third of the Australian detention centers for would-be immigrants.

The camps in all three sites are dreadful and dismal places, filled to bursting with Afghans and Tamils and Pakistanis and Syrians fleeing from war zones or from the Taliban or from ISIL or from a host of other despots and desperadoes. All the incarcerated, some held for the many years that it now takes to process their applications for Australian residency (which most likely will be denied), took months to find their way to this place. Most of them fled first to Indonesia, waiting for countless months in dreadful conditions, before taking to the sea and to what all hoped might be the sanctuary of Australian waters. But there, instead, were the ever-watchful ships from the Australian navy, determined to prevent them with force from ever reaching Australian territory and thereby being able to claim refugee status.

I was in Darwin in the November summer of 2014; and all the local talk was about the customs boats that left the little port, their crews scanning the hammered-pewter surface of the sea, looking for the tiny and barely seaworthy craft that had come down from Java or Sulawesi or the Banda Islands, and intercepting them and brusquely scooping up their passengers. They would take them on next to Manus or Nauru—or even to Christmas Island, which the Canberra government had, with Orwellian cunning, deaccessioned, ensuring that any migrant who managed to land on its shores could not claim to have landed on Australian soil. Christmas Island was indeed, legally and constitutionally, still sovereign Australian territory—except, technically, for the sole purposes of immigration, when it is deemed to be a foreign place.

Not that landing on Christmas Island was ever easy. In December 2010 a boat crashed into the cliffs of the island's Flying

Fish Cove, tossing scores of its passengers into the raging waters. Forty-eight of them died, their drownings witnessed by hundreds ashore. It was a dreadful tragedy—yet the Australian government simply used it as justification for the national policy of prohibiting boats from trying to reach land. Keeping the boats away would prevent disasters like this from happening again, the government said. Interception by armed warships was for the refugees' own good.

The Christmas Island tragedy of 2010 served to diminish still further the luster of Australia as a would-be model member of the western Pacific's Asian community. Canberra's immigration minister of the time hardly helped when he complained publicly upon learning that his own government had paid for some family members to attend the funerals of the victims. When asked if he thought it was heartless to complain that a man who had lost his wife and two young children to the sea had been given a compassionate flight to the graveside, the minister went on the radio to declare that the cost of the man's flight was unreasonable. Few thought the minister anything other than entirely pitiless.

On the one hand, pitilessness; on the other, compassion. The difference is stark, and serves as a vivid reminder of the two very separate Australias that still exist. On the one hand is the delightfully nuanced and multicultural urban Australia, with Sydney and Melbourne now among the most gorgeously admixed cities to be found anywhere in the world, representative of an Australia as a pitch-perfect member of the western Pacific community. On the other, however, stubbornly displayed by a residue of politicians and would-be politicians (Pauline Hanson a type specimen), is an Australia remarkably and woefully out of touch with and unsympathetic to the ways of the Asian world. The two sides of the argument, an argument long settled in almost all other former British colonial possessions, encapsulate a nagging and potentially serious problem.

For is this enormous, wealthy, talented, and truly fortunate country part of the Pacific, a real working component of the great engine work of Asia? Or is it still an outpost of Olde England, a place of beer, bellies, and bogans, set dustily down on the western edge of this mighty sea? It seems in part to want to be Asia, to play its role, to be a powerful component and a moral counterweight to China, to be a place of well-mixed values and of tolerance and understanding, and of a people who present, in and of themselves, a microcosm of the very ocean that washes the continent's shores.

But there is an awful undertow at work still as well, a concatenation of white-dominated, blinkered, complacent, and reactionary forces that may yet keep this once lucky place pinioned and fettered firmly in its past, and thereby not allow it to become a true member of the community in which geography has settled it, now or maybe for some long while to come.

A great place to live, as a friend in Sydney said to me one evening. A great place to live. But not a great country. Not yet.

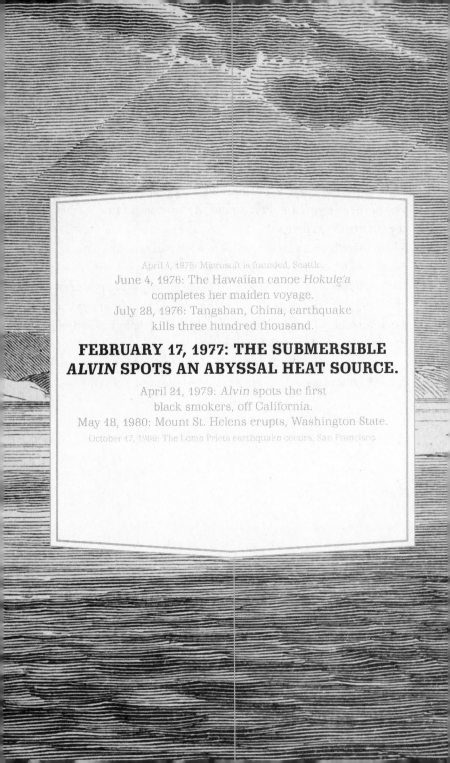

April 4, 1975: Microsoft is founded, Seattle.
June 4, 1976: The Hawaiian canoe *Hokule'a*
completes her maiden voyage.
July 28, 1976: Tangshan, China, earthquake
kills three hundred thousand.

FEBRUARY 17, 1977: THE SUBMERSIBLE *ALVIN* SPOTS AN ABYSSAL HEAT SOURCE.

April 24, 1979: *Alvin* spots the first
black smokers, off California.
May 18, 1980: Mount St. Helens erupts, Washington State.
October 17, 1989: The Loma Prieta earthquake occurs, San Francisco.

Chapter 8

THE FIRES IN THE DEEP

Below the thunders of the upper deep
Far, far beneath in the abysmal sea,
His ancient, dreamless, uninvaded sleep
The Kraken sleepeth: faintest sunlights flee
About his shadowy sides: above him swell
Huge sponges of millennial growth and height.
—ALFRED, LORD TENNYSON, *THE KRAKEN*, 1830

In 1977 the Human-Occupied Vehicle for deep-sea exploration known as HOV *Alvin* was already thirteen years old, salt-stained outside, and well worn inside. And though a snapped cable had once caused her to sink and spend half a year lying unrescued on the floor of the Atlantic, she had by now performed enough deep-sea research around the world to be thought of as quite venerable, an oceanic workhorse, up for anything and down for everything. She was (and indeed remains, still working today, after half a century) as adored as she is revered. Water Baby is how she is known to some—red and white and cheery-looking, a toylike craft, built for the U.S. Navy and operated on the sailors' behalf by the Woods Hole Oceanographic Institution, in Massachusetts.

On a Thursday morning in mid-February 1977, this doughty miniature research submarine, so precisely engineered and so heavily armored as to allow three explorers to be brought down into the ocean deeps and then driven safely back to the surface, was lowered into the warm blue waters of the eastern Pacific for the 713th logged dive of her career. What she would find later that day, in the abyssal gloom almost two miles down, would laser-etch her name into oceanography's history books as having made perhaps the greatest maritime discovery of all scientific time.

For she discovered down in the dark a whole new undersea universe, a previously unimagined dystopia of crushing pressures and scalding temperatures, of curious topography and even more curious life-forms, all gathered around a family of hitherto unknown phenomena that were immediately named for the gaseous torrents that they spewed ceaselessly out into the sea. *Alvin*, on that midwinter's day in 1977, first discovered the existence of what were to be called deep-ocean hydrothermal vents, gushings of gas and superheated water in places where all was believed to be cold and dark and dead.

For science alone, a find like this, with all its implications and possibilities for research, might perhaps have been enough. But thanks to *Alvin*, there was much more to come: further finds, also in the Pacific Ocean, that excited not just the scientific community, but the commercial brotherhood as well. For, not long after that first find, there came news of another, in which the undersea gushings were not fluid at all, but were solid and enormous and studded with minerals, and came to be known as *smokers*.

The little craft had already proved herself a boon. In 1966, when just two years old, she found a missing American hydrogen bomb, one of four that broke free from a crashing B-52 bomber over eastern Spain. Three of the weapons had been found, more or less intact, on a tomato farm. But one had parachuted into the Mediterranean, causing the Pentagon to panic that the Soviets might find it first. Twenty warships and 150 divers searched the sea for three months, but in vain. In the end, little *Alvin* was brought in, and with her Woods Hole crew of scientists sworn to secrecy, she eventually found the weapon, lying half a mile down, snagged on the edge of an undersea canyon. Crews from other navy ships, clumsily trying to haul the menacing-looking ten-foot-long silver cylinder up from the depths, managed to drop it, twice, but eventually wrestled it up, wrapped it in tarps, and flew it posthaste back to America. A Spanish fisherman who had seen it fall from the sky, and who had first shown *Alvin* where to look, was given a hefty salvor's fee.

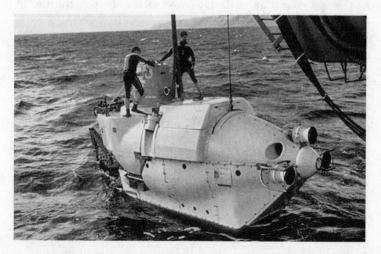

Alvin, the doughty three-person submersible, has allowed scientists from the Woods Hole Oceanographic Institution to find many of the most significant deep-sea structures, including the hydrothermal vent fields and black-and-white smokers.

Alvin would go on to make even better-known discoveries later on in her long, long career.* Her most famous find was in 1986, when she carried Robert Ballard on a dozen dives to investigate the wreck of the *Titanic* in the North Atlantic. The wreck had been found a year earlier, by the *Argo*, an unmanned Woods Hole underwater sled; but *Alvin* allowed divers to see her close up and in person, and this mission brought the tiny craft enduring fame.

Even though finding a lost hydrogen bomb in 1966 and finding a lost passenger liner twenty years later were significant accomplishments, it was the uncovering of a long-hidden natural creation in the eastern Pacific on February 17, 1977, that proved the kind of truly significant contributions the *Alvin* could make.

The finding of the first smokers was a discovery that had fourfold implications. It had a formidable impact on humankind's understanding of the workings of the planet. It introduced wholly new thinking to the understanding of the origins of life itself. It hinted at untold wealth yet to be found on the bed of the sea. And it unleashed, as a corollary, the possibilities of major environmental mayhem, even as the Pacific Ocean—this being the mid-1970s, a time of building ecological anxiety—was taking pole position in the planet's current concern over its fast-spoiling oceans.

Both discoveries (which history now logs as having been made on *Alvin* dives numbers 713 and 914) occurred on or close to the six-thousand-mile-long chain of underwater mountains known as the East Pacific Rise, a place where the seafloor spreads outward, just as the Mid-Atlantic Ridge does on the far side of the world, and which can be thought of as the true birthplace of the modern Pacific Ocean. The Rise is an underwater chain of mountains that run in a more or less north–south direction from close to the Salton Sea at the hot upper end of the Gulf of

---

* Though little of her original 1964 structure remains intact, she is still at work in 2015, much refurbished and retrofitted and more agile after half a century of work than she was in 1964, when she first left her factory in Minnesota.

California, down to a landless point in the empty wastes of the cold South Pacific, a subantarctic place of albatrosses and wandering icebergs, with the huge waves and endless storms of the Roaring Forties.

It is the relatively modest eastern Pacific section of the so-called global Mid-Oceanic Ridge system—one of the planet's biggest physical features and certainly, at forty thousand miles long, the most extensive of all the world's mountain chains, even if it is invisible, entirely covered by water. The system has numberless branches and offshoots, and were its waters to be drained from the ocean and the planet dried out, its ridges would look like a web of fibers somehow helping to hold the earth together, like the stitches on a baseball or the sutures on a skull.

Its existence was confirmed only recently, although there were suggestions as early as Victorian times that lines of unexpected shallows were to be found out in the mid-ocean deeps. HMS *Challenger*, surveying the Atlantic in 1872 to find the optimal route for an undersea cable, found that depths in the middle of the ocean were many thousands of feet less than expected, and a century later, German oceanographers noticed that this same upwelling continued around the African coast right up past Madagascar.

Yet Albert Bumstead's classic 1936 National Geographic map of the Pacific (the same one used, as mentioned, by Colonel Charles Bonesteel III to divide postwar Korea) still gives no clue that anything similarly significant had yet been noticed on the Pacific side of the world. Most of that ocean, so very large, so little explored, was depicted as blue and almost entirely blank, with just a few curving lines hinting, and probably fancifully, at the unsurveyed depths below.

If it was one thing to determine the existence of mid-ocean ridges, then it was quite another to figure out their significance. Initially, and as the ranges were found, they were thought to be undersea mountain chains, pure and simple, and that was that.

The first inklings that they actually offered clues to the origins of the planet came only in 1947, when geophysicists based in New York, dredging from a Woods Hole surface ship, found that the undersea rise in the middle of the Atlantic was made of basalt, and not the granite of which most continents are composed. The scientific community was greatly puzzled, and set about trying to determine why this might be so.

A decade later a pair of Americans, Marie Tharp from Michigan and Bruce Heezen from Iowa, both working at Columbia University in New York, decided to create a comprehensive map of the entire ridge system, surveying every ridge in every ocean. Working with the U.S. Navy, and with much of their initial research quite secret, Heezen took a survey ship, a three-masted iron-hulled schooner named *Vema*, and employing all the new sonar technology at the navy's disposal, first mapped the entirety of the Mid-Atlantic Ridge, arctic end to subantarctic other end. Miss Tharp was initially not allowed aboard the survey ship: in those unenlightened times, her sex forbade it. She had to be content to crunch the numbers back in her cartography laboratory in New York. She made her first voyage as Heezen's shipmate only in 1965, after which the mapping work accelerated near-exponentially.

They soon discovered that the ridge was not only long and sinuous, and reflective of the shapes of the continents on its two sides—curving out where Africa bulged out, and sashaying back in where South America curved in—but also that it was very much more complex an entity than the mere heaped-up pile of seafloor the early surveyors had supposed it to be. Not only was it made of basalt, but also it had a curious and quite unexpected topography. It had a deep groove, a rift valley, that ran along the summit of its entire length, and within this groove, according to the seismometers that the pair took on their expeditions, lay the epicenters of a bewildering number of earthquakes.

Tharp had an immediate epiphany. She thought the ridge-

and-valley-and-ridge feature somewhat resembled the Rift Valley in Kenya, and that its existence might help explain how the continents that ran parallel to it on either side had been formed. The submarine ridge had perhaps disgorged new volcanic material, basaltic material, which had then forced the seafloors on either side of the ridge to spread themselves outward, pushing the continents away from one another as they did so. That might be a reasonable supposition today, but even as recently as the middle of the twentieth century, it was, in some circles, still somewhat premature to imagine that the all-too-solid, all-too-fixed continents could possibly have moved. Proponents of the theory—continental drift, as it is called—had long been derided as apostates. Some of the more elderly geologists denounced such thinking as impertinently and irreligiously challenging the sacred order of the universe.

Except that, as it happened, and at almost the same time as Marie Tharp came up with her ideas about the Mid-Atlantic Ridge, other research going on in the Pacific Ocean was beginning to show that continental drift was not a sacrilegious fantasy, but in fact one of the central driving forces behind the making of the present-day world.

In a series of secret U.S. Navy experiments that were begun in the early 1960s, a flotilla of antique warships was used to drag sensitive magnetometers back and forth across those Pacific ridge summits that had been found off the Oregon coast. Scientists then analyzed the recordings, and carefully noted the traces of the magnetism that had been detected in the rocks below. What they found quite astonished them, as the submarine rocks displayed, with an elegant and instantly understandable symmetry, the record of the already known phenomenon of the earth's magnetic field reversal.

Every fifty thousand years or so, and for no certain reason, the direction of magnetism of the planet abruptly changes: compasses that point to the north suddenly point to the south, to put it simply. It had already been known for many years before this Pacific experiment began that the direction of the earth's magnetic field is always faithfully recorded in those kinds of rock that contain iron, as the millions of tiny iron crystals align themselves like miniature compasses, all pointing to wherever the pole happened to be when the rocks became solid. And if the magnetic field's direction changed, the tiny rock magnets would record that change, as it happened.

What was to be found off the Oregon coast showed that the field reversals were recorded not just in one set of submarine rocks, but in the rocks *on both sides of the ridge* over which the magnetometers were being dragged. Moreover, they were to be seen not just on both sides, but each at exactly the same distance out from the ridge itself.

It was blindingly obvious what had happened. Molten rock had emerged from the ridge's central rift and had then flowed over the two sides, dividing itself equally as it did so. The two streams had then spread outward away from each other, slowly unrolling like a pair of conveyor belts proceeding in opposite directions. As they continued to roll outward, every fifty thousand years or so, each belt of rocks recorded within itself the effects of magnetic field reversal. One field reversal was recorded in the rocks on one side of the ridge, and the very same reversal, occurring at exactly the same time, was recorded on the rocks now scores of miles away on the other side, giving the mirror image that appeared, at first so mysteriously, on all the recording traces.

The corollary was clear and, to geophysicists, intensely exciting. As the ocean floor spread outward from its ridges, and as new ocean floor was being created, the continents on either side of the ridge were being pushed away from each other—they were *drifting*, as only a short while ago the apostates and heretics had been fancifully supposing. A geological revolution was in the making.

The results from the Pacific exactly coincided with what Tharp and Heezen in the Atlantic would soon so boldly imagine: that in all the oceans, new seafloor was being created by the endless volcanic gurgitations along the mid-ocean ridges and then, as still newer material followed it, was spreading itself away from the central rift valleys.

The world's newest material was being born in such places: the ridges were the locus of the origins of today's continental geography. Western Africa stands where it does, and is shaped as it is, because of all the eruptions and movements in an invisible underwater suture line lying a thousand miles off its coastal horizon. The same is true for almost every other coastline in the world. The ridges made them all.

The ridges were also central to the construction, also in the mid-1960s, of the ideas that gave rise to the now familiar theory of plate tectonics. The theory had been built directly onto this now confirmed and fully believed idea of continental drift. It is

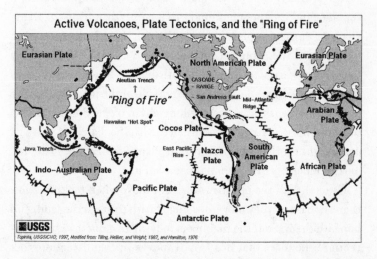

The unique tectonic architecture of the Pacific Ocean, with its major and minor plates jostling and shifting around its edges, has created an immense coastal zone displaying the most intense volcanic and seismic activity, the so-called Ring of Fire.

a theory of such logic, elegance, and beauty that we sometimes imagine it has been with us for eons past; it is in fact not much more than half a century old.

Current thinking holds that the world's outer solid crust is composed not of one continuous surface, as on an orange or a baseball, but of a number of enormous plates, each of which floats on top of the hot and relatively mobile upper mantle of the planet. There are seven major plates, eight lesser plates, and a host of other, new ones being discovered all the time. No fewer than sixty-three had been named at the time of writing.

The Pacific Plate is by far the biggest. It occupies 103 million square kilometers, thirty times the area of the continental United States. It is roughly the same shape as the island of Ireland. It has a long and quite smooth, convexly curved eastern boundary that runs southward across from the Gulf of Alaska down to the Southern Ocean.

The western side of the plate has a different appearance: a serrated and indented boundary that runs down from the Kamchatka Peninsula, past Japan and then New Guinea, turning back toward the center of the ocean, and then shifting sharply down southward, to where it casually bisects New Zealand—with the country's North Island on the outside of the plate, half of the South Island within it, and the long chain of the Southern Alps marking the dividing line between the Pacific Plate and its western neighbor, the Indo-Australian Plate. The Pacific Plate underlies much of the ocean, but not all of it.

Crucially, all the plates move. They move when magma below them swirl, and they move in concert with the swirling going on beneath them. So if the magma is moving in a northwesterly direction, the plate that lies atop it moves in that direction, too. Most plates move relatively slowly—the North American Plate, for example, is shifting westward at about twenty millimeters a year, somewhat less than the rate at which human fingernails grow. The Pacific Plate is, by contrast, something of a speed demon: it moves five times as rapidly, and in a habitual north-westerly direction, covering something like ten *centimeters* each year.

The evidence for this is plain to see. A glance at any physical map of the Pacific Ocean shows that almost all the myriad island groups on its western side are strung out in roughly elongated lines, all stretched in a generally southeast–northwest direction. This is because the plate on which they sit is moving beneath them from the southeast to the northwest, persuading them to align themselves just as boulders and debris are aligned on the surface of an ever-moving glacier. By contrast, the islands that lie beyond the plate's known borders are arranged higgledy-piggledy, with no evident pattern to their location on the planetary surface.

All that is seismically spectacular about the Pacific Ocean—and there is plenty, with earthquakes and volcanoes and tsunamis happening with what, to humans, is dismaying frequency—happens along the edges of its underlying plate, where it abuts its neighboring plates. Most famously, there is the so-called Ring of Fire, which runs for twenty-five thousand miles around the ocean's northern, eastern, and western edges. This ring—or, more suitably (since it is discontinuous, and isn't truly a ring), this belt—plays host to more than four hundred volcanoes. Mount St. Helens, Mount Pinatubo, Krakatoa, Taupo, Popocatepetl, Unzen—the majority of the planet's earthquakes occur on these same three edges of the Pacific, including the three biggest ever recorded in history, which occurred in Chile in 1960, in Alaska in 1964, and in Japan in 2011.

Yet, for all their savage spectacle, these earthquakes are not necessarily important in strictly scientific terms. It turns out that the most geophysically significant discovery of recent times was not made among the giant volcanoes or violent earth shakings of the Ring of Fire. Rather, it was made above the East Pacific Rise, which appears relatively peaceful, unspectacular, and quite lacking in the power and dangerous majesty so visible elsewhere.

For the Rise is actually where the very makings of the modern Pacific Ocean occur, the one place in the Pacific where ocean floor spreading is provably and visibly happening. This is where the present-day Pacific Ocean is being manufactured, and has been manufactured since the plate made its first appearance about one hundred eighty million years ago. Elsewhere, at all those places around the plate edges where there are volcanoes or earthquakes, the plate is either subducting beneath a neighboring plate (in Japan, the Kuril Islands, the Aleutians, and the Cascade Range in the Pacific Northwest), or else sideswiping its neighbor (most infamously along the San Andreas Fault, where it sideswipes the North America Plate, and triggers historically important earthquakes).

The East Pacific Rise is a classic mid-ocean ridge, a range of undersea mountains marking the boundary between the Pacific Plate and its three southeastern neighboring plates: the tiny Cocos Plate, the enormous Antarctic Plate, and between them, most critically, the Nazca Plate, which lies off South America's west coast and runs from Colombia to halfway down Patagonian Chile. This is the most energetic of the ridge's spreading zones. The Pacific Plate and the Nazca Plate are moving apart very fast: the crust above them moves about 7.5 centimeters a year on each side, or 15 centimeters of total spreading annually, much faster than around any other mid-oceanic ridge.

Bruce Heezen died in 1973, after which Marie Tharp took her ship alone and onward into the vastness of the Indian Ocean and then farther on east, to the Pacific. With the research from that trip, she completed in 1977 the first-ever map* of the world's entire

---

* The beauty of what was formally known as the U.S. Navy World Ocean Floor Map derives

undersea mid-ocean ridge system. And once the ridges were fully mapped, and had been accepted as the places where new material was gushing out of the earth's mantle to form the greatest features of our planet, armies of geophysicists descended on them, to determine exactly what was happening there.

The *Alvin* would give them the ability to do precisely this. So, in early 1977, the heroic and salt-stained little craft, shackled onto the deck of her mother ship, the *Lulu*, journeyed for the first time in her career through the Panama Canal, bound for her assignation with oceanographic history.

Another Woods Hole vessel, the *Knorr*, had preceded her, heading down to a spot in the ocean where curious temperature anomalies had been detected, hints of something odd, something worth divining. It was suspected that something, quite possibly hot water, was pouring out the top of the ridge, much as geyser water would gush out of solid earth at volcanically manic places such as Yellowstone and Rotorua. The site in the ocean was some four hundred miles west of the Ecuadorian coast, two hundred fifty miles northeast of the Galápagos chain, on a ridge that spun out from the eastern flank of the East Pacific Rise.

It was here that the discoveries would be made by *Alvin* on Thursday, February 17, that would startle and amaze the world.

The *Knorr* went exploring first, placing herself neatly into position above the site where a previous expedition, in 1972, had detected decisive hints of strange goings-on below. Instruments aboard a submersible device owned by the Scripps Institution of Oceanography, Woods Hole's congenially competitive opposite number on the Pacific coast, which had been towed along that year through the 8,500-foot-deep, pitch-dark, and near-ice-cold waters over the ridge, had detected two strange spikes. One was of temperature, which had inexplicably risen—no more

---

largely from its having been first painted in watercolors by a Tyrolean illustrator, Heinrich Berann, who was otherwise famed for creating a series of great mountain panoramas (Yellowstone, Yosemite, the Cascades) for the U.S. National Park Service.

than a fifth of a degree Celsius or so, but it had risen nonethe-less. Moreover, the spike was detected a hundred feet and more above the seabed, suggesting the presence of an upward gush of something hot, most likely water. The other spike was a sudden increase in dissolved iron and sulfur, and in just the place where the temperature made its own sudden rise.

The *Knorr*, using new and highly accurate maps made as part of the secret U.S. Navy magnetism researches, first sent down three sound beacons, transponders the pilots named Sleepy, Dopey, and Bashful. They would lie doggo on the seabed and emit signals to help keep on target any vehicles that the Woods Hole scientists sent down into the blackness of the deep sea.

The first vehicle was an unmanned two-ton, hundred-thousand-dollar steel-caged contraption named *ANGUS* (for Acoustically Navigated Geophysical Underwater System), which had powerful strobe lights, a collection of thermometers, and, most critically, high-definition cameras. Late on the afternoon of Tuesday, August 15, as computer-controlled propellers kept the *Knorr* above from drifting off target, a giant crane lowered the *ANGUS* downward, directly above the ridgeline. It took two hours to pay out 8,250 feet of twinned wire cables.

While the *ANGUS* was then electronically ordered to keep her position by communicating with the three transponders, the boom operator up on the mother ship was commanded to raise and lower the cables so as to keep the costly vehicle from hitting the seafloor. Then, fifteen feet above the seabed, the *ANGUS* switched on her powerful strobe lights, and then her array of cameras, and began to move, snapping one photograph of the bottom every ten seconds.

After six hours into the first watch, by which time the *ANGUS* had covered five miles, the needles on the many dials in the *Knorr*'s control room suddenly quivered upward for nearly three minutes, as the seawater became briefly hotter and hotter. It was a temperature anomaly, perhaps a rise of a fifth of a degree Celsius. Then the dials quivered back downward, as the temperature cooled just as rapidly. The *ANGUS* hovered above the Rise for another six hours, until a signal came that the film had run out. The *ANGUS* was winched carefully to the surface, her crew now agog to see what the three thousand photographs showed.

The developers worked through the morning, the pictures snatched from their hands as the sheets emerged from the fixing baths. Hundreds upon hundreds showed nothing other than rocks and darkness. But the photos from the spot where the *ANGUS* had recorded the temperature anomaly showed something very different, something quite unexpected. For down there, strobe-lit in the abyssal night, was a sudden abundance of wholly unanticipated *life*. Creatures were to be seen, living creatures, growing in the dark; oblivious to the cold, to the dark, and to the skull-crushing, hull-crushing, life-denying pressure tonnages of the two miles of seawater above.

There were just thirteen pictures of interest, but they showed something quite amazing, images that left the biologists aboard openmouthed with astonished delight: hundreds, maybe thousands of completely unexpected clams and mussels, living where no creature had the right or duty or supposed ability to be alive. The water here was blue and misty. The bivalves were apparently in good health, brightly colored, fronded, and evidently alive. How could this be? There were no nutrients. No light. No sun. And yet these creatures existed, here, on the floor of the sea— enigmatic and evidently eternal, the fact of their presence profoundly puzzling, and aching for an answer.

Just as the final pictures were being examined—and after the thirteen-image orgy of fascination, the next fifteen hundred images showed coils of glassy lavas changing to pillow piles of dull basalts, and nothing else at all—the other Woods Hole vessel, the *Lulu*, broke the horizon.

Frantic radio messages were sent out: Could the *Alvin* dive the next morning? Did she have the ability to dive that deep? Were there crewmen able and available to descend eight thousand feet, in a vessel that only recently had been upgraded with a new titanium sphere to hold the crew, to dive that deep?

To each inquiry, the answer was an unqualified yes. So the *Lulu* moved close in, and then positioned herself directly over the spot where the thirteen relevant pictures had been taken. Crane operators lifted the little *Alvin* up and over the gunwales and down onto the surface of the warm blue sea. It was Thursday, February 17. Three crewmen clambered in and strapped themselves onto the well-worn seats inside the cramped and damp little craft. Jack Donnelly was the craft's pilot; two marine scientists, Jack Corliss and Tjeerd van Andel, were the observers.

Donnelly closed the hatch and flooded the air tanks, and the water closed over their heads. The cables were released, and the craft began to head downward at a stately hundred feet a minute. Within no more than three minutes, darkness had quite enveloped them; through the porthole there was just the faintest glimmer of the pale blue of the surface; and then, with the dark loom of the mother ship's hull barely distinct, it faded away, too. The pilot switched on the powerful strobes.

He had seven thrusters with which to adjust his position, his heading, his attitude. It took an hour and a half of weaving and bobbing to reach bottom—where, to Donnelly's delight, he found they were a mere five hundred feet from the target. He gunned his motors, adjusted his thrusters, and according to an official account of the expedition, "they entered another world."

The lava fields below them were crisscrossed by cracks, and billowing up from the cracks in shimmering clouds were endless gushings of what the sensor probes showed to be very hot water. The shimmering itself was mesmerizing—but just a few feet away, the hot water mixed with the bitterly cold seawater, precipitating certain chemicals that turned the color to a powdery blue as they settled heavily on the seafloor, staining the surrounding rocks with crystals of deep umber.

This was spectacular in itself, and Jack Corliss, a geologist, was seeing before his very eyes confirmation of his theory that hydrothermal vents clearly did exist, which further supported the presence of spreading ridges beneath the sea, and which would lead to the creation of new ocean floor.

Then he cried out in astonishment, and asked a young woman named Debra Stakes, in the *Lulu*'s control room two miles above, "Wait—isn't the deep ocean supposed to be like a desert?" Stakes patiently replied that, yes, this was what was believed. To ask her so basic a question strained credulity—it was as if an astronaut had asked if it was true there was no oxygen in space. "But," spluttered an evidently flabbergasted Corliss, "there's all these animals down here!"

They had stumbled onto a huge and densely populated biological community, in a part of the planet where life was previously thought to be entirely impossible. This turned out to be only one of four such fields they found that session, each different, each pullulating with robust displays of living existence. There were enormous clams and crabs, and creatures on long stalks, like dandelions. There was an octopus of a kind never seen before, and scores of eyeless shrimp. There were forests of waving tube worms, some of them seven feet tall, licking hungrily at the waters, seeming to suck nutrients from it.

The three crewmen were quite stunned, and noted that these creatures, illuminated for the first time by the strobes, did not run for cover or dive for shelter. They just sat there, pulsating with life.

Back in the 1970s the *Alvin*, though not as technologically sophisticated as today, had grappling arms and sample bottles, so while the crew's air supply remained intact, and the pilot kept his craft on station, the two scientists delicately plucked clutches of living specimens from this newfound world, and sucked water into bottles for analysis up above. They had to get answers to a series of hitherto unimagined questions: What were these creatures? What were they doing here? How were they living? What were they eating? The Pacific Ocean swiftly became the nexus for a set of quite fundamental inquiries that had never been either imagined or supposed before.

When the trio broke the surface hours later, they had animals with them, among them a huge white clam bigger than Corliss's two hands. They had scores of new photographs. And they had water samples. When they opened the bottles, they were hit by the unmistakable odor of rotten eggs. The water was clearly heavy with dissolved solids, its odor suggesting the presence of the element normally seen as a yellow powder around volcanic vents: sulfur.

John Edmond, the young Scots geochemist who had come out from MIT to be aboard the *Lulu* for this expedition, remembered the ecstatic moment of realization that was born of this particular chemical find. For he realized that whatever the importance of the vents in the story of the formation of the ocean floor (whatever the geological significance, in other words), the presence of animals and plants and, crucially, of sulfur, was more significant still: for this told him and his colleagues vastly important things about the origins of life itself.

The biology team immediately knew that the relatively complex creatures they had found below must be feeding on *something*. Logic told them that whatever that something was, something lower down the food chain, was likely to be more primitive than the creatures that were doing the feeding. Most probably the foodstuff consisted of bacteria, of some kind. So, somewhere in these hot streams of water, logic said, there had to exist some very primitive living creatures that could somehow reproduce themselves and so serve as the very base, and an endlessly replenished base, of the planetary food chain. Whatever these creatures were, they had no apparent need for sunlight, or for oxygen, or for any of the other chemical or physical components commonly connected with the endowment of the vital force. Such bacteria, if that is what they were, probably originated in places and circumstances like these newfound hydrothermal vents of the Pacific.

"A whole lot of things sort of fell into place," Edmond said. "We realized that regular seawater was mixing with something. It was a unique solution I had never seen before. We all started jumping up and down. We were dancing off the walls. It was chaos. It was so completely new and unexpected that everyone was fighting to dive in *Alvin*. There was so much to learn. It was a discovery cruise. It was like Columbus."

The finds made that day confirmed one aspect of tectonic plate theory. But they also entirely upended the hitherto comfortable assumptions concerning the origins of life. For no longer were sunlight, chlorophyll, oxygen, and warmth considered necessarily essential for life's beginnings. Another, and quite new, option had now revealed itself here in the Pacific Ocean. For whatever it was that lay at the base of this East Pacific Rise food chain (still undiscovered at this point, but surely to be found someday soon) had somehow come into being in this most inhospitable of environments. It had been born in what was, essentially, another version of the already much-vaunted primeval soup: liquid that in this case would soon be shown to be ferociously hot, was already known to be eternally dark and chemically rich, and was of the kind of sulfur-rich composition that was suspected to have existed at the volcanic dawn of the earth's story.

The notion that life itself, that living cells prodded into life from simple amalgams of chemistry into some primitive beginnings of sentience, did begin in the hydrothermal vent ecosystems would swiftly set the biological world afire. The curious were about to have a field day.

A young woman named Colleen Cavanaugh was one of them. She was a biology student from Michigan, and she happened to be at Woods Hole just before the institution's little submarine was discovering the vents. She had arrived there, innocently enough, to take a summer course on the mating habits of horse-shoe crabs. But at the course end, her car broke down, and she never made it home to Michigan. Instead, she decided to complete her undergraduate degree in Boston, and was then invited back to Woods Hole (an hour away, and on the sea) in the summer of 1977. That was the so-called discovery year—except that Ms. Cavanaugh's work was unrelated, and was concerned as it had been before with the love lives of horseshoe crabs.

But everyone in the sprawling campus of the Woods Hole laboratories seemed now to be talking about events five thousand miles away in the Pacific, and about the sensational finds that the *Alvin* had made the previous winter. People were talking about the geology, true. They were talking about the ocean ridges' mineral potential, true. But they were talking most energetically about just how it was that clams, tube worms, subsea dandelions, and, yes, crabs (relatives of Cavanaugh's crabs) managed to flourish as they did in the high-pressure darkness right beside scalding-hot vents.

Colleen Cavanaugh became a passionate believer that bacteria of some kind must provide the key to the story. It was she who would then go on to discover both the actual nature of the bacteria in these hydrothermal vents and, perhaps most crucially, the chemical process that they undertake to provide nourishment for the creatures that cluster beside them. She most famously enjoyed her epiphany by interrupting a classroom discussion on the biology of vent-living tube worms: the moment she heard the lecturer casually remark that the worms had crystals of sulfur inside them, she stood up and loudly asserted that it was "perfectly clear" that the creatures had to have sulfur-oxidizing bacteria inside them. Somehow these bacteria were manufacturing organic material (sustenance for the tube worms) out of inorganic building blocks. Making life, in other words, out of purely elemental whole cloth.

It is a process known as chemosynthesis. Not a new process—a remarkably prescient Russian musician turned chemist named Sergei Winogradsky, working in Saint Petersburg in the 1890s, had proposed the theory that some specialist bacteria could produce energy from purely inorganic materials, could then employ that energy to obtain carbon, and with that could produce sugars: organic material, in other words, the basis of life.

Cavanaugh, who would in time become a tenured professor at Harvard with her own eponymous laboratory, was eventually able to demonstrate that chemosynthesis was exactly what was happening in deep-sea hydrothermal vents. The tiny globules of sulfur found in the gut* of a giant tube worm (one of the dauntingly large, six-foot-long, red-tipped creatures brought up from the deep) indicated to her that bacteria living inside the worm were able to create energy from the hydrogen sulfide that was dissolved in the vent's hot-water gushes. They then would use that energy, just as Winogradsky so presciently suggested, to capture carbon from the methane and carbon dioxide also found in the water, and manufacture food on which the tube worms could feed.

It was a truly edge-of-your-seat scientific advance. Before this, the scientific community believed that all life ultimately demanded energy radiated from the sun, that the process of photosynthesis, in which light is an absolutely essential component, lay at the basis of all living existence. Winogradsky's theorizing, and Cavanaugh's impertinent epiphany and her work on the deep Pacific tube worm (*Riftia pachyptila*), showed beyond doubt that energy could be derived from within the earth itself, with no need whatsoever for any contribution from a distant star.

Colleen Cavanaugh announced the find in 1981, with a paper in *Science*, "Prokaryotic Cells in the Hydrothermal Vent Tube Worm *Riftia pachyptila* Jones: Possible Chemoautotrophic Sym-

---

* *Gut* is a misnomer, since tube worms do not have a digestive tract; instead, there is an organ called a trophosome, which is inhabited by the bacteria that provide the tube worm's energy.

bionts." It remains one of the milestones of modern science. And that it was derived from discoveries made by the *Alvin* in the Pacific Ocean underlines the formidable importance of the planet's mightiest of seas.

All the other oceans have since been discovered to house hydrothermal vents. More than three hundred fifty clusters of vents have been found and seen since that first *Alvin* dive. It was swiftly realized that the water gushing up from them was seawater that had seeped down through cracks in the ridges, had been heated, and like a geyser out on the dry surface of the world, had erupted back outward again. This is not newly created water—rather, it is existing seawater recirculated through the ridges so the total volume of water in the seas remains constant. The recirculation is a massive planetary engine: all the world's oceanic water is thought to circulate through these chains and clusters of vents about every ten years, and to leach out immense amounts of crustal chemistry into the deep sea as it does so.

Most of (but not all) the vents have been found along the rifts at the top of their various spreading ridges. Most have been given rather prosaic names, like those given to obscure stars or small asteroids. But some vents are so large and powerful that they have been given appropriately memorable titles: White City, Loki's Castle, Bubbylon, Magic Mountain, Mounds and Microbes, Neptune's Beard, Nibelungen, Salty Dawg. Not surprisingly, numerous international bodies have been established to coordinate and regulate ridge research—one of them born in 1992, when two ships arrived at the same mid-ocean site at the same time with plans to send submarines down to the very same ridge to look for the very same vent fields.

Though the role these vents play in the search for life's origins is fascinating, another motivation for today's activity over the deep-sea ridge lines is more economic—and that commercial interest was spawned by a second discovery that was made two years later, also by the *Alvin*, also in the Pacific, on the craft's dive number 914. This was the dive that found, at the tops of the most active ridges, almighty "submarine towers," the massive solid and semimetallic consequences of all the fluid gushings beneath. If dive 713 has become part of scientific legend, then dive 914 is best remembered for revealing the commercial possibilities behind that legend—and for offering the alluring hint of treasure, there for the picking, down in the world's deep waters.

The *Alvin* had been kept busy in the months following her first vent discoveries. She performed twenty more dives north of the Galápagos, and then headed back through the Panama Canal to spend the rest of the year in the Caribbean, before heading home to Woods Hole. In 1978 she had more nip-and-tuck work done (much steel was replaced by titanium), to prolong her working life and enable her to probe ever deeper and for longer periods of time. A second grabber arm was added so the scientists could seize more samples of the marvels waiting at the vents. And there were new cameras, new lights, and a basket at the bow to hold ever more samples.

Thus kitted out, she then performed a scattering of more workmanlike tasks (investigating nuclear waste dump sites off the New Jersey coast, for example) before heading back south to warmer waters, and then through the Panama Canal once again to the Pacific's exceptionally active rift zones. She made two dozen further dives northeast of the Galápagos in the winter of 1979—during which time her crew discovered that the dandelion-like creatures below were actually specially adapted colonies of thousands of even tinier creatures known (when clumped together) as siphonophores, related to the jellyfish lookalike called a Portuguese man-o'-war. These beasts did not handle the pressure at the surface well, and exploded on deck or otherwise vanished, another indication of the vast amount of entirely new science that was being uncovered in this new hydrothermal universe.

It was in April that the *Alvin* headed north. Her mother ship, the *Lulu*, voyaged for eighteen hundred miles, carrying the *Alvin* up and onto the crest of the northern sector of the East Pacific Rise. She was assigned to a spot in the tropical seas at twenty-one degrees north latitude, within sight of the cliffs at the very tip of Baja California. Surfers were riding the waves here, oblivious to the work that was about to begin far offshore.

The *Lulu*'s cranes hoisted the fifteen tons of *Alvin* over the side. Dudley Foster, a thirty-three-year-old former navy pilot (who once said that the *Alvin*'s arm was an extension of his own, and that he wore the sub as part of his body), was in command. An American geologist and a French volcanologist were the observers. It was Saturday, April 21, 1979.

The Frenchman, Thierry Juteau, was aboard because of a curious discovery made nearby the year before by the French minisubmarine *Cyana*. Her crew had not encountered a vent, but they had dredged up a great number of tantalizing rock samples, including one more bizarre than most, which seemed to have assumed the form of a long, hollow metal tube glistening with crystals. These turned out to be precipitates of a zinc ore called sphalerite;* there were traces of iron, copper, lead, and silver on the sides of the tube as well, suggesting both a vast trove of deep-ocean chemistry below and also the presence of the exceptionally hot water needed to dissolve it.

The *Alvin* descended quickly into the darkness, switching on her powerful new lights as the sunlight vanished. By mid-morning she was close to the bottom, and almost immediately spotted the white clams that are the most visible signature of

---

* As a teenage would-be geologist, I collected the beautiful yellow or brown sphalerite crystals that I once found littering a secret cleft in the moors in Cumbria, in northern England. I would trade them for other specimens (and occasionally for pocket money) to a rock dealer in London.

a Pacific vent field. Dudley Foster turned the craft to follow the ever-thickening concentration of shells until, suddenly, quite without warning, he had to slam on his brakes.

Before him, staggering to see, was something that no human had ever witnessed before. Rising directly in front was a tall spire of dark rock, with what looked like a jagged crystal-fringed mouth at its top, from which gushed, without cease, a torrent of thick, black, coiling fluid, looking just like dark and oily smoke, belching upward as if from a ship charging full ahead or from a railway train racing down the line.

Foster nudged his craft closer—and found its considerable tonnage bucking and rearing under the immense water turbulence beside the edge of the fountain. For a moment he lost control, and the sub was knocked into the pillar, breaking it and widening the hole even more, and filling his viewing screen with pitch-black fluids that briefly blinded him. He steered frantically back into clear water and then turned to view what they had found. The three sat entirely mesmerized by the show. It was like watching a leak from a rogue oil well, with tens of thousands of gallons of pitch-black fuel coursing upward without end into the pristine sea.

After a few moments, the crewmen's courage regained, their breathing rates stabilized, they advanced the sub slowly back toward the tower, this time with an electronic thermometer gripped tightly in the manipulator arm. Using the thrusters to keep the platform horizontal and moving in a straight line, they pushed a sensor gingerly into the liquid uprush—whereupon it promptly shot off scale, showing a temperature of more than 90 degrees Fahrenheit. No such figure had ever been experienced in such deeps. It had to be wrong. They tried again—the needle banged hard up against the end of the range again, and this time the instrument went dead.

Only when they got to the surface did they look at the thermometer and realize why. Its sensor tip had melted. The temperature of the smoker fluid—which they then determined on their next dive, with a thermometer capable of working in a blast furnace—was some 662 degrees. It was an incredible figure. If this was water, then it was hot enough to melt lead. Magnesium would soften, too, and so would tin. Sulfur would be almost at its boiling point.

This column of fluid clearly was not composed of water, at least not principally. The fluids were so intensely hot, the pressure so insanely high, that metals or their compounds had first been dissolved and extracted wholesale from within the material of the earth's crust deep below. The upward-gushing fluids were most probably made up of considerable concentrations of dissolved compounds of gold, silver, iron, magnesium, lead, zinc, and tin. They could almost be thought of as a molten alloy of all these base metals, mixed with sulfur and seawater. And when this chemical-laden torrent suddenly confronted the ice-cold deep-sea waters, the base metals and the sulfur compounds almost instantaneously precipitated out of solution and created a bewildering smorgasbord of solids—either compounds of metals or else, in rare cases, clanging, shining shards of the wholly pure metals themselves.

These gathering masses of solids would then fold themselves out of the uprush and, as they fell to one side, still half-molten, would pile ever upward around the circumference of the gushing liquid column, like metallic stalagmites. As the gush continued, so the towers became ever taller and taller, until they could not stand under their own weight—or else until careless passing research submarines knocked into them. Towers like these could be built in a matter of hours, climbing skyward in the dark, and yet never (like the trials of Tantalus) quite making it upward beyond the limits imposed by physics and gravity, but instead slumping back onto the seafloor, for the ever-hopeful gusher to try again.

These towers on the East Pacific Rise were called black smokers, for obvious reasons—although this smoke is a precipitate of particulate metal rather than, as is traditional, combusted ash. Other huge underwater chimneys—soon to be discovered in the Atlantic Ocean and exhaling, by contrast, torrents of paler-colored fluids, and emerging at lower temperatures—turned out to be laden with calcium and barium. These quite reasonably came to be known as white smokers.

When heavy metal sulfides are caught up in the geysers of superheated water gushing from a submarine vent, huge but fragile towers of metallic minerals—black smokers—form and rise dozens of feet from the seabed, until they eventually collapse under their own weight.

Hundreds of smoker towers have been found in the years since. They have been found in all the expected places: along the summit traces of the ridges in the Indian and Atlantic Oceans and, most profligate of all, along the immense tracery of ridges and beside the ocean trenches and island arcs that mark the boundaries of the great Pacific tectonic plate.

Not surprisingly, it took no time at all for those in the mining business to realize the value of these gleaming pipes of crystalline metal compounds. And some of the towers are truly immense. One black smoker, named Godzilla, lying deep off the Pacific coast of Canada, could be watched as it grew fully one hundred fifty feet up from the seafloor, before it finally collapsed under its own mighty weight. And that mighty weight was made up not of coral or clams or crabs or tube worms, but of metallic compounds: sulfides, most commonly, of exploitable, minable, potentially salable metal.

Surveys have recently identified an alluring number of collapsed smoker pipes and vent crystals deposits, making what are now called seafloor massive sulfide (SMS) deposit fields or sites. Eye-watering tonnages of copper, lead, and zinc sulfides, and ores of gold and silver, have been assayed in the glittering, meteorite-heavy, and metal-like SMS fragments that have already been dredged up or brought to the surface by *Alvin*-type submarines.

For years following the first finds, the world's mining industry became intrigued with trying to work out how best to win these SMS deposits from the sea. But the companies' enthusiasm was tempered by caution, and understandably so. The industry was already somewhat gun-shy, since it had been badly burned back in the 1960s by the commercial failure of the much-vaunted manganese nodule boom, in which billions of tons of mineral-rich pellets lying on the ultra-deep seafloor turned out to be far too costly to bring to the surface. SMS fields, by contrast, were richer in minerals and, as they lay on or beside mid-ocean ridges, were in much shallower waters. If only the technological challenges of getting at them could be met, and if the world price of these various metals remained high enough, then there was an absolute fortune waiting down there in the deep.

A Canadian firm called Nautilus Minerals has become the first in line to try to make a business out of the metal-laden ruins of old black smokers. It has identified two sites, both in the Pacific, both on the Ring of Fire. One is in the Bismarck Sea, just north of Papua New Guinea, halfway between the islands of New Ireland and New Britain, and thirty miles north of the infamous and unfortunate city of Rabaul.* The other is on a ridge to the west of the Kingdom of Tonga. The means Nautilus has contrived to extract from these sites the hundreds of thousands of tons of metal ores suspected to lie two miles below are clever, cunning, and, in the view of some environmental groups, deserving of wide condemnation.

The United Nations has set up a regulatory body, the International Seabed Authority, based in Jamaica, which lays down rules for deep-sea mid-ocean mining. The first two undersea sites chosen by Nautilus fall within the territorial jurisdiction of their neighboring states, Papua New Guinea and Tonga, and so are not subject to UN rules. However, another Pacific Ocean site that Nautilus believes could be exploitable lies in what is known as the Clarion-Clipperton Fracture Zone, a two-million-square-mile expanse of sea that stretches from a point some five hundred miles southeast of Hawaii right across to the coast of Mexico. The ISA does wield authority here, though the United States refuses to accept it, not being a signatory to the Law of the Sea convention that set up the ISA in the first place. In any case, the ISA's control over seabed mining in the Clarion-Clipperton Fracture Zone is at present more academic than actual, since no one has yet come up with an affordable technology that will allow mining under the frigid and crush-

---

* Rabaul is not a happy town. It had an exceptionally unhappy war—it was once a huge Japanese naval base, but after being essentially isolated by Allied air raids and almost unable to defend itself from attack, it was regularly pummeled by Australian planes and then totally overrun shortly before the Japanese surrender. Then, in 1994, the two volcanoes close by (Ring of Fire volcanoes) erupted, with lightning strikes killing residents and forcing the entire town to be evacuated and then abandoned after almost every building was destroyed or covered by ash. Though the volcanoes have been quiet in recent years, little economic activity has resurfaced in the ruined city.

ing environments that exist five miles down. That is for the future.

But technology is now being created to exploit the shallower parts of the seas, such as those unregulated inshore waters off Papua and Tonga. Nautilus is planning to deploy a small armada of three massive, powerful (and highly waterproofed) new machines known collectively as seafloor production tools, remotely controlled robotic crawler miners that will be lowered gingerly down five thousand feet directly onto the site where the sulfides are known to be. The machines are made, uniquely, by a firm in Newcastle upon Tyne, in northern England, known as Soil Machine Dynamics, a company that specializes in building "remote intervention equipment, operating in hazardous environments worldwide."

Sitting on the factory floor, each of the white-painted engines towers over the gaggle of Geordie workers laboring to assemble it. They are working in a factory where for years steam turbines were assembled, back when this part of Newcastle was a shipbuilding city and made destroyers for the Royal Navy and oil tankers for the merchant fleets of the world. Nowadays the shipyards are mostly silent, but the new machines being manufactured in their old assembly buildings seem just as huge, just as heavy—and instead of floating on the sea, they are being designed to work far beneath it, carving cargoes out of the seabed rather than transporting them on the ocean surface.

The three machines initially delivered are truly monstrous, both in size and in appearance. They have long iron arms and huge spinning blades; gouging devices and giant claws and buckets that could hold whole cars, and lay down tracks, as if they were tanks or bulldozers, and that allow the vehicles to crawl and lumber at will over any of the steep hillsides and through any of the canyons they might encounter deep below.

The auxiliary cutter moves in first, thrashing its mighty knives and dozer blades, cutting wide benches into the rock-faces and scarifying and otherwise preparing the ground for the arrival of the suboceanic big boy, the unromantically named bulk cutter. This fearsome creature, two hundred tons of raw blade power, then grinds its way along the benches and cuts and slices and hauls and finally crushes the sulfides out of the cliffs, leaving them in many-tonned piles scattered in rows along the seafloor, waiting to be collected by the last member of this iron-bound trinity: the collecting machine.

This is much like a robotic dump truck, only much larger than any ever seen in the world's biggest opencast mines. Down below, it is obliged to run on tracks, rather than on the mansion-size tires seen up on the surface. It scoops up the sulfide litter piles and, responding always to commands from its remote driver sitting like a drone pilot in the mother ship two miles above, takes them across to the slurry pump and riser. This is a heat-hardened vertical rubber tube fully two miles long that, like an elephantine vacuum trunk, then sucks the material up to the surface and onto the deck of an enormous mining control vessel.

This ship, of a kind never before made, is being manufactured in China for a Dubai-based chandlery, and it will by rented by Nautilus for the first five years of the project. It will cost the not inconsiderable sum of $199,910 a day. The vessel will act as a controlling guardian angel for the three machines growling away below. It is also being built to receive through its two miles of hard-rubber umbilicus the thousands of tons of sulfide-and-water mixture that the three monsters manage to claw out of the seafloor. Once enough has been piled up into wells on deck, this ore will be strained through an immense net and, to use the miners' term, dewatered, before the surplus and de-ored waste-water is then sent back down to the ocean bottom.

The solid sulfide ore will finally be swished by conveyer belt across to a flotilla of waiting barges, and after each barge is filled to its brim, it will leave for a metals processing plant on the

Yangtze River, three thousand miles across the open Pacific and the East China Sea.

Out of every thousand tons of ore, Nautilus expects to get seventy tons of solid copper, and sizable poundages of gold and silver. The mine will pay for itself, the firm's Canadian shareholders are assured. The Pacific will begin to yield up its bounty from about 2018 onward, and the bounty, for a copper-starved world in particular, will prove an immense boon to all.

This is the plan, and the company prints attractive brochures and makes slickly produced films to underline the point that it is doing all this with unalloyed concern for the Pacific's fragile environment. Of course, the firm has released an environmental impact statement, and it has acknowledged that two types of deep-water snail might have their habitats briefly disturbed. Nautilus says the regional environment, however, will escape unscathed. Others are far from sure.

As one might expect, the usual protective agencies (the World Wildlife Fund, Friends of the Earth, and Greenpeace among them) are concerned that the seabed is going to be ruined in the name of profit and greed. But in this case an indigenous foe has arisen, too: Papua New Guinea Mine Watch. At the time of writing, it is producing an energetic, intelligent, and highly coherent argument against seabed mining generally and in the coastal waters off Papua New Guinea in particular. The arguments are both principled and technical. In summary, though, the question that dominates is simple: why place at risk the sanctity of our oceans briefly to sate our endless appetite for planetary growth?

The affair is of deep significance to the Pacific story. How this single debate plays out over the coming years will offer some indication of just how the Pacific Ocean is going to be regarded in the future—by outsiders who see it mainly as a major resource to be exploited, and by those who live there and have long drawn their sustenance from it and wish to see it treated with proper reverence and care.

The arguments are complex on many levels. To limit the professed worldwide need for copper, say (the kind of copper that Nautilus plans to claw up from one of its chosen undersea fields), the most acceptable solution seems always to be: to lean on the BRIC countries (Brazil, Russia, India, and China) and all other such developing countries to limit their use of the metal, to lower their populations' expectations, and to wind back such standards of living as depend on the use of copper—and in today's high-technology consumer world, that is a huge number of uses.

Not unsurprisingly, the citizens of these countries cry foul. They want to know why they should not enjoy the standards that Westerners have long taken for granted. Why should they have to bear the consequences of the environmental damage that our past wanton overconsumption has caused? Why should they not have copper, for example, and acquire it from wherever it may be lying?

If such an argument wins the day—and it most probably will—the first submarine mines in the South Pacific will almost certainly be developed. The bulk cutter and the auxiliary cutter and the collecting machine will, in due course, be lowered into the ocean and will start to crunch, grind, and tear their unheard and invisible ways through the undersea ranges of the Bismarck Sea, and will turn the seafloor into a moonscape of unutterable ugliness—or, it would be ugly, were anyone able to see it. But since the sea is so deep, and the seabed so dark, and once scoured of its riches, it need never be seen again, probably few will care. Nautilus and its shareholders will do well, will sleep happy, and the firm will move on to other projects in the same great ocean.

Meanwhile the *Alvin*, now well into her fifties, will no doubt continue to dive ever deeper, and will make still more spectacular breakthroughs in submarine science. Whether mankind then makes responsible use of the ever-widening knowledge that the busy little craft brings back to the surface is, however, another matter altogether.

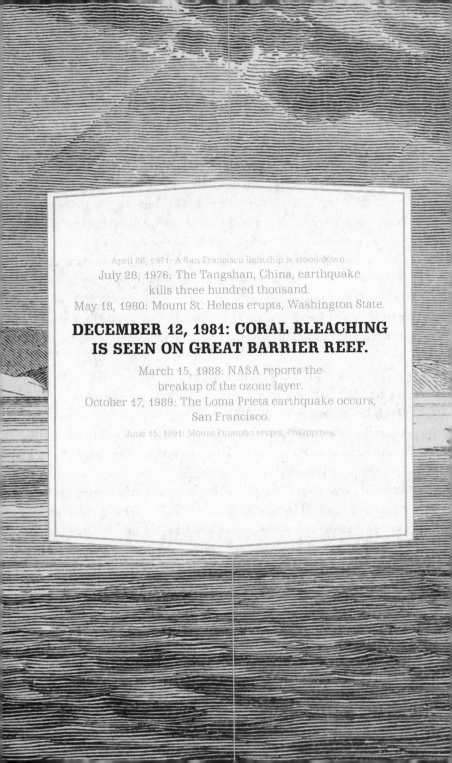

April 26, 1971: A San Francisco lightship is stood down.
July 28, 1976: The Tangshan, China, earthquake
kills three hundred thousand.
May 18, 1980: Mount St. Helens erupts, Washington State.

DECEMBER 12, 1981: CORAL BLEACHING IS SEEN ON GREAT BARRIER REEF.

March 15, 1988: NASA reports the
breakup of the ozone layer.
October 17, 1989: The Loma Prieta earthquake occurs,
San Francisco.
June 15, 1991: Mount Pinatubo erupts, Philippines.

Chapter 9

A FRAGILE AND
UNCERTAIN SEA

———⟨∘⟩———

> *Flowers turned to stone! Not all the botany*
> *Of Joseph Banks, hung pensive in a porthole,*
> *Could find the Latin for this loveliness . . .*
> —KENNETH SLESSOR, *FIVE VISIONS OF CAPTAIN COOK*, 1931

Charlie Veron was diving contentedly in the warm shallows off
the central Queensland coast's Pandora Reef on a perfect sky-
blue, early summer Saturday when something highly alarming
suddenly caught his eye. He kicked his way down through darting blizzards of tiny fish to inspect the banks of multicolored
corals below, in particular a cluster of branching specimens that
were somewhat uncommon except on this one small island. In

due time, Veron himself would give this unique coral the species name *Goniopora pandoraensis*.

These clusters were mainly a rich brown and yellow, colors that contrasted brilliantly with the pinks, ochers, blues, and vivid greens of the other coral species that were later to make Pandora a favorite inshore site for divers and tourists (these days, mainly Japanese). But what alarmed Veron that Saturday was a highly unusual patch of pure white that he saw in the center of one of the *Goniopora* crowns. It was circular and maybe six inches across.

He reached down with his bare hand and gently touched one of the white coral clusters. Most of the columns were still firm and alive—and sharp, as anyone who has touched a sliver of coral with a bare hand knows. Had they been dead, fronds of the skeleton would have snapped, the pressure of the merest touch causing fragments to tumble like snowflakes down to the seafloor. Veron was reassured that they were still alive, but at the same time he was highly alarmed, since their sickly appearance suggested they might be starting to die. He reached for his waterproof camera and snapped a single image: the first time he had seen on Australia's Great Barrier Reef, long the pride of Pacific biology, this unhappy harbinger of a potentially lethal phenomenon now known as coral bleaching.

It was one of the first indications that the sea (and in this case, the Pacific Ocean's portion of the world's universal sea) was in serious trouble.

Veron had been a naturalist, a scientist, a coral expert, and a student of the Barrier Reef's fantastic coralline loveliness since 1972. In later life, he would discover, describe, and catalogue an immense proportion of the world's 845 known species of hard, reef-building corals, and would write the definitive encyclopedias of the coral universe. So that day, he was uniquely qualified to realize that what he was seeing was a foretaste of something quite awful: a mass bleaching event that would spread around the entire tropical planet in the months and years to come. "It is horrible to see," he remarked later. "Corals that are four, five, six hundred years old, they turn white and die. It is a very recent thing."

Often called the largest living organism on the planet, Australia's Great Barrier Reef stretches for nearly fourteen hundred miles along the Queensland coast. But it is under threat from a rise in sea temperature and acidity, and its corals and inhabitants are in peril, their fate a potent symbol of global warming's impact.

This last was significant because it meant that some recently generated external force, at the time still unknown or uncertain or not wholly admitted, was causing these lovely and highly sensitive animals to wither, whiten, and in many cases pass away.

Veron's friend and colleague in Brisbane, Ove Hoegh-Guldberg, had already put forward a convincing case that corals, uniquely, foretold with great accuracy the coming of global climate change—they could be considered the mine shaft canaries of approaching climate problems. He had long claimed that corals (being animals that look like plants, and which build castles for themselves of stone, thereby becoming a confusing trinity of biology, botany, and mineralogy) were vastly more important than as mere ocean-side decoration. They are among the most acutely sensitive of nature's early-warning devices. They react very quickly to the minutest changes in their environments, a facility that allows them to serve as predictors of any number of the earth's environmental troubles.

And this is exactly what they appeared to be doing on that midsummer's day on Pandora Reef: they were sending out an alarm to alert the world. Charlie Veron, acutely sensitized to even the subtlest changes in a coral population, was the first to notice it. So sweeping were the implications of his find that he has since devoted his life to promoting the importance of coral reefs—of their beauty, fragility, and impermanence, and of their abiding capacity to warn us of dangers ahead.

The Pacific Ocean has a formidable amount of coral within its borders. It has twice as many species of coral as the Atlantic does. It sports thousands of coral atolls, numberless fringing reefs, and above all else, it has at its far southwestern edge, where the Coral Sea deeps meet the upwelling seabed off the beaches of eastern Australia, the three thousand reefs and nine hundred islands that make up the fourteen hundred miles of the Great Barrier Reef.

And just as the panda and the blue whale have come to symbolize both the beauty and the impermanence of mammalian life; and just as creatures such as the bluefin tuna, the Grand Banks codfish, the dodo and the great auk, and the Japanese flowering cherry blossom have come to stand for the precious fragility of nature, the Barrier Reef has come to stand for the earth's delicate and finely balanced frailty. Not just the frailty of corals. Not just of ocean life. But of the planet's life, in total. For as goes the Great Barrier Reef, science is able to claim, so goes the natural world.

Australia's reef is certainly massive, the biggest in the world, far larger in length and area than any reef structure in the Bahamas; in the Red Sea; off the coasts of Belize, Yucatán, and Guatemala; off Florida or China; or among the growling dangers of the half-invisible Chagos Bank, in the middle of the Indian Ocean. It is so large that astronauts can see the pale green of its shallowings as it spears its ragged way northwestward off the coastline, all the way from the glittering tropic seas near Gladstone* up to where it fizzles out and dies in the estuarine-muddied waters of the Torres Strait.

Reefs take up a tiny proportion (just one-fifth of 1 percent) of the world's surface, yet because they are home to such an astonishing diversity of marine life, they are of far greater significance than their size alone suggests. The Great Barrier Reef has almost four hundred types of hard and soft coral: brain corals, staghorn corals, pillar corals, plate corals. A quarter of all marine life is then supported by the gigantic reefs that corals such as these manufacture. The simple existence of reefs like these protects coastlines, nurtures fish, and contributes untold treasures to

---

* Off Gladstone, the reef lies rather more than 125 miles from shore, and when in May 1770, Captain James Cook swept his tough little barque HMS *Endeavour* into the temptingly smooth waters inshore, he little knew he was entering a nearly fatal ship trap. For the reef crept closer and closer to shore as he and his crew slid ever farther north—until his ship slammed hard into a coral spike and stuck fast, holed and half wrecked. Superhuman efforts managed to warp the vessel off the razor-sharp coralline rocks; and Cook limped into a river mouth where the settlement of Cooktown now stands. Aboriginals on the hills watched him and his wounded vessel; probably they would not care that he named the headland off which he nearly foundered Cape Tribulation.

those who live by or close to them. The limestone of which the coral skeletons are made draws carbon dioxide from the atmosphere and plays a crucial role in the planet's carbon cycle.

A coral reef is the marine equivalent of a rain forest: full to bursting with life in all its glory, yet fragile, vulnerable, and presently in the gravest danger. Anyone with a face mask and a snorkel who slips over the side of a boat and into the warm and limpid shallows above the outer reef will be readily amazed. Just inches away is a polychrome feast of life—there are corals of all kinds and colors: green and yellow, red and pink, pale blue and rich brown; in every crevice waft the perpetually hungry cilia fronds of anemones or else the slow-opening and -closing mouths of clams; on each smooth slope of coral, hand-size crabs scuttle slowly sideways back and forth; and darting between the coral pillars, like yellow-striped moments of iridescence, are brilliant electric-blue pulses of sudden light, each one a tiny fish bent on its mysterious business. Larger fish, silver and sedate, weave their polite ways slowly through the currents; and below, in the coral sands, tiny creatures bury themselves in a whirlwind of particles, or emerge blinking into the green sea light. This immense tableau seems perpetually to rock and sway under the press of the tides, the currents, and the swells, all its imagery refracted down from the silver-sided surface above.

No fewer than fifteen hundred species of fish have been counted in the Great Barrier Reef. There are also, in huge variety and abundance, all manner of turtles, dolphins, manatees, skates, sharks, stingrays, small whales, and porpoises. There are numberless varieties of snails, anemones, clams, sea slugs, sea grasses, sea horses, and seaweeds. On the sand islands that peek above the sea surface, there are ferociously dangerous saltwater crocodiles, as well as generally benign flora and fauna and a significant percentage of the world's shorebirds and waders. There is an entire spectrum of seabird types, from the quite commonplace flocks of gulls and terns, shearwaters and tropic birds, frigate birds and boobies to the majestically solitary white-bellied sea eagle, a creature revered for thousands of years by the aboriginal peoples who live on the coasts nearby.

Charlie Veron, most recently chief scientist with the Australian Institute of Marine Science, knew exactly what was causing the bleaching he first spotted in 1981. A coral polyp usually contains an abundance of algal plant cells, called zooxanthellae, which live in symbiotic harmony with the coral animals, and which perform the photosynthesis that allows the coral to flourish. The coral provides a protected home to the algae. The algae, when secure and happy, then produce the oxygen that the corals need in order to make, among other things, the calcium carbonate for their skeletons—skeletons that are in essence the building blocks of a coral reef. The algae also give the coral its color—but only when, once again, they are secure and happy.

But once in a while, and for a variety of reasons, the corals suddenly instruct the algae to leave home. Their previously symbiotic relationship is brought to a rapid end, whereupon the coral polyp equally rapidly finds itself in the awkward position of being unable to get the oxygen it needs, or to get the glucose and amino acids that are the by-products of the algae's photosynthetic efforts—and also unable to sport the decorative colors for which zooxanthellae are renowned. To the outsider, the coral's color is suddenly shed; it becomes deathly white. It has been bleached. And if the condition persists, without oxygen and glucose and proteins, the coral polyp dies.

Veron also suspected why his *Gonioperae* were so brusquely expelling their algae and ending the once friendly and necessary relationship with the coral. It was entirely due to stress. Coral is an animal that has taken skittishness to an art form. On every occasion when the warm and shallow sunlit waters of its home are seriously disturbed, it goes into a state of shock, with the zooxanthellae being the first victims of what might be called the coral's new and distressed state of mind.

There are two prime reasons that corals become distressed: the sea's temperature may rise, and its chemistry may become more acidic. This was precisely what was happening back in the 1980s. For reasons that were then neither fully accepted nor agreed upon, the average ocean temperature had risen by about one degree Celsius—meaning that the very hottest days on the reef were even hotter than a coral could stand, and it reacted with disastrous and very visible drama. And worse was to come.

That initial bleach was to be suddenly spread just weeks later. The affliction cannonaded not just around the Pacific—an early episode of bleaching had been seen, at about the same time as that on Pandora Reef, among corals eight thousand miles away on the Galápagos Islands—but around the world. This first-ever worldwide mass bleaching event of 1981–82 indicated that the problem in Australia was undeniably the planet's problem, too.

The cruel irony for Australia was that only a matter of weeks before the Pandora Reef bleaching, UNESCO had placed the Great Barrier Reef on the list of World Heritage Sites. This was a great honor, but it also focused the world's attention on Australia's being the custodian of an entity that was now officially revered by all humanity—such that any troubles that might befall it could be Australia's to bear alone. Australia had won the honor, but now Australia would bear the blame.

There have been further global assaults on corals in the years since. Most notoriously, in 1997–98, in a ferocious bleaching disaster, a species of hot-water corals, which exist mainly in the Red Sea and around the Andaman Islands, were all being killed off, since the waters became so warm that even this coral's special adaptation to high temperatures proved useless. The year 2001 saw an even more massively widespread attack, such that many well-respected marine biologists began to predict that all coral reefs could well vanish from the world in another fifty years.

Already Australia's reef has lost half its corals, with most of its loss occurring since the disastrous 1998 season. Frantic efforts are now being made to reverse the situation. Laboratories in Hawaii have enjoyed some success, most notably with bold experiments to breed new varieties of heat-resistant corals. Whether this is merely the postponement of disaster remains to be seen. Most scientists[*] believe this assault on reefs is caused by ocean warming and ocean acidification, and though arguments still rage over humankind's culpability in this, apprehension continues to grow that matters have now gone too far, that the dire situation is irreversible, and that corals may soon be fossils.

Australia's particular role is rather more complex. Local pollution is proving a major menace: the runoff from the many Queensland rivers, now heavy with fertilizing chemicals, pesticides, herbicides, and animal waste, is especially harmful to the reef. Tourists are blamed for causing it casual damage (by being careless in their boats, clumsy in their diving, greedy in their souvenir hunting), and overeager fishermen, foraging in protected areas, are blamed for triggering even more extensive ruin.

For many years the reef also battled the infamous crown-of-thorns starfish. This spectacularly unpleasant carnivore, which has as many as twenty-one arms, thousands of needle-like spines, and the ability to secrete both venom and a foul-tasting detergent-like foam, settles itself atop a coral, extrudes its enormous stomach over the entire creature, and begins digesting it on the spot: a single starfish can kill sixty-five square feet of reef a year.

Ingenious means were devised to kill these starfish off: One involved increasing the population of a marine snail called a Triton's trumpet, which is known to tear the starfish to death

---

[*] But not all. A few atmospheric scientists continue to question the link between climate change and reef destruction. They point in particular to Cuba's pristine and well-managed, highly regulated coral reefs, and suggest that elsewhere local threats, especially in countries with less regulation, are the more likely culprits.

with its razor-plated tongue. Another plan had divers inject-
ing the starfish—skilled divers boast they can take care of two
a minute—with an otherwise harmless chemical that causes the
creatures to blister and ulcerate, lose their bodies' turgor (their
plumpness), and finally wither away and die. Concerted assaults
like this have caused the starfish to fall back, for now at least.

Meanwhile, other threats appear to be gaining ground. The
Marine Park Authority, the government body that since its
founding in 1975 regulates and protects the reef (and is funded
largely by the fees it charges tourists), quickly points out that
climate change does the greatest damage. But it has been widely
criticized for allowing developers and, most especially, mining
companies to undertake projects that place the reef in even
greater and more immediate peril.

The most classic example of such a development is a proposed
huge new coal port to be built at Abbot Point, near the central
Queensland town of Bowen; this port will be neighbor to the
Whitsunday Islands and Hayman Island and Airlie Beach and
Lindeman Island and a host of other of the most lyrically beautiful of the reef's best-known treasures. The argument over the
building of this port, and over the millions of tons of seabed that
will need to be dredged and dumped to make way for its loading
wharves (smothering with dust the clear waters that are so essential for the life of corals and sea grasses), has pitted Australia's perceived economic needs against the long-term hopes of
the much wider world community.

This argument serves as a reminder that Australia possesses a
formidable reserve of the minerals needed by the ever-growing
economies of East Asia—China, most especially. A large number
of Australians have profited hugely from this trade. The country's economy has largely managed to insulate itself from the
world's recent economic storms, with the Australian people en-
joying and preserving an enviable standard of living even while
much of the world beyond has been tightening belts and cutting

spending. More than a third of Australia's exports go to China; coal and iron ore account for 70 percent of that total.

Not surprisingly, Canberra is doing all it can to help keep it that way. In 2013 the government's ministry in charge of the environment approved the plan for the dredging of Abbot Point, and the government-run Marine Park Authority then issued the necessary permit. Two Australian bodies notionally charged with the protection of nature have in this instance found it more expedient to protect coal mining interests—and have come in for widespread condemnation as a result.

One might think that even the most myopic and misinformed would regard mixing coal with coral as self-evidently unwise. The entire country was gripped by a drama in April 2010 when a fully laden Chinese-registered coal carrier, the *Shen Neng 1*, cut a corner in coming out of a coal staithe in Rockhampton, grounded on the reef, gouged a two-mile-long scar in the reef, and left behind it a two-square-mile oil spill. Fatigue was offered as an excuse. The master and his deck officer were arrested, and they appeared, quite bewildered, in a local court—where they were promptly given bail and allowed to go back to the ship and head on home to China. The reef that was hit by the ship will need at least thirty years to recover, if it ever does. The Marine Park Authority solemnly promised it would issue sterner rules for dealing with such navigation errors. Yet three years later (memories being short when Chinese money is at stake) this same government body gave permission for yet more coal docks to be built nearby.

Charlie Veron, already grieving at what he sees as the reef's impending ruin, is angry and skeptical. On hearing about the Marine Park Authority's approval of the Abbot Point dredging scheme, he remarked that the reef's sole official local guardian and protector was "committing suicide." In early 2015 the World Wildlife Fund further condemned the government for its indifferent stewardship, for bowing to commercial interests, and for placing the economy and the China trade before the wider needs of the planet.

And ten thousand miles away, in Geneva, officials at UNESCO have been similarly exercised. They have the ability and the right to declare the reef's World Heritage Site status under threat; and they also have the right, in extremis, to withdraw that status altogether. Australia, needless to say, would be publicly humiliated. But commerce being what it is, and reefs to some being merely pretty, Australia would undoubtedly also continue to sell its coal to China in ever-swelling tonnages. Business, in the new Pacific, is a powerful trump.

The object I have long revered most in the world is contained inside a seven-foot-tall box made of mahogany and glass that stands toward the back of the Pitt Rivers Museum in Oxford. Heavy velvet curtains shield its contents from the fading effects of sunlight; a button at the side allows artificial illumination, carefully timed, to bathe it briefly in a harmless glow.

The object inside is very fragile, very old, and vivid with an almost unbelievable torrent of color: slashes of bright scarlet and yellow, and scimitar curves of the richest black. It is a ceremonial cloak from Hawaii, originally worn by a member of the islands' royal family. Such cloaks, known as *ahu'ula*, are often worn with a curved, close-fitting feather helmet known as a *mahiole*. What makes the few that survive today so memorable and priceless is that each was hand-stitched from hundreds of thousands of exquisite bird feathers, and yet without a single bird dying, or so it is said, in the process of manufacture.

The cloak was acquired more than a century and a half ago by the Pitt Rivers, a much-beloved Oxford institution currently housing half a million items of ethnological fascination, from bagpipes to shrunken heads, totem poles to war canoes, the tools of ancient Papuan dentists to the scalping devices from Amer-

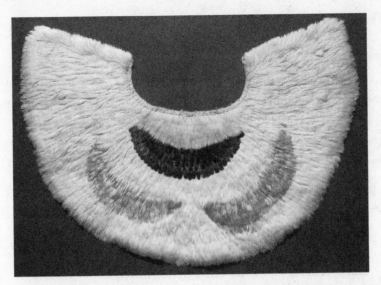

The gorgeously colored ceremonial Hawaiian feather cloaks, or *ahu'ula*, are treasured relics of a time when birdlife on the islands was much richer than today. The red, yellow, and black feathers were taken without harming the birds; the demise of certain species is the fault of newcomers, not of the Hawaiians.

ica's Great Plains. It was brought to Britain by one Sir George Simpson, who for most of his career had been governor of the Hudson's Bay Company, based in Montreal. On his retirement, this majestic and autocratic figure decided to travel the world with a secretary, and in a leisurely crossing of the Pacific in 1842, he chanced upon the Sandwich Islands, soon to become Hawaii.

Here he met the Hawaiian king's daughter Kekauluohi, the young woman who, as the constitution of the day allowed, was running the country in her sickly father's stead. Simpson was well aware that, at the time, and with the Pacific in a frenzy of imperial seizure, outsiders were busily trying to win the hand of the Hawaiian leadership—with a view, ultimately, to becoming their colonial masters. The Americans were most especially jockeying for influence over the islands, as were the French, the Belgians, and, naturally, the universal imperialists, the British.

Simpson wanted none of it. From his viewpoint, and considering the commercial trading interests of the Hudson's Bay Company, he thought it essential that Hawaii remain wholly independent of all foreign powers. He told the young princess this, and she, in a fit of gratitude, realizing that he shared views that she already had, promptly asked a servant to bring down from her private chambers the most magnificent *ahu'ula* cloak. She presented it to Simpson, though cautioning him that he was to give it to his wife.

We do not know if Frances Simpson ever wore it. On her husband's death, though, the cloak somehow passed to his acquaintance General Augustus Pitt Rivers, joining a vast amount of South Seas material already given by the family of Captain Cook. It was eventually mounted in its glass-fronted case in Oxford, and soon established its reputation as one of the best-known items in one of the best-loved museums in England.

As an icon of early Hawaiian independence,* the Pitt Rivers *ahu'ula* cloak plays its role with silent gusto. As a warning of the environmental perils that face the modern Pacific, it plays a role also—though mostly as a reminder of how mankind's relationship with the natural world used to be, and how that relationship has deteriorated over time. Much as the aboriginals of Australia had long revered the Great Barrier Reef and the creatures it harbors—the white-bellied sea eagle, now threatened with extinction, was a powerful totem to many groups—so the feather

---

* There are still stirrings of life to be found in Hawaii's many independence movements, and once in a while the leaders of these groups, all firmly committed to peaceful demonstration, take control of a government building or two to remind outsiders who it is that properly owns the islands. Lately the Chinese have shown a vague interest in offering them support. Each time the American government offers aid or weaponry to Taiwan—which China dearly wants back in its own fold—suggestions are offered in Beijing that China should similarly help the Hawaiian nationalists. Just now the exchanges are lighthearted, few taking them seriously. Just Chinese mischief-making, most likely.

cloak here recalls the reverence in which many Hawaiians held the natural riches around them, most notably the birds.

The eight volcanic islands of the Hawaiian group, drifting above their hot spot, above their local upwelling of magma in the middle of the tropical North Pacific, are the most isolated archipelago in the world. For this reason Hawaii, much like the Galápagos, offers biologists a near-perfect laboratory (little contaminated by the influence of outside biologies) both for the study of evolution and for an examination of the flourishing or otherwise of endemic species. A survey in the late 1990s by Honolulu's Bishop Museum showed the islands' amazing fecundity: more than twenty-one thousand species of living creatures have been counted there—with fifteen thousand creatures living on the land, three hundred in the rivers, fifty-five hundred in the surrounding seas.

Hawaii's bird population, however, has suffered in recent years, with no fewer than sixteen of its three hundred island species rendered extinct since Captain Cook arrived and since Europeans began to settle amid this gorgeous assemblage of tropical flora and fauna. This loss was not due entirely to the poor stewardship of the newcomers, and it would be idle to pretend that Hawaii's earlier inhabitants never had any adverse effect on the wildlife—in any contest between human beings and the natural world, humans invariably win, at least at first. But the Hawaiians' assault on the local birds happens to have been very much more limited than that of the newcomers, and it is a most curious irony that it is the very existence of the famous feather cloaks that shows this to have been so. Because, to put it most succinctly, the making of these cloaks demonstrates just how assiduously most early Hawaiians actually cared for their wildlife.

Maybe it was no more than enlightened self-interest: a recognition that to produce ever more elaborate cloaks, thousands upon thousands of feathers were needed, and it was vital to keep the bird population thriving to produce them. Or maybe early

Hawaiians did indeed have a real reverence for nature. Whatever the reason, the cloaks display a way in which humans managed to deal with nature in a kindly fashion. To do so in a sustainable way.

The cloak makers traditionally gathered feathers from two main types of birds: the o'o and the i'iwi. The o'o (the name coming from its distinctively melancholy, bell-like cry) provided the cloaks' black and yellow feathers; while the i'iwi offered up the scarlet. In both cases the methods of catching the birds were astute, quick, and entirely humane.

Skilled bird catchers were sent up into the valleys where the birds were known to congregate. To catch the o'o, which have special tongues enabling them to prise nectar from flowers, the hunter would quietly extend up into the branches a stick smeared with very adhesive honey. An o'o would be tempted, would settle on the stick to dine—the birds are of a family generally known as honeyeaters—and it would be snared, its feathers firmly glued to the stick. The hunter would then gently reel his stick in, would carefully relieve the bird of perhaps a dozen of its most brilliantly yellow feathers (usually from the thighs, which were the best colored), and would take also a handful of the most richly black chest feathers. He would then clean the bird and release it, to flutter and soar back into the forest.

It would be much the same with the i'iwi, except these tiny birds were not so keen on honey. This made it more difficult for the catcher, who had to climb into the tree canopy, wait patiently for a bird to perch nearby, and then snatch it quickly from the branch, taking care not be gored by its tiny, curved needle of a bill. This time the catcher took a few of the reddest feathers from its underbelly, and then opened his hands and sent the bird away, too, its rush of scarlet vanishing swiftly into the woods.

There are four species of o'o and all are now believed to be extinct—but, and pertinently, they became extinct more than a century ago, long after the last feathered cloak was made. Disease is said to be one reason: *o'o* are particularly prone to avian malaria, and are dead within a day after being bitten by a mosquito.

The last true Hawaiian *o'o* was seen in 1934, and the last Kauai version was believed to have been sighted in 1987. I, no bird-watcher of any kind, was fortunate enough to see one of the very last of the Kauai birds on a blistering summer's day in the late 1970s. I was up in the Alakai Swamp, writing an essay on rain (comparing the feel of intense rain up on the island summit with what was then the other contender for the world's wettest place, Cherrapunji, in Assam), when, suddenly, my guide shushed me, his hand on my arm.

Through the dripping silence of the rain forest came a melodious, flutelike call, clear and loud. It sounded for maybe a full minute, then stopped—whereupon from directly ahead there swooped a compact black bird with a white-speckled chest and a short, uptilted tail and with, so distinctive that even one as bird blind as I could spot them, a pair of brilliantly yellow-feathered upper thighs. My guide nodded enthusiastically, barely able to speak, he was so enthralled. It was indeed one of the last living examples of *Moho braccatus*, of which back in those days maybe just a dozen pairs survived.

Ten years later, all have gone, and the only feathers of an *o'o* bird are those stitched on the costumes of Hawaiian monarchs, who are quite extinct as well; and some of the choicest are to be found in this eccentric old museum in Oxford, as far from Hawaii and the Pacific as it is possible to be. A reminder of the sadness of all extinctions, and of the ways in which Pacific peoples, the royalty as well as the commonalty, once took far better care of their natural world than we have ever managed to do.

The ocean has some recent success stories, though, and some noble attempts to turn back the trend toward oblivion. One was achieved by a Japanese academic ornithologist named Hiroshi Hasegawa, who had been working for more than thirty years to head off the potential loss of the North Pacific's largest seabird, the short-tailed albatross, *Phoebastria albatrus*. In 2012 he announced that the crisis had finally been averted. A breeding colony of the birds had now been fully reestablished and was proving to be quite stable. The bird was, in consequence, tumbling off the world's endangered lists, and was to be seen instead in graceful soaring flight in almost all its old northern habitats, from Alaska to Kamchatka, the Aleutians to Midway, Japan to Mexico. Hasegawa made his announcement at an albatross conference in Tasmania. Visibly moved by what many had already decided was his achievement alone, he allowed that the bird's recovery had been nothing less than "dreamlike."

The albatross has a near-mythic standing among the world's seabirds, a creature of poetry and painting, of legend and superstition and a thousand bosuns' tales. Coleridge is to blame, in part; his Ancient Mariner was tortured and plagued by fiends, according to the poet's famous eighteenth-century *Rime*, simply because he had used his crossbow to shoot an albatross, contrary to all the rules of seamanship—the birds were commonly believed to embody the souls of dead sailors; a terrible fate awaited a man who ever killed one.

The biggest of all the birds lives in the Southern Ocean: *Diomedea exulans*, the wandering albatross, which can follow a ship for weeks at a time, an ever-present guardian-companion, but always out of reach. The vessel may be wallowing uncomfortably through the waves and troughs of the Roaring Forties, a sink of illness and misery, but all the while the great white and gray bird, as much as a dozen feet from wingtip to wingtip and with a body the size of a large child, will be gliding effortlessly alongside, riding on the gales, at one moment rising as high as the mainmast, the next with its wings almost grazing the sea. Once seen, an albatross is never forgotten. Anyone sailing by night into the Southern Ocean, and then waking to find a wandering albatross outside the porthole, has a story to tell the grandchildren.

The North Pacific's version of this magnificent bird is of the genus *Phoebastria*: narrower in its wings and with a more compact body maybe, but still the largest seabird north of the equator. It is also the longest-lived—one of the genus recently found on Midway Island had been tagged sixty years before and was thought to be the oldest wild bird on the planet. The bird is easily recognizable: it is alabaster white, with a golden streak on its head and a bill of bubblegum pink with a pale blue tip.

The difference between these two creatures, south and north, *Diomedea* and *Phoebastria*, is that the first belongs to a population that is adequately healthy, and is respected and unbothered by those few humans who make it into the cold of the southern seas. On the other hand, the short-tailed albatross, in the north, has been on the virtual edge of extinction for much of the last half century. Its principal predator, making a regrettable contrast to the bird-loving Polynesians of Hawaii, has been the Japanese hunter.

The breeding grounds* of the short-tailed albatross have always been on the volcanic Izu Islands, a chain that forms a long, straggling line running for four hundred miles out into the Pacific due south from Tokyo Bay. It forms the northern part of the so-called IBM Line, the Izu-Bonin-Mariana Line, marking the western edge of the Pacific Plate, where it subducts beneath the small Philippine Plate. The islands, stretching between Tokyo

---

* The Hawaiian Islands were for many years the extinction capital of the Pacific, since mankind had imported so many cats, dogs, rats, and mongooses that land and seabird populations were being savagely reduced. A recent experiment in throwing a half-mile-long rat-proof fence across the neck of Ka'ena Point, at the very western tip of Oahu, has resulted in an explosion of new life, and the return in particular of large numbers of breeding pairs of the once-rare Laysan albatross. As with the short-tailed albatross in Japan, the experiment has shown that with effort and imagination, some of the damage done by man can occasionally be reversed.

and Guam, are marked by near-ceaseless seismic activity, and are thinly populated.

Torishima, an island a mile wide and dominated by its twelve-hundred-foot volcano, lies almost at the southern end of the line. For reasons best known to the species, the dangerously active slopes of the volcano have been long favored by the short-tailed albatross as its primary breeding ground. There were once hundreds of nests, each one built on a small pillar of tussock grass; the albatross habit of mating for life would result in the production of a single egg per couple each year. The parents would teach the fledged youngster how to take off (at which albatrosses are notoriously inept), after which each would be on its own, bent on scouring the seas for the rest of its days, flying from Alaska to China on a perpetual hunt for food.

Until the 1860s, Torishima was quite unpeopled: only shipwrecked sailors found sanctuary there.[*] But then, with the surge in economic activity all across Japan that followed Admiral Perry's arrival and the full restoration of the emperor, the little island became home to a population of hardy fishermen who collected guano from the enormous albatross population. A volcanic eruption in 1902 then killed every last one of them (an event serious enough and rare enough to be reported on the front page of the *New York Times*), and no one ever came to live there again.

But word of the island's huge albatross population spread quickly, and for the next several decades, hunters would come out on expeditions to Torishima, bent on clubbing the young birds to death to collect their feathers. For the next half century, there was an immense export of feathers from Japan—the quills

---

[*] One of the first Japanese to live in America, Nakahama Manjiro, was wrecked on Torishima in 1841, only to be rescued by an American whaling boat and taken to New Bedford, Massachusetts—where he went to an American school, learned English, and eventually returned to Japan to act as interpreter during the country's reluctant opening up to the West. He was probably the first Japanese to take a train or ride in a steam-powered ship, or to take part in the California gold rush.

were used as writing instruments, the wings and tail feathers for decoration, the downy underbelly feathers for mattress stuffing. And the birds were so easy to kill: the Japanese called them *ahodori*, "fool birds," because they stayed patiently at their nests without moving while hunters clubbed one after the other. There followed years of casual and very profitable slaughter. The U.S. government calculated that during the early decades of the last century, five million birds were clubbed to death on Torishima—with the hunters stopping only when they had reduced the albatross population almost to zero.

In 1940 there were only fifty birds left on the island; in 1951, just ten. Five years later, it was widely believed that the bird was just about extinct. The Japanese government, shocked by the realization and shamed by what had happened, promptly declared the bird to be a Natural Treasure. Though it seemed the avian equivalent of shutting the barn door after the horse had bolted, feather hunting was swiftly made illegal and all further landings on Torishima were strictly banned. A few birds were still to be seen, clinging on, but their situation seemed dire, nearly hopeless.

It was then that the young ornithologist Hiroshi Hasegawa, at the time a graduate student in the biology department of Toho University in Tokyo, decided he would devote the remainder of his days to saving these magnificent birds from total ruin. He won permission to travel to Torishima, and he spent the next thirty years working out cunning ways to protect, to sustain, and finally to expand the breeding ground, and to bring it back to normal.

Phoebastria, it turns out, were victims of their own fool birdishness. Unlike the wandering albatrosses of the southern seas, who build their nests on flattish and relatively protected islands off the coast of South Georgia, the short-tails of the north have the perverse habit of nesting on the steep slopes of an active volcano. The couples mate, build a ramshackle nest, lay an egg—and in all too many cases it promptly rolls down the slope and drops into the sea. Infant mortality isn't much of a factor, because there are so few infants hatched in the first place.

Hasegawa tried to ease this situation by first planting grasses on the nesting slopes, and in a single season—grasses grow furiously fast on Torishima, thanks to rich volcanic soils and warm subtropical weather—the egg survival rate doubled. But only for one season; in the next, there was a torrential rainstorm, and a mudslide carried everything away. Without a feather hunter in sight, the albatross population had suffered a serious setback. The first, as it turned out, of many.

Still, Hasegawa persisted. The Japanese government gave him some money to help terrace the slopes, to make them more congenial for the birds. The numbers then crept up again; and as more terraces were added, matters got better. Still, the site was hardly ideal, and it required constant and costly human intervention. So Hasegawa came up with a more radical plan: he would lure the island's courting albatrosses to a better, flatter, less vulnerable breeding site on the other side of the island. He would make these foolish birds change their minds—or if not their minds, then their habits.

He found a new, flatter site. He hand-painted dozens of life-size decoy birds and placed them on the grasses. He set up hidden loudspeakers and tape recorders and played the sounds of albatross mating cries and courtship rituals. Overnight, he turned himself into a kind of circus barker, a street salesman, performing a birdman version of three-card monte, to attract a crowd and to get matters moving.

And it worked. In only a matter of hours, clusters of young males appeared in the skies above the new site. Some of them settled, and some began performing the curiously choreographed mating dance rituals (heads to the sky, beaks clacking furiously) for which all albatrosses are renowned. And then the females started to arrive. Friendships were initiated, mates

The short-tailed, or Steller's, albatross, *Phoebastria albatrus*, almost vanished, the victim of Japanese commercial feather hunters. But a lone Tokyo academic, Hiroshi Hasegawa, has managed to rescue the birds, and there are now healthy populations, no longer threatened with extinction.

were selected, matings occurred—eggs were laid, they remained where they were, the broodings began—and with the volcano obligingly quiet, the hatchings started. Chicks were fledged, taught how to fly, and pushed off from the tussock grasses and into the air—to wander around the skies of the North Pacific in ever-increasing numbers, year after year after year.

Nesting sites then expanded to other islands in and around Japan. Down in the Bonin Islands, one new nest has been recently seen. And in the Senkaku Islands—claimed by the Chinese, a new hot spot in the coming collision between the world's superpowers (since the United States is pledged to come to Japan's side if the latter's territorial sovereignty is impugned)—there are albatross nests, too. Good reason, one would have thought, to make sure the region remains at peace.

In August 2012, Hiroshi Hasegawa, mild and modest like so

many in the field, was able to stand before the audience at the Fifth International Albatross and Petrel Conference, in Wellington, New Zealand, and to declare that three thousand birds were now living on Torishima. Albatross have been seen soaring above the waves from Skagway to Shanghai, from Midway to Nome. By 2018 the numbers will have risen to eight thousand, by which time the bird will be officially declared no longer at imminent risk. The paper Hasegawa presented was headed "Success!" and the conference attendees stood to offer their applause—to one piece of good news brought via the work of one man dedicated to the natural life of the Pacific Ocean.

There are still threats to this and many other Pacific birds, of course, threats mostly from the human community. The birds get entangled in long-line fishing gear, they are fouled by spilled oil, they are assaulted by mammals (rats, mainly) that are carelessly introduced to breeding grounds. One of the more spectacularly displeasing kinds of suffering comes from the creatures' ingestion of plastic ocean debris, of which the Pacific has lately been found to contain more than its fair share, all of it swirling around in what is popularly and nightmarishly known as the Great Pacific Garbage Gyre.

On the Kamchatka Peninsula in the fall of 2014 I was walking along a beach, a sandbar a mile long with the Sea of Okhotsk on one side, a lazily flowing estuary on the other, and behind, the grassland and low marshy scrub and pine forests that are typical of this remote corner of Far Eastern Russia. It was a brisk sunny day, the water boisterous with white horses, the wind so bracing a walk became essential to keeping warm.

But everyone in our group kept stopping: to collect the abundant flotsam we found on the tide line. There were amusingly shaped lengths of driftwood, of course; and there were old coils of frayed ropes and bottles abraded until they were smooth and opaque. But the most sought-after items were the small hollow spheres, each four inches in diameter, that seemed to be sprinkled randomly over the beach. They were floats that had once held up the nets of a Japanese fishing boat. Some were black and plastic; some were machine-cast glass, with a seam from the mold. But the choicest were the hand-blown floats, each different, all slightly imperfect, that had bobbed about on the ocean surface for the many decades since they were first made. They had survived a thousand storms and had covered untold distances from where some trawler captain had cut them loose. Each time one of us found one, he or she cried out with pleasure, like a child on an outing.

Ocean flotsam and jetsam were long thought of as mainly charming. Once in a while—it happened once when I was sailing, though in a different sea—you hit a half-submerged shipping container, and it damages your boat, and you curse its presence in the ocean. But more often, if you find a glass float or a rubber duck or, best of all, a bottle with a message inside, there is a moment of quiet delight: the sea has yielded up a small treasure, part of what we suppose the sea is meant to do.

All this changed in 1999 when a sailor, passing between Hawaii and the coast of Northern California, claimed to have passed through an enormous field of floating plastic garbage. He reported that he had seen bottles, tires, milk crates, the hulls of broken boats, toys, old fishing nets, and all in concentrations that suggested one might well be able to walk across the sea on top of it. Three U.S. National Oceanic and Atmospheric Administration scientists in Auke Bay, Alaska, said they were not in the least surprised: they had written a paper a decade before suggesting that if careless dumping from ships and thoughtless overuse of indissoluble plastic on land did not end swiftly, then the patterns of currents in the Pacific would assemble the entire mess into one great vortex of discarded rubbish. What had been found, evidently, was what they had warned about.

The story from the mariner gained traction from that moment on. Quite swiftly the word *gyre*—all oceans have gyres, gigantic mid-sea swirls around which all the currents circulate—became, along with *flotsam* and *jetsam*, part of a new lexicon of world-affecting horrors, part of the immense matrix of anthropogenic global warming accusations that have dominated human dialogue ever since. But thanks to the popular press, and the difficulty most people experience in visiting the gyre to check, the story has gotten slightly out of hand.

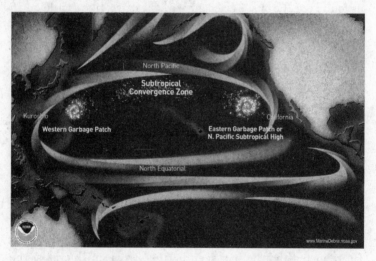

"The Pacific Garbage Patch" is the imaginative name given to an area in the northeast Pacific, between San Francisco and Hawaii, where there is a high concentration of barely visible particles of man-made plastic. Its discovery has triggered major moves to limit ocean pollution.

The garbage gyre certainly exists. Samples of seawater have proved incontrovertibly that there are concentrations of unwanted material floating in an area of the sea once known as the

horse latitudes, suspended in the calm waters* lying between
the westerlies and the trade winds; these waters lie north of the
Tropic of Cancer and run from just east of the Hawaiian chain to
five hundred miles off the American and Canadian west coasts.
But the gyre, the vortex, the patch—whatever it is called—is much
less dramatic than first reported.

It consists mainly of tiny confetti-size particles suspended in
the upper few feet of the sea, with around twelve pounds of garbage in every square kilometer of ocean. It is generally quite invisible to all, although it is no less dangerous for being unseen.
You may not collide with it in your boat, you may not be able to see
it from space, you may not be able to take photographs of it—but
the particles are indeed made of plastic, they are small enough
to be ingested by surface-swimming sea creatures, and they can
well enter the food chain, disruptive to all who live along it. And
where an island exists close by, its beaches are often badly polluted by the larger pieces of material that have been sifted out by
time and tide. Photographs of a dead Laysan albatross on Midway
Island, its stomach filled to lethal bursting with plastic rubbish
of one kind and another, have brought widespread public sympathy and attention.

So a small industry has lately arisen to deal with it. Research is
being undertaken with great expenditure of time and treasure.
Ships large and small—the JUNK Project, a sailing raft made of
garbage, being a small one; and two much larger ships, called
the *New Horizon* and the *Kaisei*—are involved in laudable projects in the gyre. Sea skimmers take off to try to collect rubbish,
often with little result. Websites exist (5gyres.com is one) that
describe the ocean's woes. As a result, the public has become
increasingly aware that plastics (in particular those that are

---

* The calms here so slow down ships that, on passage through them, many sailors worked
out what was called the "dead horse," the period for which they had been paid wages in
advance, so they celebrated by hauling a piñata-like stuffed horse up the mast and then
casting it out to sea. Polluting this part of the ocean has a long history.

nonbiodegradable, nonphotodegradable, and indissoluble) are highly dangerous, do great damage, and leave harmful legacies. This was born of a NOAA prediction and one sailor's observation in the Pacific Ocean, and it may yet result in a significant change in the habits of the whole world's population.

Because of the world ocean's sheer size, and the difficulty of policing what were once the high seas but are now known as ABNJs (Areas Beyond National Jurisdiction), big and powerful bodies, ideally ocean-aware and supranational bodies, are vitally necessary to confront environmental challenges of great magnitude, and to try to give protection to the evidently fragile, ever-changing seas. The lack of such entities in the past has had dire consequences. In 1992, for example, when savage overfishing caused the collapse of the Newfoundland cod fishery, there was simply no interested, responsible, or influential organization around to help prevent it. Then again, the tuna, sharks, turtles, and whales that nowadays are so greedily sought in the southern Pacific have long swum similarly unprotected—with all their populations now facing crises and with the diminishment of certain species, the possible extinction of others, a looming likelihood.

For too long, impotence was pervasive. The nongovernmental bodies (Greenpeace and the World Wildlife Fund most notable among scores of others) always had the voice, and they had a good measure of influence, but what they never did have was power. And in the past, few national institutions had the political will to wield such power as they had over the open ocean waters. Happily, and late in the day, this now appears to be changing.

As the realization grows of the very real threats to ocean life (warming, acidification, overfishing, pollution, mining), there does seem just now to be a fresh abundance of good intentions—maybe genuine, maybe merely the fashion of the moment, maybe an ambition among world leaders for personal legacy building, or maybe attempts to assuage the guilt for past environmental sins. Whatever the motivation, though, a number of what are known as marine-protected areas are currently popping up all over the place, and the Pacific in particular has recently seen the huge expansion of existing protected zones and the establishment of new ones.

The American government is currently attempting to exert its influence on the central Pacific's environment by declaring vast square mileages of it protected, and maintaining them as strictly policed sanctuaries for the creatures that live on, in, and beneath them. During President George W. Bush's administration, and using the curious and wide powers of the 1906 Antiquities Act—which President Theodore Roosevelt first signed into law, allowing him and his successors to give protection to national treasures—the U.S. government created four large protected areas in the Pacific. The first, with a name of appropriate size, was the 138,000-square-mile Papahanaumokuakea Marine National Monument in the northwestern Hawaiian Islands. The other three, named just two weeks before Bush left office, protect from mining, drilling, or commercial fishing the almost 200,000 square miles of waters that wash around a dozen lonely and unpeopled mid-ocean islands, which include Rose Atoll in American Samoa, seven tiny morsels of land in the Line Islands group (declared the Pacific Remote Islands Marine National Monument), and three more in the Marianas. All the fish, crabs, sharks, mud-nesting birds, volcanoes, and hydrothermal vents that lie within Bush's Spain-size tracts of sea will be protected, for now at least.

In 2014, President Barack Obama went even further, using the same century-old act to place under American protection seven hundred thousand further square miles, all to be added to President Bush's Remote Islands Monument. And all the specific bans declared by Mr. Bush would operate in these new areas, too, with the addition of a prohibition on tuna fishing, which provoked predictable howls of rage from fishermen.

Other countries have since weighed in. The British are considering placing under protection the waters around Pitcairn Island and its three uninhabited neighbors—though a Royal Navy ship would need to be nearby to enforce the zone, and the British have little money to pay for it.

Kiribati, the former British colony, has already placed strict limits on commercial fishing in an area within its own immense jurisdiction. It is the only nation that occupies all four hemispheres, its islands and atolls being scattered north and south of the equator, and east and west of the International Date Line, so it has ample room from which to select a suitable area—1.35 million square miles, with just 100,000 inhabitants.

But Kiribati has another problem that makes overfishing pale in comparison, since it may well not exist in a few decades. The island appears to be drowning. Steadily rising sea levels are doing it, so the island could be the first true sovereign victim of climate change. The rising waters are already causing regular and frequent floodings, and considering that its atolls have an average height above the sea of only seven feet, it could well founder by the century's end.

Happily, the Kiribatians have an ark. They have lately bought six thousand acres on the Fijian island of Vanua Levu, and the current Fijian president says they will be most welcome to come south and bring their entire population there. For now, the Kiribati government is building sandbag seawalls, and hopes the ocean will perhaps calm itself and drop back to what the landlubbers regard as its proper level.*

---

* The Kiribatians are not alone. There are calculated to be 12,983 habitable islands in the Pacific Ocean. Were the sea level to rise by three feet, 15 percent of them would be effectively drowned, according to the Intergovernmental Panel on Climate Change. Worldwide, a million and change would be displaced.

Poised between those unsung and solitary figures who have
wrought some measure of beneficial change (the Australian
coral scientist, the Japanese albatross protector, the generations
of nameless Hawaiian cloak makers) and the might and power of
governments who hope to do so, too, is a scattering of others who
occupy a different category altogether: a *corps d'élite* of the extremely rich, usually men, who entertain the wish to leave their
own substantial mark on the planet. One man who has publicly
declared his wish to make such an impression on the Pacific
Ocean is Larry Ellison, the billionaire sea-loving founder of the
Oracle Corporation. His ambition for immortality stems from
his purchase in 2012 of nearly all of the somewhat played-out but
delightfully shabby little Hawaiian island of Lana'i, which once
was the Dole Food Company's largest pineapple grove, and the
largest such plantation in the world.

Ellison plans to turn the island into a sustainable version of
Paradise. His point man on the island, Kurt Matsumoto, has declared that Lana'i will be reengineered into "the first economically viable, 100 percent green community." If Ellison succeeds,
then the ocean surrounding Lana'i (to say nothing of the three
thousand people who live there) will have ample reason for gratitude. But in his first few years of ownership, Ellison has inspired
less than perfect confidence that he will leave the kind of legacy
for which he is hoping.

Ellison's wealth is quite staggering, with *Forbes* magazine
placing it at $56 billion in 2014. If the figure is correct, then this
seventy-year-old college-dropout son of an unmarried teenager
from the Lower East Side of Manhattan has more money than
the individual GDPs of 120 of the world's sovereign states. He is
worth more, for instance, than the 3.4 million people of Uruguay
make together in a single year, four times as much as the 15 million people of Cambodia.

It was computers, both the hardware from which they are made and the software that makes them tick, that brought Ellison his fortune. Until he stepped down as CEO in 2014, he ran the Oracle Corporation, the company he founded with a twelve-hundred-dollar loan in 1977. Like so many computer companies—from Microsoft up in Seattle to Apple down in Cupertino and Google in Mountain View—his Oracle in Santa Clara can rightly be said to be a Pacific Ocean company: though he is by birth a New Yorker and by self-interrupted education a Midwesterner, he moved to Pacific California when he was twenty-two and has effectively never left.

Once there was much ostentation. He sails enormous boats; he flies (he has two ex-military jets); he owns gargantuan oceanside estates. And yet more recently, and displaying something of the kind of marriage of East and West that the Pacific encourages, he prefers people to know that he is becoming ever more considerate of his circumstances and his surroundings. He is wanting to become known as a friend of the ocean, as a man of great generosity, a man who gives away, or plans to give away, much of his fortune. When in 2004 he donated $105 million, the sum represented just 1 percent of his wealth. Now he says he expects to give away 95 percent of it before he passes on.

He lives principally in a Japanese house with a Japanese garden, eats Japanese food, and is said to marinate himself in Japanese culture and espouse some Buddhist beliefs and practices. He bought almost all the island of Lana'i for a relatively modest and, for him, easily affordable $300 million. In total, the island—its name is the Hawaiian word for "hump," since its weary volcano was the only thing visible across the five miles of sea from the vastly more interesting island of Maui—occupies 90,000 acres, 141 square miles. Ellison would have taken all of it, except that a few islanders had long ago acquired title to their homes and some lands had been traditionally Hawaiian-owned. So he had to settle for just 87,000 of the acreage—all the pineapple groves; all the beaches; all the mountains, the docks, the churches, and two very large luxury hotels (that happen to lose a great deal of money each year). He also bought Island Air, the tiny airline that flies to Lana'i from Honolulu, with ambitions to bring an inrush of wealthy visitors.

At first sight, Lana'i currently does not radiate much interest or attraction—and it never did. Before the Europeans came, its legends involved man-eating monsters, ghouls, and a mountain god given to provoking nightmares; its shores were littered with shipwrecks; its landscape spotted with the graves of unfavored members of the Hawaiian aristocracy who had been sent there in exile. The first European to settle was a Mormon missionary who turned out to be a confidence trickster; and after he had been excommunicated by the Salt Lake City elders and his island put on the block, James Dole bought it, for a song, to grow pineapples.

It still sports much of the agricultural detritus of his now abandoned monoculture. For seventy years, its sixteen thousand cultivated acres were given over entirely to pineapples. (Most of the rest is steep forest upland, hiding pronghorn antelope, mouflon sheep, feral cats, and deer, and with thousands of the Norfolk Island pines that the Royal Navy in its sailing days favored felling for mainmasts; or else it is ringed with steep coastal canyons, spectacular but unusable.)

Lana'i City, laid down in a perfect grid in the cool and shaded upland center of the teardrop-shaped island, was very much a company town, with small tin-roof houses that were home to the Filipino plantation workers, and larger establishments for the Japanese and *haole* overseers who kept them slaving away in the fields. H. Broomfield Brown was the best remembered and least liked of Jim Dole's overseers. He spent his days astride his horse up on the Munroe Trail, scanning the fields below with his telescope: if any one of the broad-hatted, goggle-wearing Filipinos seemed to him to be slacking, he'd gallop down the network of laterite paths and give the hapless fellow a tongue-lashing, maybe cut his Dole Company pay; and if a persistent lay-about, the man would be tossed out of his Dole Company house, transported to the Dole Company dock, and sent packing back to the mainland and likely deportation home to Manila.

For those who didn't slack, life was quite agreeable. "Have happy workers, grow better pineapples" was the Dole mantra for many years—and with free Dole schools and free Dole doctors, and men who mowed the lawns and weeded the gardens around your house so that Lana'i City looked as pretty a company town as possible, the island had a utopian quality to it. A million "pines," as they were called, shipped out of Lana'i every day. But the fruit became increasingly expensive to grow: the workers were unionized; fresh irrigation water was scarce; and electricity, imported by underwater cable from Maui, was costly. By the 1980s, Ecuador and the Philippines were beginning to grow cheaper fruit, and the Dole pineapple empire started to struggle.

The company's entire business was eventually bought by an elderly Californian named David Murdock, who built himself on Lana'i a handsome home and a fine Victorian-style orchid house for his prodigious collection; threw up two enormous hotels, one in the cool of Lana'i City and the other down by the ocean; and when he arrived on-island, progressed through town in a horse-drawn carriage expecting displays of adoring fealty from his plantation workers, whom he called "my children."

But Murdock soon tired of the island, as most unsentimental businessmen tend to tire of an entity that doesn't make them serious money. So it was Murdock who did the eventual deal with Larry Ellison: for the three hundred million dollars he paid, Oracle's boss got just about everything available on the island, except for the orchids—he did get the empty orchid house, though, which has since been refilled—and a prescriptive right that is still retained by Murdock, who at the time was energy-obsessed, to build a wind farm on a headland up in the island's remote northeast.

Ellison has had more or less free rein to do as he will since the summer of 2012. He formally announced his intentions in 2013: his island, he said, would become "an experiment in sustainability." According to his company, the plan was "clear in concept, vast in scope and complex in implementation."

Lana'i City would be tripled in size, to twelve thousand people. Electricity, currently generated in Maui, would now be made locally, by a solar power station. A soi-disant "solar soothsayer" named Byron Washom was hired from the University of California's San Diego campus to design a series of microgrids for the island, with computers allowing the community rapidly to switch the source of power generation among three sustainable sources—solar, hydro, and wind—depending on need.* Fresh water, admitted by all to be the key to success of any such venture, would be manufactured from the sea with a large desalination plant. The dried-out old pineapple groves would be turned into organic vegetable farms, with drip-feed irrigation, and would produce such volumes of vegetables that the island could earn immense sums from an export trade with Japan. Grapes would be grown, and there would be an island winery. The islanders would drive electric cars. There would be a branch university established on the island, clusters of more luxury homes, a film studio, a second runway at the little airport, a third resort hotel, a tennis academy.

Though hardly anyone ever anticipated a sudden and swift

---

* The self-contained electric microgrid that has made UC San Diego a poster child for this fashionable new technology has not been an unqualified success. After a statewide blackout in 2011, it took the microgrid five hours to recover—the city itself took thirteen. Early suggestions that, at a moment like this, a microgrid would prove "an island of light in sea of darkness" were not borne out. Washom is hoping that a similar test on Lana'i, after a decade of further fine-tuning, will be more of a triumph.

turnaround for the island—some did, and for a while Ellison was being publicly touted as the island's long-awaited savior, almost a messiah—some on the island now note that after three years, less has been accomplished than had been hoped. True, the two hotels are being renovated, and some grand new houses have been built. A public pool, closed by David Murdock, was reopened in 2014, to widespread approval. And the tiny island movie theater has been wholly renovated, now gets first-run films, and is mobbed by patrons.

But the first drip-fed organic garden set up to supply food for one of the hotel restaurants was overrun with wild turkeys, apparently as keen on organic food as the hotel clients, and no other such agricultural experiments have yet been started. The islanders are fighting with Ellison's local Lana'i management company, named Pulama Lanai, over plans for the desalination plant, which many fear will harm their own very limited aquifer—and that has delayed the start of construction. The most recent design statement for the airport now has only a single runway. There are no plans for wind generators.

"Mr. Ellison appears to have gone to ground," the editor of the local newspaper remarked. "He has other things on his plate. It looks as though he has gone off us."

More than a few islanders were publicly grumbling that they were being told little, and at the beginning of 2015 an apprehension was growing among some of the initially hopeful islanders that Lana'i was perhaps becoming little more than a costly plaything for this faraway mogul, a chattel, disposable and forgettable.

Ellison, however, still has supporters aplenty—not least his employees and hired hands, who are fiercely loyal and, with strictly enforced nondisclosure deals, unwilling to talk about plans. There is also a retired French Canadian psychotherapist and former Hare Krishna zealot named Henry Jolicoeur, now living in Vancouver, who regularly broadcasts filmed paeans to

Ellison and his efforts, and mounts spirited attacks on anyone with the temerity to criticize him. A newspaper story that once spoke of Ellison's having a man in a speedboat follow his yacht and collect any basketballs that he lost while shooting hoops on deck especially piqued him. Why shouldn't Mr. Ellison employ such a man? Jolicoeur demanded. He is rich. He can do as he likes. The reporter who wrote the story, he told viewers, was merely fourth-rate, and worse.

The only other privately held island in the Hawaiian chain, Niihau (bought from the Hawaiian king in 1864, and owned ever since by the purchaser's descendants, a family named Robinson) is extremely impoverished. Many of its population of some one hundred thirty native Hawaiians are on welfare. It has no radio or TV, no electricity mains or telephone, though its primary school is solar powered. The Robinsons' local farming and beekeeping businesses have done poorly, and most islanders are compelled to live by subsistence farming. Niihau's owners earn most of their income from allowing a small unmanned U.S. Navy radar station to be sited atop a cliff: it monitors the U.S. military's own high seas missile tests, being conveniently downrange from the navy launch center on Kauai island nearby.

By contrast, Lana'i's abundance of land and loveliness will be likely to keep it from suffering the dismal fate of dusty, barren, treeless Niihau. Whether it will ever achieve the sustainable success that has been forecast for it is hard to imagine. But at least an effort is being made. Also, to the extent that one further corner of a hitherto environmentally neglected ocean is now attracting attention, the island of Lana'i joins the Great Barrier Reef, the Japanese volcano of Torishima, and scores of tiny atolls in the central Pacific as threatened and overlooked places in which we are at last expressing a keen interest, and about which, finally, we are seemingly starting to care.

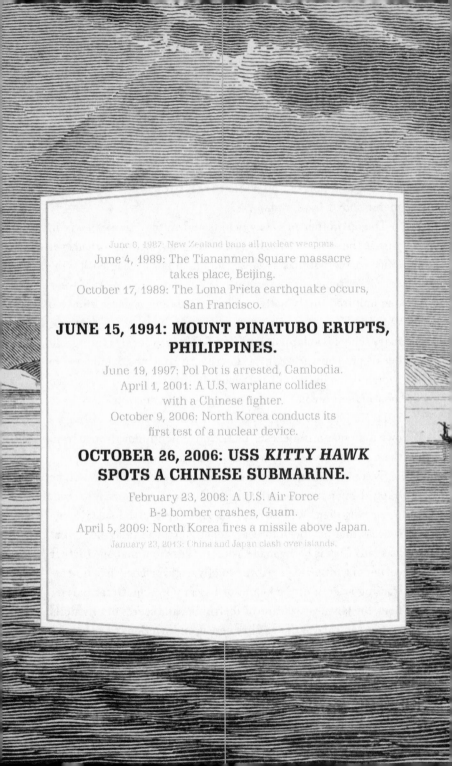

June 8, 1987: New Zealand bans all nuclear weapons.
June 4, 1989: The Tiananmen Square massacre
takes place, Beijing.
October 17, 1989: The Loma Prieta earthquake occurs,
San Francisco.

JUNE 15, 1991: MOUNT PINATUBO ERUPTS, PHILIPPINES.

June 19, 1997: Pol Pot is arrested, Cambodia.
April 1, 2001: A U.S. warplane collides
with a Chinese fighter.
October 9, 2006: North Korea conducts its
first test of a nuclear device.

OCTOBER 26, 2006: USS *KITTY HAWK* SPOTS A CHINESE SUBMARINE.

February 23, 2008: A U.S. Air Force
B-2 bomber crashes, Guam.
April 5, 2009: North Korea fires a missile above Japan.
January 23, 2013: China and Japan clash over islands.

Chapter 10

OF MASTERS AND COMMANDERS

不知彼，不知己，每戰必殆
If you do not know your enemies, nor yourself,
you will be imperiled in every battle.
—Sun Tzu, *The Art of War*, 2600 BP

It is essential to the welfare of the whole country that . . .
the enemy must be kept not only out of our ports, but
far away from our coasts.
—Alfred Thayer Mahan,
The Influence of Sea Power upon History, 1890

More than fifteen years separate a pair of highly unanticipated
events that took place in the western Pacific. On the face of it,

nothing connects the first and the subsequent one. Yet the first, the almighty and highly damaging volcanic eruption in the northern Philippine islands in 1991, with a devastating typhoon thrown in for good measure, was so catastrophic that it certainly played a role in the second, the sighting of a small steel periscope slicing its way through the international waters of the Philippine Sea.

The periscope belonged to a Chinese diesel attack submarine that had crept, stealthily and undetected, to within torpedo range of an American aircraft carrier battle group on routine patrol in the area. Though the official Pentagon inquiries would later concentrate on the alarming embarrassment of the U.S. Navy, with all its sophisticated underwater detection apparatus, having failed to notice the incoming attack vessel, the deeper and initially unanswered question was why this Chinese submarine was in this corner of the Pacific, in what were long assumed to be American waters, by use if not by right.

Before long, Chinese ships of all shapes and sizes began popping up around the western Pacific waters. The sighting of the first periscope on Thursday, October 26, 2006, became a reference marker, the Year Zero for the Chinese navy's steady and relentless expansion that has continued to this day. And though the link may not be immediately obvious, the expansion originated during the months of volcanic mayhem that got under way in the Philippines fifteen years earlier, on the wet and windy tropical dawn of Saturday, June 15, 1991.

As a global spectacle, the Mount Pinatubo eruption has generally faded from the West's memory. It was a Ring of Fire event, true. But not that memorable to most—who, instead, more likely call to mind the blast of Mount St. Helens, in May 1980. When asked about earthquakes, they will speak of the ripping apart of Anchorage on Good Friday 1964 or the Loma Prieta quake of

1989, which collapsed a freeway and took down a bridge in San
Francisco. And when tsunamis are mentioned, most will remember the unimaginable death toll that spread across the sea
from Banda Aceh in Sumatra, on the day after Christmas 2004,
or the nuclear-tainted horror show at the Fukushima power station following northern Japan's powerful undersea earthquake
of 2011.

Though Pinatubo was the second-biggest eruption of the
century,* it is generally now no more than the stuff of archives
and reference tables. Perhaps this had something to do with the
times, which were filled with a general weariness over the all-
too-slow exit of President Ferdinand Marcos, his wife, and their
cronies, and exasperation that the regimes that followed were
unable to achieve much to improve the dire state of Philippine
politics, back then quite dominated by corruption and ineffi-
ciency. Whatever the reason, the Philippines had become associated more with frustration than fondness—and the eruption
of a giant volcano there won less interest and sympathy than it
should have.

Yet, as a hinge of history, this one eruption was of rather
greater moment than the many other, more lethal seismic events
that have lately occurred around the Pacific, because it closed
down, abruptly, two of America's biggest military bases, snatch-
ing away a good portion of America's military authority in the
region. At least in this immense part of the world, this knocked
the United States very much onto its back foot.

Pinatubo was never a particularly noticeable or interesting
mountain. In a country where some of the volcanoes are true
classics of the kind—Mount Mayon, a perfectly symmetrical
Fuji-like cone a hundred miles farther south, ranks as one of the

* The biggest of the century occurred in 1912, on the Aleutian Range on the Alaska
Peninsula, when the stratovolcano Novarupta began a five-month series of enormous
eruptions that expelled more than three cubic miles of ash into the sky. In human
memory, only Krakatoa and Tambora, which erupted in the previous century, were larger.

world's most beautiful*—Pinatubo was low, lumpy, covered with enough jungle to render it almost invisible, and barely rated as a volcano at all. It had never been known to erupt, and it was generally associated with ill-fortune: the much-loved Philippine president Ramón Magsaysay named his official plane *Mt. Pinatubo*, and in 1957 it crashed into a hillside, killing him and everyone else on board.

An early clue to possible fresh trouble on the mountain was a big earthquake that rocked central Luzon in July 1990. Locals, members of a hunter-gatherer group called the Ayta, reported gouts of steam pouring from the jungled flanks of Pinatubo; though when the seismologists arrived, they found nothing amiss. The real trouble began in March 1991. The volcano's ridge seemed suddenly to unzip itself, and scores of small steam vents opened up, hurling curtains of scalding water and ash up into the sky. At the same time, seismographs at the U.S. Geological Survey, which had a formal partnership with the volcanological community in the Philippines, recorded dozens of small tremors in the earth below. A magma body was forming, the geologists concluded: a volcano that had erupted only three times in the previous six thousand years was about to do it again. On June 7, when a cloud of dust and ash mushroomed four miles into the atmosphere, the Philippine government issued a formal warning to evacuate. Pinatubo was about to go mad.

Although the volcano was a scant ninety miles north of Manila, with six million people living cheek by jowl beside it, the Pentagon was fretting because it was dangerously close to

---

* I climbed it in the 1990s, with a friend and two guides. As we neared the summit, a full gale blew up and the guides ran away in terror. The two of us pressed on, clambering the final few hundred feet to the crater lip on our hands and knees, drenched by rain and pummeled by high winds. We later found our guides huddled in a cave, quite incapacitated by smoking so much marijuana that we were obliged to reverse roles and guide them down to safety.

two critically important U.S. bases. The U.S. Subic Bay Naval Base was twenty miles away as the crow flies, and Clark Air Base was less than nine miles distant—close enough to be the location of the U.S. Geological Survey's immense array of seismographs. Should the volcano truly blow its top, then both bases, and the USGS offices, would be directly in its line of fire.

The military stakes were enormous. Clark was the most populous of all American overseas bases: two hundred thirty square miles in extent, with fifteen thousand people living there, and with runways perpetually busy with fighters, huge transport aircraft, bombers carrying nuclear weapons, troop carriers, helicopters. It had been especially busy during the Vietnam War, as the main hub for men and matériel on their way to and from Saigon. Once that conflict had wound down, the base served as a forward operating headquarters for the closing years of the Cold War. Although planes were sent off to existing bases in Alaska and California once the tensions with Moscow abated, the Pentagon still expected to keep the base active as China started to assert its role in the region. But nature had other plans.

The Subic Bay Naval Base was an even more significant operation. Together with its adjoining naval air station, which in its own way was quite as busy as Clark, Subic Bay was about the same size, at more than two hundred sixty square miles. It was principally used as a forward operating base for the U.S. Seventh Fleet, and during the Vietnam War it repaired and resupplied hundreds of vessels and their aircraft.

In 1987 the RAND Corporation, which advises on long-term military planning for the American government, had little doubt as to the strategic value of the two bases: "It would be devastating to regional security if the USA's relationship with the government in Manila deteriorated to the point where it lost its base access. Without the naval and air facilities at Clark Field and Subic Bay, the ability of the United States to support the defense interests of the ASEAN [Association of Southeast Asian Nations] countries and others who depend on the security of the sea-lanes would be seriously jeopardized."

RAND was focused on the frosty relationship that existed then between Washington and Manila. But politics would not be what did in the bases. Instead, it was the power of fast-rising magma and the eruptive explosions that finished off everything.

The official government warnings that the Pinatubo volcano was about to go critical went out on June 7. The Ayta people were the first to move away, a communal decision that served as a reminder that traditional peoples often have folk memories of impending catastrophe. For example, the seldom-contacted Andamanese, aboriginal inhabitants of a chain of islands in the Bay of Bengal, similarly moved to high ground long before the arrival of the 2004 tsunami that killed so many of their less-informed neighbors.

Despite the gathering seismic crisis, about three hundred thousand local inhabitants who lived within a twenty-five-mile radius of the peak stayed put. The commanders of the two bases decided to follow suit, and keep their crews and ships and aircraft where they were, batten down the hatches, and hunker down—at first. But this volcano was not going to settle itself, it seemed, and three days later the USGS officers monitoring their instruments declared that a true cataclysm was about to begin: Clark Base had to be evacuated, and all personnel moved across the country to Subic, where they should be readied to get on warships to be taken as far away as possible as quickly as possible.

On the morning of June 12 (by chance, Philippine Independence Day), seismic matters took their predicted serious turn. The half-solid lava dome, which geologists suspected had built up over the previous months, was suddenly ruptured by the accumulation of millions of tons of gas-charged magma beneath it—and a gigantic cloud of material rocketed up into the sky. It was a crystal clear day; images of the miles-high cloud boiling gray against the vivid blue of the sky were both awe-inspiring and deeply troubling.

The Americans got going. The Pentagon had already named its planned rescue Operation Fiery Vigil, and that got fully under way. All the aircraft had flown safely away from the air base and the naval air station. Streams of military trucks and private cars were taking airmen and civilians down to Subic. Traffic jams extended for miles. Heavily armed U.S. Marines lined the choked highway, in case Marxist rebels operating in the region took the opportunity to stage ambushes. The noontime temperature was well over one hundred degrees Fahrenheit, and the combination of fear, urgency, and the ever-present sight of the erupting volcano nearby made for a volatile couple of days.

But all progressed smoothly enough. By the time the major eruptions began, Subic Bay had fifteen thousand extra people to feed and house, and thousands of vehicles to park on its docksides. The Seventh Fleet commanders were busily ordering in vessels from wherever nearby they were on patrol: seventeen warships, including the carriers USS *Midway* and USS *Abraham Lincoln*, with their respective battle groups, had already been diverted and turned into the bay, ready to collect their unanticipated passengers.

Pinatubo's final eruption started during the predawn hours of June 15. By this time, the roars of the exploding mountain were almost drowned out by the howl of the gale-force winds from Typhoon Yunya, which happened to strike the island of Luzon at the very moment its great volcano was erupting. The cannonade of millions of tons of ejected material from Pinatubo was believed to have shot twenty miles up into the sky, but no one could see the spectacle because of the thick rain clouds and the swirling thunderstorms. All they saw were lightning storms of an unimaginable intensity, and then thousands of tons of liquid ash pouring down on top of everyone and everything for the entire day, collapsing buildings, making roads impassable, turning rivers into torrents of near-solid mud.

The unexpected eruption of the hitherto peaceable Mount Pinatubo smothered two critical U.S. bases nearby—one of them a navy headquarters—with enough ash and mud to cause their abandonment. The Chinese navy, abhorring the maritime vacuum thus caused, entered the adjoining waters with enthusiasm.

All the seismographs went dead. Everything was pitch dark, Hadean, and terrifying. The skeleton crew who had stayed at Clark to guard sensitive equipment were finally ordered to smash it all and leave as best they could in the remaining heavy vehicles. A very few of the most foolhardy decided to ride the storm out, secure in the central core of the base's strongest buildings. Everything shook, the noise was frightful, the air almost impossible to breathe.

Then, at about ten thirty that night, the atmospheric pressure returned to normal, the rains stopped, the clouds cleared away, and the eruption gurgled to a full stop. It was over. Nine hundred feet had been erased from the mountaintop,* hurled out and into the upper atmosphere and then back down to earth, which meant the two bases below had been wrecked, utterly. They were covered in a foot and more of heavy, gray, and greasy mud. It would take millions of dollars to clean up the bases, and they would be inoperable for years.

The Philippines suffered terribly from the volcano and the typhoon. Ash extended over forty-eight thousand square miles of the islands, wrecking entire villages. Floods had devastated whole valleys. Eight hundred people died, two million were directly affected, eight thousand houses were destroyed. Ten billion tons of magma had been ejected, twenty million tons of acid gases. Pollution was widespread and profound. The republic's economy, faltering at the best of times, went promptly into a steep nosedive.

As far as the American military bases were concerned, though, the evacuation was a success, a model of efficiency and good organization. And the fast-advancing science of volcanology had also scored its first impressive triumph: the USGS scientists had predicted the eruption of Pinatubo long before it occurred; they got the time of the eruption right and correctly estimated its ultimate severity. Their work had undoubtedly saved lives, possibly in the thousands. Volcanologists could now be counted among the ranks of forecasters, specialists fully able to predict at least one aspect of the earth's behavior with quite as much reliability as meteorologists predict the atmosphere's behavior.

---

* Some weeks later, when all residual activity had quieted, a photographer and I took a helicopter up from Manila to see the devastation. We brought along kayaks so we could explore the crater lake that now replaced the missing summit. As I made my way toward the lake's center, the waters became hotter and hotter until, directly above the volcano's mouth, the water bubbled and boiled, and the plastic of the boat began to soften and to bend under my weight. There were a nasty few moments of furious paddling back out into cool water. The kayak's owner later complained about the distortion of his craft, though he could never rightly understand how it got that way.

Within a month, President George H. W. Bush's secretary of defense, Dick Cheney, announced that the damage at Clark was so severe that the U.S. forces would abandon it. American troops had been housed there for almost ninety years. There was brief hope that Subic Bay, which Theodore Roosevelt had placed there in 1901, could be cleaned up and reoccupied. But Philippine domestic politics then intervened. There was a sudden spasm of anti-American sentiment among the more radical members of the Philippine Senate, and that hope, too, became a chimera, and soon vanished. The Stars and Stripes were then lowered at the two ruined bases—at Clark in November 1991, and at Subic on November 24, 1992.

For the first time in a hundred years, there were no Americans at any base in the Philippines. Moreover, for the first time in five centuries, there were no foreign fighting forces in the Philippines. The island republic (which itself had only minuscule deterrent forces: a tin-pot navy of fifty elderly U.S. Navy surplus vessels and only five working jets in its air force) was now essentially defenseless. Overnight, the western Pacific had become a vacuum.

One that the Chinese military was only too ready to fill.

Confirmation that the Chinese were building themselves a deep-water navy came on that October day in 2006, in what the U.S. Navy still calls "the *Kitty Hawk* incident." The details remain classified, but so far as can be established, the venerable eighty-thousand-ton carrier *Kitty Hawk*, at the time the oldest ship in America's active fleet, was halfway between Okinawa and Palau. She and her escort group (a cruiser, two destroyers, seventy aircraft, perhaps a submarine or two, and a distant oiler for long-range operations) had left their home port—Yokosuka, in Tokyo Bay—some weeks before. They were performing routine exercises as Carrier Strike Group Five, reporting directly to the

In October 2006, an aircraft from the elderly carrier USS *Kitty Hawk* operating in the Philippine Sea spotted, to general astonishment, a Chinese Song-class diesel attack submarine surfacing a mere five miles from the American strike group. This was the first indication of the Chinese ambitions for a "blue-water navy" of their own.

commander of the U.S. Seventh Fleet, a thousand miles away in Japan, and then to the four-star admiral commanding the Pacific Fleet, five thousand miles away in Pearl Harbor.

On this particular Thursday, the carrier commander ordered the launch of one of his Lockheed S-3 Viking submarine tracker aircraft, for no better reason than to practice looking around the immediate sea area where the strike group was operating. To the astonishment of the four crewmen aboard, a plane affectionately known as the *Hoover* (because of its engines' low growl) received a message from one of the FA-18C Hornet fighters already in the sky: the white-water trail of a possible speeding periscope had been sighted. The *Hoover* flew swiftly overhead to check, and promptly spotted what was unmistakably (and in this case, unforgettably) a submarine surfacing from deep-water patrol, and not five miles from the carrier. It was a Chinese-made Song-class diesel attack submarine, brand new, deadly, almost exactly beneath the American carrier group, and clearly not there by accident.

News quickly reached Washington. "As big a shock as the Soviets launching Sputnik," one Pentagon official in the Office of Naval Intelligence was quoted as saying. Memories of Pearl Harbor and the World Trade Center attacks were promptly resurrected; seasoned naval officers said that those dreadful casualty figures would be thought of as minimal compared with the devastation of an American supercarrier being attacked while on patrol in international waters. An accurate torpedo strike could cause the mighty ship to roll over and sink, drowning more than five thousand sailors. The loss would be punishing on all kinds of levels.

This nightmarish possibility, however slender, has haunted America and its naval strategists in the Pacific Ocean ever since. A full-blown seaborne confrontation between the naval forces of China and the United States—the steps leading to such a hitherto-improbable scenario were initiated as far back as February 1992, eight months after the Pinatubo eruption closed down Clark Air Base and nine months before the Americans abandoned Subic Bay. America was easing herself out of the region; China was taking her place.

It began with Beijing's formal announcement in February that 80 percent of the South China Sea was now to be formally considered Chinese territorial waters. Beijing had promulgated a sweeping and brand-new decree, its "Law on the Territorial Waters and Their Contiguous Areas." Any objections to this unilateral nationalization of the area would be dismissed out of hand; any incursions by uninvited and hostile vessels would be certain to have consequences. It was a vivid demonstration of the way that a newly emboldened, and newly rich, China was about to begin behaving.

The inference in Washington was stark: Beijing had clearly decided that with the closure of the Philippine bases, and the relative lack of American hardware in the area, the United States was now either unwilling or unable fully to protect the waters lying between the Philippines and the Chinese coast.

China's thinking here was part of a much larger regional doctrine, one dominated by the country's abiding historical and philosophical faith in what it likes to call the First Island Chain. This chain, which one can see on any map and is constructed by the tectonic forces of the Ring of Fire, is a ragged line of islands that runs from the tip of the Kamchatkan Peninsula in the north to Borneo in the south. China sees this as an inalienably pan-Asian feature of geography, but also as a protective outer cloak, a carapace, that shields China from its mostly non-Asian adversaries to its east.* Secure the chain, goes China's current belief,

---

* There is an outer group, the Second Island Chain, passing from Japan through Guam and to the western tip of New Guinea, toward which China also entertains ambitions. And in

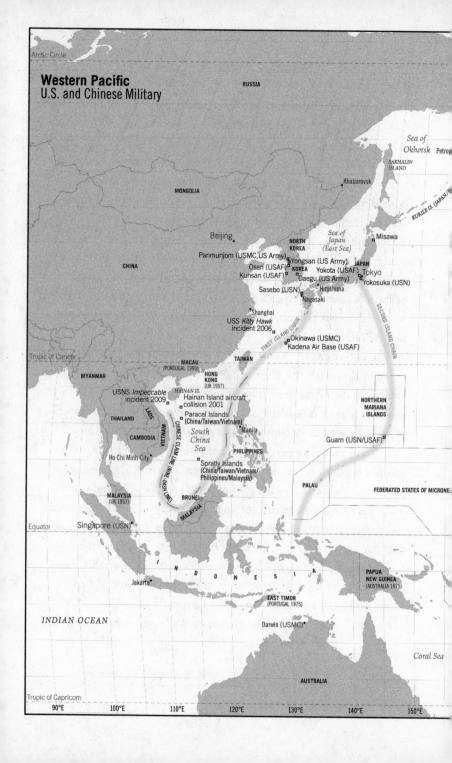

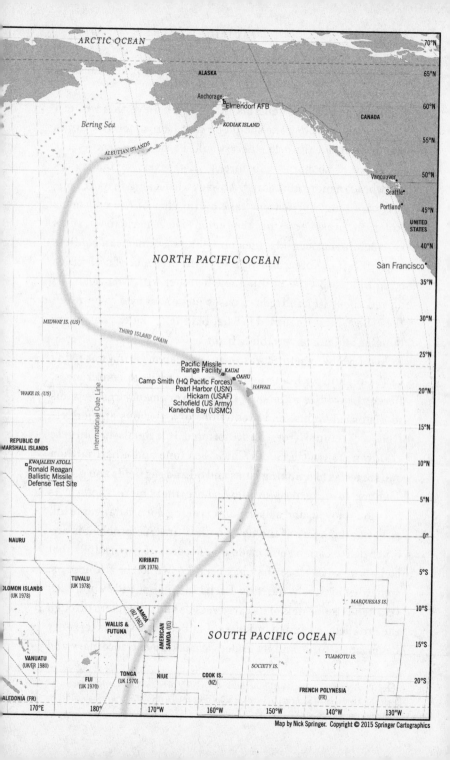

ARCTIC OCEAN

70°N
65°N

ALASKA

Anchorage
Elmendorf AFB
60°N

Bering Sea
KODIAK ISLAND
CANADA
55°N

ALEUTIAN ISLANDS
50°N

Vancouver
Seattle
Portland
45°N

UNITED
STATES
40°N

NORTH PACIFIC OCEAN
San Francisco
35°N
30°N

MIDWAY IS. (US)
THIRD ISLAND CHAIN
25°N

Pacific Missile
Range Facility KAUAI
Camp Smith (HQ Pacific Forces) OAHU
WAKE IS. (US)
Pearl Harbor (USN) HAWAII
Hickam (USAF)
Schofield (US Army)
Kaneohe Bay (USMC)
20°N
15°N

REPUBLIC OF
MARSHALL ISLANDS

KWAJALEIN ATOLL
Ronald Reagan
Ballistic Missile
Defense Test Site
10°N
5°N

NAURU
0°

KIRIBATI
(UK 1976)
5°S

TUVALU
(UK 1978)
SOLOMON ISLANDS
(UK 1978)
10°S
MARQUESAS IS.

SAMOA
(NZ 1962)
WALLIS &
FUTUNA
AMERICAN
SAMOA (US)
SOUTH PACIFIC OCEAN
15°S

VANUATU
(UK/FR 1980)
TUAMOTU IS.
SOCIETY IS.

FIJI
(UK 1970)
TONGA
(UK 1970)
NIUE
COOK IS.
(NZ)
20°S

CALEDONIA (FR)
FRENCH POLYNESIA
(FR)

International Date Line

170°E   180°   170°W   160°W   150°W   140°W   130°W

and you secure Asia. Make the green waters within it your own, and you are kept safe.

The admirals and generals in Beijing believe this western Pacific demarcation zone should be kept sacrosanct. In China's vision of an ideal world, there would be no American forces whatsoever, no non-Asian forces whatsoever, in the waters anywhere inshore of the line. Already China and its more conservative neighbor nations take a dim view of the enormous American presence within the region. The American warships in Japan; American Marines on Okinawa; American soldiers and aircraft in Korea; and until 1991, the presence of both kinds of equipment in the Philippines; together with American GIs dotted about in hundreds of bases just about everywhere—all these have long amounted to what Beijing considers an unpardonable and unsustainable affront.

Just for now, rhetoric was the only weapon China was using in the region, with hostile declarations made at conferences, fierce rhapsodies published in the government press, claims to historical rights, and predictions of a glorious and *laowei*-free—a foreigner-free—future offered in schools to Chinese children. Yet in the South China Sea (one and a half million square miles of shipping lanes and oil and gas fields and untold seafloor mining possibilities), more than mere rhetoric is being employed, and a flash point is being born, a danger to all.

For at the heart of the new South China Sea problem is another demarcation zone made up of scores of tiny islands that constitute what since 1947 has been known to the Chinese as, somewhat bizarrely, the Nine-Dash Line. The outside world each year sends thousands of its cargo vessels and oil tankers through the crowded shipping lanes here. But Beijing says this territory belongs to China, and its leaders are taking extraordinary and

the still longer term, there is on Chinese maps even a Third Island Chain, which passes from the Aleutians to New Zealand and includes the western tip of the array of islands that leads down to Hawaii.

dangerous-looking steps to make sure this is so. If a flash point is in the process of being made, then its axis is the Nine-Dash Line.*

This dashed, or dotted, line was first drawn onto maps in 1947, by officials of what was then the Chinese Nationalist government, the Kuomintang. The intention was to show the world that, under the terms of the Cairo and Potsdam declarations, China was taking charge of those islands in the South China Sea that were once seized by Japan and had now been surrendered. The line joining them all, drawn by some long-forgotten Chinese bureau-crat, was udder-shaped, dewlap-shaped, cow's-tongue-shaped—the name depended on the local language used to describe it—and first extended southward from the waters to the east of Taiwan. It proceeded down inside the main Philippine territory of Luzon, farther down inshore of the island of Borneo, before recurving northward, steering well clear of any islands claimed by Indo-nesia, heading up along the Vietnamese coast toward the Gulf of Tonkin, and finally stopping some two hundred miles shy of the large Chinese-owned island of Hainan. (There were eleven dashes on the original map, with two more drawn in the gulf be-tween Hainan and Vietnam; these two were later erased.)

Informal though that first Kuomintang map may have been, its boundaries were immediately adopted by the Communist government when the People's Republic was declared in 1949. The Nine-Dash Line became, by the stated policy of the Polit-buro, the outer limit of Chinese sovereignty, and it has remained so through successive Beijing regimes, no matter what others might say.

---

* In the late 1990s the British container vessel *Cardigan Bay* would sweep through the Paracel Islands every few months on routine passage between Singapore and Hong Kong, and the master was invariably asked to take photographs from his bridge of any new structures he saw. On one journey that I made with him in 1995, we were met at the container port by a Royal Navy intelligence officer, who took the film from the captain. "Most helpful," he said. "Always good to know what our friends up there"—he jerked his thumb back toward the hills of China—"are up to."

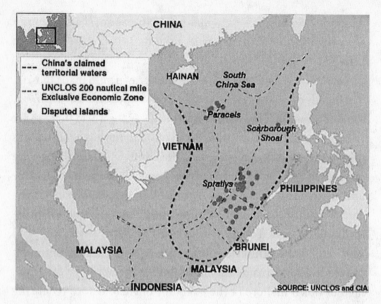

The Nine-Dash Line, drawn by China as a series of penciled lines on a postwar map of the South China Sea, showed the area of ocean still claimed by the Chinese as their own. The People's Liberation Army (PLA) Navy insists it can act with impunity within the line; others, the United States included, dispute the claim, hotly.

And they say a great deal. There are scores of islands within the line, most notably the Spratly Islands down south, close to the Philippines, and the Paracel Islands up north, near Hainan. The governments of Malaysia, the Philippines, Brunei, Vietnam, and even Taiwan all insist they have historic claim to the Spratlys. Vietnam and Taiwan say the same about various Paracel Islands. China, however, dismisses all these claims with an airy wave of the hand: it wants them all. And however long it takes, and whatever resistance Beijing might encounter, it intends to get them all.

The first Chinese land seizure within the Nine-Dash Line was of a low horseshoe-shaped bank of uninhabited coral lying in shallow waters off the Spratlys, little more than a hundred miles from the Philippine coast. It lay well inside that country's exclusive economic zone. It was known as Mischief Reef,* and only very rarely did anyone make visits there.

But in January 1995, a Philippine fishing boat did sail to Mischief Reef, and discovered something unimaginable: four enormous platforms had been built in the shallows behind the reef, and on top of them, eight large prefabricated octagonal huts, dozens of feet above the water, and supported by steel stilts. Each hut had a satellite dish on the roof.

The fishermen were even more astonished when eight armed vessels suddenly appeared over the horizon, and a posse of angry Chinese sailors boarded their little boat and declared the crew prisoners, charged with trespassing. It did no good for the fishermen to protest that the reef was actually Philippine territory; the Chinese laughed at them. The men were detained for a week. When eventually they were released, it was only on the condition that they swear never to reveal what they had seen.

Once they were safely back home, they naturally told all. The Manila government was stunned and furious. The Chinese ambassador was hauled in to explain himself, and he angered the Filipinos further by cheekily suggesting that the huts were no more than shelters for itinerant Chinese fishermen, an explanation that no one in Manila believed.

But all the angry talk did nothing. Nor did the softer approach,

---

* Though China's systematic expansion began after Pinatubo, a scattering of earlier confrontations had occurred between China and the Vietnamese, most notably on reefs and islets in the western Paracels in 1974 and in the western Spratlys in 1988. In both cases, the Chinese, employing considerable violence, drove the occupying Vietnamese away. The Americans lent some early support to the South Vietnamese, including the loan in 1974 of a CIA officer, who was briefly detained by the Chinese. But once the Vietnam War had been lost, and all Vietnam became Communist, American policy reverted to "a plague on both your houses," and steered clear of involvement, most notably during the Johnson South Reef Skirmish of 1988, which gave the Chinese their first real foothold on the Spratlys, three years before the Pinatubo eruption.

which had the Chinese president Jiang Zemin visiting Manila in 1996 and singing a duet of "Love Me Tender" with Philippine president Fidel Ramos during a karaoke cruise around Manila Bay. Still, the structures on Mischief Reef not only remained but were reinforced with concrete and searchlights—and a few years later, they even acquired a helipad.

One of the region's wiseacres, the Singaporean founder-premier, the late Lee Kuan Yew, who had a fondness for sardonic wisdom, offered an adroit summary of what was taking place: China, he said, was behaving "like a big dog going up to a tree and marking its presence, so that the smaller dogs in the region will know that the big dog has been past and will come back."

The dog tries to name its trees, too. In the years since its annexation of Mischief Reef, China has given Chinese names to all the various specks of land in the region—and to further confuse navigators, Vietnam, Malaysia, and the Philippines have given them their own names, too. Most seasoned mariners who are passing through along the long-sea routes stick to the islands' older chart names, nearly all of them in English: Scarborough Reef, Truro Shoal, Helen Reef, Robert Island, the Crescent Group, Macclesfield Bank, the Reed Tablemount, Bombay Castle, the Royal Charlotte Reef, Second Thomas Shoal, Rifleman Bank. The list is vast, the words conjuring up images of Victorian seamen, with their sextants and telescopes, spotting yet another low rise of coral and palm, or of shallowing green seas, and marking in indelible blue ink names from home, or of famous ships, or of long-dead sailors of renown.

Whatever the names, whatever the claims, China is now bulldozing her way through all objections, and literally. She plans to build airstrips, barracks, schools, even whole towns on these plainly titled islands, making each as permanent a part of China as appears symbolically possible. Buried in the middle of the Paracel group, for example, is Woody Island, the scene of much recent activity. It is among the largest islands in the South China

Sea, though occupying a mere six hundred forty acres, and is low and flat, sandy, palm-covered and tropical, and quite undistinguished.

The onetime significance of Woody Island derived mainly from the fact that so many countries have wanted to plant their flags on it, for reasons more symbolic than strategic. It is in theoretically international waters, close enough to the Chinese coast to appear on old Chinese maps, including those of the famous Muslim eunuch-sailor Zheng He, who went off to Africa in the fifteenth century and brought back a giraffe for the imperial court in Peking. Woody Island was close enough to Japan for it to be occupied in the 1940s; then the Nationalist and later the Communist Chinese took it over in the late 1940s. The French abruptly seized it on behalf of the Vietnamese in the mid-1950s, and then the Vietnamese claimed it for themselves.

A few fishermen's huts, steles, and beacons were thrown up by the temporary Woody Island tenants, and from time to time soldiers were garrisoned there. Someone built an airstrip, initially quite basic, and it was given a wash and brush-up by the Chinese in 1990. But little truly substantial or permanent-looking occurred there, until that time when the previously ever-watchful Americans turned their backs, or left the area after the eruption of Pinatubo. Then the visits began, and the construction—all pretty timid at first, but culminating in the arrival of a party of Nine-Dash Line–minded Chinese who moved there surreptitiously in 2006 and started to turn Woody Island into something quite different and, from the American perspective, dangerously so, and almost overnight.

Low, sandy, and useless, Woody Island is now the city of Sansha, and it serves proudly as the administrative capital of all China's supposed properties in the entirety of the South China Sea. This means it directs all the day-to-day businesses of a vast empire of claimed real estate—and ironically, though it is the smallest of all Chinese regional capital cities by population, it wields power over by far the largest administrative area. It has 630 permanent residents, mostly fishermen, but its reach now extends across a notional seven hundred thousand square miles of ship-busy, oil- and gas-rich, island-speckled ocean, a territory the same size as Alaska or Algeria, and very much bigger than Texas.

Sansha is not merely modern and clean, and designed to be a magnet (if just now a little improbably) for tourists of the future; it is also rapidly gathering military muscle. It has a modern airfield with a runway now almost two miles long, and it can handle the newest fighters and transport aircraft. There is a large and well-protected modern dock, easily able to accommodate both frigates or destroyers, and which will likely be used as a forward operating base for the scores of new patrol corvettes that China's naval dockyards are now said to be turning out "like sausages." The island, in short, appears to have been turned into the beginnings of a significant military headquarters, one that augments the large strategic submarine base on Hainan Island to the north, and that is well able to police and control the seas between the Chinese mainland and faraway Borneo.

And as if that were not enough, Sansha already seems to have spawned a subsidiary down south in the much more widely contested Spratly Islands. In 2014, Chinese sappers were seen to be dredging on the western edge of this huge area of shallows and reefs and banks, beside what is generally known as Fiery Cross Reef. Within weeks, satellite images showed a dock and a two-mile-long airstrip; within months, the Chinese navy corvettes were to be seen on every horizon. Fishermen on the nearby islands in the Philippines were scared of being harassed and told by Chinese soldiers they could no longer work their traditional fishing grounds.

The change in status of the sea and its islands and islets and reefs appears to be relentless, unstoppable. Much the same thing as took place on Fiery Cross is now happening nearby, on John-

Coral reefs and uninhabited islands all across the South China Sea are currently being seized and claimed by China, whose construction teams are blasting and dredging and pouring cement to create new "weather observation stations" or, rather, the United States believes, military staging posts.

son South Reef, where there was a deadly confrontation with Vietnamese sailors in 1988—there, too, dredging and concreting are under way, and yet another new island is being manufactured. One by one, strings of newly engineered, artificially expanded islands are sprouting up: seven so far, seventeen habitable others still to go, and yet more on Beijing's radar.

And it is all being done most cleverly, with Machiavellian stealth. The steps that China is taking are individually small enough to make retaliation difficult—an American carrier strike force going after the occupation of a lonely mid-ocean reef would seem ludicrously disproportionate. But when these steps are viewed collectively, they become deliberate strides and, with so many already taken, well-nigh impossible to reverse. It is a typically Chinese strategy: infuriating and clever, and right out of Sun Tzu's handbook.

Occasionally there is defiance from other claimants. Much attention has been given to the sorry state of a group of Filipino marines ordered to live aboard a rusting, lethally unsafe seventy-year-old former American tank landing ship that in 1999 was deliberately grounded on an island in Second Thomas Shoal: the marines serve for four months at a time, their food and water dropped by parachute, their discomfort watched by Chinese patrol vessels that lie, patiently hove to, in the deeps outside the reef.

The marines are there to keep the flag flying, to maintain the formalities of sovereignty. But there are more Chinese flags in the region now, a new red Chinese flag fluttering from every one of the newly made, newly settled morsels of Middle Kingdom real estate. The fresh-formed Chinese lands are all being sternly protected by naval guns. And Chinese claims are creeping slowly outward, too: on Scarborough Reef, startlingly close to the Philippine coast, Chinese naval vessels have lately blockaded the reef entrance, denying access to all. Yet another piece of Chinese sovereign territory is being manufactured in full public view, and one day, in all likelihood, it will be defended at all costs, by all and any means.*

It was once assumed that China's steady encroachments were predicated on the oil and gas reserves that lie beneath the South China Sea. Such is no longer the principal reason. Nowadays it is much simpler: the Chinese want all the green waters, all of what they call "the near sea" within the First Island Chain, to be free of interlopers. They know also that "might is right," and with stealth and determination, they can keep away those who are not welcome, and thus protect themselves, creating, as Shakespeare remarked of Richard II's England, a "fortress built by Nature for herself Against infection and the hand of war."

The Americans dislike the way the South China Sea is currently being carved up, with so much to China's unilateral benefit. They dispute the argument that the Chinese are doing no more in the western Pacific today than America, armed with the Monroe Doctrine of the 1820s, spent the last century doing in the rest of the world. Yet the Americans, the only other force with the military power to keep the sea-lanes secure and the international rules obeyed and adhered to, do little more than splutter their outrage. Not one country that claims territory in the sea has received anything more than lip-service support from Washington.

The White House does not want to alienate the Chinese any more than is necessary. But three episodes of head butting between the two superpowers show what kind of mishaps, misun-

---

* Right in the middle of the Spratlys is the island of Taiping, occupied since 1956 by the Republic of China and administered as a subdivision of a Taiwanese city. Six hundred officials live there, with power, water, public telephones, and Internet service, and with a cell phone signal inadvertently supplied from a Beijing-occupied reef nearby. Despite its political opposition, Beijing has made no moves to acquire the island, assuming that, as with Taiwan, it will revert naturally in good time. Its existence as an oasis is similar to that of Guantánamo, in Cuba, or of Russia's Kaliningrad, on the Baltic.

derstandings, and misreadings could bring them into serious trouble. All three episodes involved U.S. efforts to gather intelligence about various aspects of the Chinese military. And the Americans insist that all three occurred in international waters or airspace, making them entirely legal. China disagrees.

The first incident occurred on April Fools' Day 2001, and it took place in the air, four miles above the sea, seventy miles south of Hainan Island. To both sides, the location is important. The Americans insisted they were over international waters and in international airspace, and were an appropriate distance both from the Chinese mainland and Hainan and from the bases on the Chinese-claimed Paracel Islands. The Chinese accept this, and agree that the plane was where the Americans say it was. But they don't agree that the water or airspace is international. They believe this area is the sovereign airspace of China, and that permission was needed for any military operations to take place inside it. And the slow, lumbering, propeller-driven signals-intelligence-gathering EP-3 aircraft that the U.S. Navy sent out that day from a base on Okinawa had manifestly not asked for permission. As a matter of policy, the Americans never seek permission to operate in the area. Hence the problem.

The aircraft was tracking slowly back home, having been in the air for some six hours. It was on autopilot, flying straight and level. The twenty technicians abaft were performing their routine tasks, listening to Chinese transmissions from down below. No one on the flight deck was especially surprised when two sleek Chinese navy fighter interceptors suddenly appeared alongside. This had happened before; the Chinese pilots had been quite friendly, acting almost as escorts, making sure the Americans didn't stray off course. On one previous encounter, one of the pilots, a lieutenant commander named Wang Wei, had held up a greeting card, and came close enough for the Americans to be able to read his e-mail address through the cockpit.

On this April occasion Wang was back, and as before he made

foolhardy attempts to approach the Americans' aircraft—twice successfully and then, on a third try, coming up from beneath. Fatally, metal struck metal.

The top of his craft smashed into the underside of the American EP-3, damaging both planes and rendering them simultaneously well-nigh unflyable. The cockpit of Wang's interceptor was crushed, and though he ejected, his parachute didn't deploy properly and, as his broken jet spiraled down to crash into the sea, he hurtled to his death. His body was never found.

Lieutenant Shane Osborn, the American pilot, and the twenty-three crew with him (three were women) were rather more fortunate. Their plane had lost a propeller, there was a hole in the fuselage that caused it to lose pressure, its radar dome was broken off, part of the left wing was damaged, and its radio antenna wires were wrapped tightly around the tail, but as the aircraft screamed downward, losing three full miles of altitude, Osborn managed to wrestle it back under control, leveled out, slowed, and ordered everyone to prepare to bail out. But then he realized just how close he was to Hainan, and to the very air base from which Wang and his colleague must have departed. He decided to risk landing there, though no one at the base would answer his distress calls or give permission.

Nonetheless, he managed to get his plane and crew safely to the ground, albeit at a secret and seldom-visited Chinese air base. Once he managed to stop, the plane was immediately surrounded by dozens of heavily armed Chinese troops. They battered frantically on the sides of the fuselage while the crew inside did their best to destroy as much sensitive material as possible, and to wreck all their radios and radar. Once the Chinese jimmied open the end doors, the crew members were arrested, interrogated, and held for ten days—until the United States signed an apology and an admission of guilt and paid a $34,000 accommodation bill. Only then were the men and women released and flown by chartered passenger plane to Guam, and then by the

U.S. Air Force to Hawaii. The Chinese held on to the EP-3, dis-assembled it, stripped it of any remaining secret apparatus, and finally returned it in pieces—minus its flight recorders. It came ignominiously back to America in wooden boxes, aboard a Russian cargo plane.

Lieutenant Osborn retired, and in due course became a politician, ending up as Nebraska's state treasurer. Commander Wang's widow received a personal letter of condolence from President George W. Bush. The incident, alarming to all, could have been very much worse. For the Americans, the event marked this particular corner of the Pacific as a place where their warplanes, on whatever kind of mission, should henceforward be exceptionally wary, a place in which to take the greatest of care.

Yet the flights went on, the sea patrols continued, most of them uninterrupted.* And eight years later a similar event occurred, in a similar place, though this time it was on the sea.

It involved another American intelligence-gathering vehicle, working in almost exactly the same spot on the sea as the EP-3 had been operating above it: seventy miles southeast of Hainan Island. It was Thursday, March 5, 2009, and the curiously shaped ship USNS *Impeccable*, slow, unarmed, but new and highly sophisticated, was steadily towing a mile-long submarine-detecting aerial through the seas. *Impeccable* looks like an ungainly box of gray steel, built to be stable in the fiercest of seas, and her principal job is finding and tracking submarines and their weapons. Since she was, on this occasion, just a few hours' sailing from China's nuclear submarine base on Hainan, her purpose that spring was clear. The Chinese

---

* But not all. On August 19, 2014, a Chinese fighter plane came to within twenty feet of a large American Poseidon surveillance aircraft operating in almost exactly the same area. The Pentagon said that the Chinese pilot had even passed the Poseidon's nose with his plane's belly in full view, as if to show the weapons he had aboard. Washington made an official complaint.

were in the process of testing their own brand-new nuclear-powered attack submarine, the type 093 Shang-class boat. *Impeccable*, civilian-manned and operated by a clandestine division of the giant Maersk container shipping line, was on station to find out about it.

The USNS *Impeccable*, a curiously unseaworthy-looking five-thousand-ton unarmed surveillance vessel, which normally tows submarine-detecting sonar arrays, was confronted by a number of Chinese warships in 2009, prompting protests from Washington.

Not surprisingly, the Chinese were less than amused when they discovered what was going on. A high-speed frigate suddenly appeared ahead, passing unannounced, and within yards of the American ship's bow. From the north, a small Chinese aircraft flew across and, at an altitude of little more than five hundred feet, began flying back and forth above the ship. A radio message was then broadcast to the Americans: leave the area immediately or suffer the consequences. *Impeccable* ignored it, continued with her mission, playing out her aerials and listening for the signatures of any submerged Chinese boats.

Two days later, matters got nastier and more risible. Five hitherto unseen Chinese vessels, three of them armed, the others rusting trawlers, came alongside and began harassing the *Impeccable*. The Chinese sailors aboard used grappling hooks to try to snag the *Impeccable*'s 155-ton towed array and dropped balks of timber into the sea directly beneath the American's bows, forcing this most clumsy of ships to try to swerve and compelling her to dodge and then come to an emergency stop when one of the trawlers, not at all worried about being hit herself, sliced in front.

Being unarmed, the *Impeccable* had as her only available response her fire hoses, which the crew turned on full force whenever a Chinese ship came too close. But it was a hot day, and flat calm, and the Chinese men simply stripped to their undershorts and pranced around on deck, luxuriating in the streams of cool water.

By now the American captain was boxed in and stopped, essentially imprisoned. He radioed Hawaii for instructions. Should he destroy all his sophisticated equipment, burn all the records of intercepts, smash his hard drives? Hawaii said no—just leave. Ask permission to go. So the *Impeccable*, tail between her legs, asked the Chinese to back away and give her clear passage, and within moments, she was lumbering squarely off in an easterly direction. In an hour's time, she was just a smudge of smoke on the eastern horizon, the Chinese ships exulting in having tweaked the American eagle's tail feathers. It was left to the politicians in Washington to do the complaining, and to the diplomats to undertake the explaining.

But the lines had been drawn: the international waters that once had been, now no longer were. The Americans were going to have to get used to the new reality that the Chinese were not truly recognizing these as international waters. Washington could complain all it wanted, but this was not the South China Sea anymore; rather, it was China's Sea, in the south. The White House swiftly rejected the impertinence, and followed up by ordering the navy to send out a sentry vessel, a powerful destroyer, to stand for a while alongside the *Impeccable* and her sister ships whenever they operated in this sensitive area of the ocean. The escorts remained on station for some months, until the situation appeared to have calmed itself. Eventually they were brought home, and the *Impeccable* has worked unbothered ever since.

A third incident, which took place in this general area of the South China Sea, would reveal another aspect of the story: the extraordinary sensitivity that China is currently displaying, with regard not just to the use of her littoral waters, but also to the release of any information about the growth of her fast-emerging blue-water navy. Anyone paying too much attention to that particular phenomenon, to just how China might be expanding her navy, is these days told with abundant clarity that their interest is most unwelcome.

When word came that China's first-ever aircraft carrier, the newly named fifty-three-thousand-ton *Liaoning*, was about to set sail on her first deployment, the highly nimble guided-missile destroyer USS *Cowpens* jumped at the chance to follow her.

The *Liaoning* had had a tortured and celebrated past. She was originally built for the Soviet navy, then transferred to Ukraine, but was never finished or taken to sea. China then bought her, though under cover of refashioning her into a casino, to be moored in the docks at the gamblers' paradise of Macau. She was towed almost the entire way, and briefly broke free of her tow in a storm off Greece.

However, once safely in Chinese waters, and to the surprise of no one in U.S. Naval Intelligence, she was not taken to Macau, but sailed instead up to the North Pacific port of Dalian. She would spend years in dry dock there, being fitted out, given powerful engines, kitted out with weapons and electronics and, then, in good time, with the squadrons of aircraft that are essential to a carrier's function. She was named *Liaoning* for the province in which she was completed, and was commissioned and handed over to what is formally called the People's Liberation Army Navy in 2012, twenty-seven years after her keel was first laid down.

Now, on November 26, 2013, she left the Pacific port of Qingdao fully laden with aircraft and escorted by two destroyers and two missile frigates, heading south. She would be part of China's first-ever carrier strike group, an essential component of naval power projection. Three days later, USS *Cowpens* slipped quietly out of Yokosuka and hurried to lock radar with and then tag well behind the Chinese strike group. They passed in convoy through the Strait of Taiwan, sailing out of the East China Sea and into the much more contentious waters of the South China Sea. Here the *Cowpens* came smack up against exactly the same kind of Chinese hostility that the EP-3 and the *Impeccable* had encountered before.

At first all was polite. On December 5, one of the two Chinese frigates broke away from the strike group and approached the American vessel, evidently trying to keep *Cowpens* well away from the *Liaoning*, which, with her massively tilted bow rendering her immediately recognizable, was sailing serenely along the horizon. Then the Chinese came on the radio: would the American ship, now well within the twenty-five-mile "inner defense layer" of the carrier strike group, kindly move away and let the PLA Navy conduct the "scientific research, tests and military drills" she had come to these waters to do?

The *Cowpens*'s captain, following his orders, replied with studied courtesy, that no, he certainly would not move away. These were international waters, open to all, and the *Cowpens* had innocent purposes only; he had every right to stay put. The Chinese frigate moved off, and yet, as she did so, a huge amphibious dock ship that had been standing by suddenly gunned its motors and sped directly into the path of *Cowpens*, which was forced to throw all her engines into full reverse and swerve to port in order to avoid what could have been a disastrous collision.

The two ships' commanders then spoke to one another and agreed that good seamanship was desirable, and the affair was not repeated. Whether *Cowpens* gathered any useful information about *Liaoning* on her first outing is not known; all that is certain is that the question "Whose seas are these?"—the like of which had not been known until the Pinatubo eruption twenty years before—had caused alarm bells to ring again, as they have been ringing ever since.

And still the contagion spreads, and becomes ever broader. In recent years, China's dominance of the South China Sea has been followed by attempts to impose similar hegemonic control over the East China Sea. A long-standing claim made by the Chinese to the disputed Diaoyu Islands, an uninhabited cluster northeast of Taiwan that the Japanese have long called the Senkaku Islands, was suddenly backed up in 2013 when the Beijing government declared the airspace overhead a restricted area, and demanded that all aircraft, civilian and military, report and seek permission before entering it.

A stunned Washington saw this as quite preposterous, said it would ignore it, and promptly flew a pair of huge, highly visible, and deliberately chosen B-52 strategic bombers from a base in Guam into and out of the area, refusing to tell anyone in advance, not filing flight plans, and not registering radio frequencies. There was no reaction, though China drily said it had "observed the violation." Japanese Airlines initially said its passengers jets would obey the new rules, but backed down once the Americans showed their resolve. Korea said it would defy the ruling, though Taiwan said it would agree. And then the world piled on: Australia and Germany were the first to call in their Chinese ambassadors and complain, and in time the steady roar of opposition made the Chinese look foolish and petulant. But their rule remains in place, if honored more in the breach. An eventual armed incident, warned the German foreign ministry, was only made more probable by decisions such as this.

Meanwhile, the Chinese navy in particular keeps on getting larger and larger, its area of operations wider and wider; and from a Western perspective, its territorial claims ever more egregious. The concern spawned by this increase in China's military ability has led to what is claimed to be a profoundly changed new American policy toward the Pacific region, one that has been, or is in the process of being, put in place.

America is presently tightening her focus on the Pacific, with a policy that enjoys two names, neither of them especially inspiring: there is the pivot, and there is the rebalance. Moreover, the proposed style of military response enjoys two names as well: From when it was first formulated in 2009 until its rebranding at the end of 2014, it was known as the Air-Sea Battle concept. But since a lengthy policy memorandum from the Joint Chiefs of Staff was circulated on January 8, 2015, it has become known as the Joint Concept for Access and Maneuver in the Global Commons. This last came with a supposedly helpful explanation from the Joint Chiefs of Staff at the Pentagon, telling users of the idea's official pronunciation: it is to be referred to in conversation as the JAM-Gee-Cee.

Whatever the semantics, all that the name suggests is that America's military has, since 2009, demanded the authority and the budget to do things differently in this very different new world. According to this memo, it plans to tilt its strategic attention away from Europe, away from worrying about Ukraine and the Baltic, away from the intractable tar baby problems of the Middle East. It plans instead to gaze toward the formidable and ever-gathering challenge of China and the swiftly evolving realities of the new Pacific. The Atlantic and the Mediterranean present no more than a thankless quagmire; the Pacific presents challenges to be overcome in order to offer America and China a high-speed excursion into a brighter future.

But what does China want, and what is it planning? Thus far, efforts to come up with coherent answers to either of these two basic questions have met with only limited success.

Two names, however, keep surfacing. On the Chinese side, there is the late Admiral Liu Huaqing, the revered architect of the country's long-term naval strategy, the Chinese equivalent of Alfred Mahan or of Teddy Roosevelt.* The plan China appears to be undertaking today was essentially laid down and promoted by the admiral and his political superior Deng Xiaoping, in 1985. At the time, both men were well on in years: the admiral was

Admiral Liu Huaqing is the principal architect of the rapid expansion of the Chinese navy's presence in the western Pacific. He believed that by 2040 the nation should have an active blue-water force in the region, with several aircraft carriers at its disposal.

---

* In one crucial sense Liu was very different from these two: he was the overall commander of forces during the Tiananmen Square massacre of June 4, 1989, and had a reputation for ruthlessness that survives him to this day.

seventy, the Chinese leader eighty-one. They were old friends, die-hard Communist revolutionaries, Long March veterans, and as it happens, true visionaries, men whose thinking has had a major impact on the warp-speed development of China in recent decades.

Liu Huaqing's view of his navy's future role came about in the early 1980s, after the sudden realization that the Soviet Union, an entity soon to disintegrate, no longer posed a consequential threat to China's future. The Chinese armed forces (all formally known under the rubric of the People's Liberation Army, but with the "Navy" and "Air Force" elements appended to the title) could afford to change their focus, and quickly.

Hitherto the Chinese navy's policy had been based on coastal defense: on harrying and intercepting any enemy forces that might appear in the coastal cities, and helping the army to dislodge them and drive them back to their lairs. Since this was now not likely to happen, the naval effort could well be spent in the future on offshore defense: on broadening China's ability to defend itself in the three great seas that surround it, the South China Sea, the Yellow Sea, and the East China Sea.

So a new concept was born. China's navy, if it were provided with enough ships and with a sufficient budget commitment for the long term, could now project the boundaries of the nation's influence outward and ever outward into the Pacific Ocean as the years went on. The concept of the First Island Chain was born during these strategy sessions: the need for China to secure the "green waters" within the Kamchatka-to-Borneo line, and to do her best to deny access to any foreign military that might wish to be there.

There were discussions of the allurements of the Second Island Chain and even of the Third Island Chain. The possibility, never before imagined, never even imaginable, that China could one day extend her blue-water power as far as Guam, maybe even as far as Hawaii, seemed suddenly within the realm of the possible.

Admiral Liu had almost overnight given China a new dream. And as the powerhouse of China's new industrial might cranked itself under way in the late 1980s—as the factories began to hum, and the exports to thunder eastward, and the dollars and the gold began to flow at an unstoppable rate into the bank vaults of Beijing and of Shanghai and, after 1997, of the handed-back territory of Hong Kong—the revenues increased and the public funds became more generously available, and serious new defense spending became a possibility, then a necessity, then an absolute essential. In the People's Liberation Army Navy's shipyards, scores of new keels were laid down, scores of new vessels launched, and all the most-up-to-date war-fighting accessories and ammunitions were added in the fitting-out basins, and the coastal waters became white-waked with sea trials and gunfire tests and missile launches, and the new Chinese naval policy was formally implemented in acres of gray steel and black smoke and the flying of battle pennants. And all this has been accelerating across the western Pacific ever since.

There is a timetable, too. By 2000, Liu and Deng had seemingly agreed, phase one of the PLA Navy's plan would see China in control of the green waters within the First Island Chain. This has not yet been achieved, but China's influence inside the chain, from the Senkaku Islands in the north to the Spratly Islands in the south, and with the building of all the bunkers and airstrips and radar sites on atolls and skerries all across the region, is considerable, is intimidating, and is growing.

The presence of a Chinese attack submarine in the Philippine Sea in 2006 shows that phase two is now under way: China is making its presence felt within the Second Chain, in the waters between Luzon and the Marianas, between Cebu and Palau, between Borneo and Vanuatu. There is even serious talk these days in Beijing and Shanghai of a Third Island Chain, which includes the Hawaiian Islands, and of China seeking rights and freedoms there as well.

No red-blooded American admiral is going to look too kindly on an aspirant power from Asia bringing its vessels regularly into the seas around Hawaii. Yet Hawaiian waters are temptingly highlighted on Chinese naval planning maps these days—most beguilingly, on a map in the 2005 Chinese translation of the biblically regarded *Influence of Sea Power upon History*, by Alfred Mahan. Admiral Liu was still in service at the time (he died only in 2011), and his stamp would have been on the decision to draw the map this way, together with his advocacy of phase three of the master plan: that the Chinese exercise some good measure of control of the seas to the west of the Third Island Chain, and become a truly global navy, to boot, by the year 2049.

By the centenary, in other words, of the founding the People's Republic, and of the founding, in August 1949, of the People's Liberation Army. That is the key culminating moment for all current Chinese regional ambitions. By then, all that China believes to be China's is expected to be back in China's hands. The recent past has been rather kind to China's ambitions for the slow and steady recouping of its territory. Manchuria is back. Port Arthur and Port Edward have come back. The British colonial outposts of Weihai-wei and Hong Kong are back; Hainan is back from the French, Macau from the Portuguese. Now all that remains is Formosa, the once Portuguese "beautiful island" lying between China and Japan, the recovery of which all know will provoke, unless great diplomatic skill is wielded by all parties, an almighty fight.

The rapid rise in Chinese defense spending ($166 billion in 2012, up 12 percent from the year before, and heading skyward) is part of the plan to help achieve all this, and to secure the seas beyond. The hardware is being amassed: the seventy-seven principal surface combatant ships China had in its 2014 navy (the U.S. Pacific Fleet had ninety-six), the sixty-seven submarines (the United States has seventy-one), the fifty-five amphibious ships (there are thirty in the U.S. Navy), and its eighty-five small fighting vessels (the United States has twenty-six) are all part of the plan. Only in one area, the possession of aircraft carriers, is Beijing well behind Washington, with the *Liaoning* and her strike group ranged against the ten enormous carriers fielded by the Pentagon. Yet even here the Chinese are catching up: three new carriers bought from Australia and Russia are being fitted out, and two more are being built in Shanghai. And while the *Liaoning* has a relatively antique "ski-slope" launching system, the new ships will have catapults, and a new carrier-suitable fighter is soon to be ready to join the fleet.

All this gives muscle to the Chinese plan, a design that has firm goals, set dates, political will, and a commitment of the money and machinery to undertake and achieve it. The American counter-scheme, its current "offset strategy," its so-called revolution in military affairs,* certainly starts from a superior position: the U.S. Navy has an overall total of 284 ships, 3,700 aircraft, and 325,00 active-duty personnel, while the Chinese have 495 ships, 650 aircraft, and 255,000 crew members. Yet it appears to be based very much on a series of back-foot assumptions about China's intentions, and about how to respond to them, rather than on any kind of forceful taking of the initiative, the kind of bull-by-the-horns approach of a figure such as Teddy Roosevelt. The four-year cycle of the American presidential election system hardly helps: whatever the Pentagon comes up with manages all too often to enjoy little long-term validity. Thanks to the realities of modern American politics, the playing of the long game is not the most prominent feature of many of Washington's policies.

Yet a counterplan exists, for now, and though like all complicated military strategies, it is a child of many fathers, it was

---

* This "RMA" concept stems from the belief that warfare styles evolve in quick bursts that are the direct consequence of the introduction of new technologies (whether chariots or longbows, the blitzkrieg or nuclear weapons, EMDs or drones), each new development prompting a new kind of fighting. Proponents of the theory say that new technologies now available to China have significantly changed the metrics in the western Pacific theater, and that America needs to change in lockstep.

designed in essence by one man, the second of the two much-revered graybeards of transpacific planning: the man who was generally regarded as Admiral Liu's intellectual opposite number in Washington, Andrew Marshall.

This most remarkable figure retired in 2015 after forty-two years as director of the Pentagon's Office of Net Assessment, to which he had been appointed by Richard Nixon back in 1973. When he stepped down he was ninety-four years old, and for decades was known familiarly in the American press as Yoda, after the Jedi Grand Master in the *Star Wars* movies, the Pentagon's preeminent keeper of peace and justice. His job was to plan for future wars, "to look at not very happy futures," as he once put it. Among the world's war makers, most especially those in Russia and China, his fame is legendary. He was "our hero" to the PLA's General Chen Zhou, one of modern China's most powerful strategists. America's RMA theory, said General Chen, was something he and his staff had studied "exhaustively." The Chinese "translated every word that Marshall wrote."

Andrew Marshall ran the little-known Pentagon Office of Net Assessment for forty-two years, following his appointment as a defense strategist by Richard Nixon in 1973. His Air-Sea Battle concept, a plan for preparing the United States for any coming confrontation with China, is a key element of today's strategy.

And what he wrote most recently (in 2009, when he was eighty-eight years old) formed the basis for the new military policy for the Pacific, what was first called by the Pentagon the Air-Sea Battle concept.

This concept assumes that China is now fully bent on keeping the Americans at bay, keeping them out of their immediate maritime hinterland, and stopping them from getting where they need to go in order to fight. Admiral Liu's now time-honored navy strategy of sanitizing the green waters inside the First Island Chain is part of it, part of what is known as A2/AD, the cumbersome military acronym for China's new anti-access/area denial capabilities. China's naval vessels already have the ability to frustrate and harass American attempts to patrol in the South China Sea: soon they may be able to forbid it altogether. In time this may spread to the East China Sea and then into the Yellow Sea as well.

Naval hardware and policy are just one aspect of China's plan. Deeper inside Chinese territory is now a growing number of missile bases and radar tracking stations and heavy artillery positions specifically designed to keep enemies away from the country's coastlines, and further deny them access to the areas in which they might need to wage war. The combination (with newly built islands in the sea and coastal missile batteries on land) seeks to hobble American power projection, to make it both riskier and costlier than before. In other words: to make any military expedition well-nigh unacceptable.

No Chinese planner has any doubt about the might of America's military, and China's strategy is not to try to match it, not for now. It seeks instead to practice an asymmetrical version of war planning, to aim to exploit America's military vulnerabilities rather than attempt to fight them head-on—to keep America apprehensive and at bay, unwilling or unable to make a first move.

One Chinese weapon in particular is currently generating much worry in Washington. It is a large new rocket, the East

Wind DF-21D antiship ballistic missile. Batteries of what is believed to be a formidable piece of musketry have been deployed
along the coasts since 2010. The missile has been dubbed the
"carrier killer": with its range of some nine hundred miles, it
would be able to stop in its tracks, if accurate enough, any American aircraft carrier that dared to push its bow through the seas
anywhere west of Okinawa. It has already been tested on a carrier mock-up, albeit in the relatively benign conditions of the
Gobi Desert, and it works like a charm.

Once weapons like this are fully operational, once all the radar
systems are in place, once patrolling submarines (which China
is building at a rate four times that of America) are on station
and the newly made cyberweapons (already highly effectively
tested on American assets) and antisatellite missiles are locked
on, and once the U.S. bases in Japan, on Guam, in South Korea
(but, courtesy of Pinatubo, not in the Philippines since 1992)
have been placed squarely in the Chinese gun sights, one thing
becomes certain: no American expeditionary force would ever
be allowed, or would ever wish, to land in China. There could be
nothing like the airborne invasion of Iraq or Afghanistan. There
could be no seaborne softening-up of any Chinese beaches. Nor
could any Tomahawk missile strikes rain down on Chinese cities
from flotillas of American Aegis-type destroyers positioned
close by in offshore waters—because such waters would already
have been closed off, proactively.

The American doctrine of shock and awe would be in trouble.
Before the massive battlements of the new Great Wall being constructed by China, any such tactic would simply wither and die.
Such is the gloomy realization that convinced Andrew Marshall
and his teams to construct what they felt were the necessary
countermeasures.

The resulting concept of Air-Sea Battle, which reduces any
army component in the coming contest and concentrates instead
on the abilities of the navy and air forces, would have at its core

two aims: first, to weather and survive operationally undimin-
ished any initial Chinese attack; and second, and more important, to launch from submarines and stealth aircraft a series of
devastating pinpoint strikes against the very radar and missiles
that would keep Americans away from the Chinese coast. The
huge American bases that dominate the western Pacific today
(Kadena, Yokosuka, Sasebo, Kunsan, Osan) would be supple-
mented by nimbler, better-protected forces on distant islands
such as Tinian and Palau, from where the counterattacks would
come.

All U.S. policy would thus be predicated on the absolute need
for the surviving American forces to secure access to Chinese waters, to defeat China's A2/AD abilities, to get in, to get
stuck in, and to stay. Once having secured the vitally necessary
access, American forces would bring to their knees any Chinese
who dared oppose them. More than two hundred specific components, most of them secret, pepper the pages of the Air-Sea
Battle manuals. And the People's Liberation Army is mentioned
in the document almost four hundred times—lest anyone doubt
the document's central aim. It clearly states the need for more
ships, more submarines, a second aircraft carrier strike group
for the Seventh Fleet (which manages the region), better and
stealthier aircraft, better missiles, revitalized small bases, and
smarter bombs.

Small wonder that Andrew Marshall's critics have proclaimed
that he didn't run the Office of Net Assessment, but instead the
Office of Threat Inflation. The U.S. Marine Corps, notably left
out of the plans, said the concept would be "preposterously expensive to build" and would result in "incalculable human and
economic destruction" if employed in a major confrontation.
The Brookings Institution insists the scenarios presented by
Marshall's office are fraudulent, the notion that China might
ever attack American assets just unimaginable.

Even with the Pentagon's assurances that no specific regime

was targeted in its various assessments, the Chinese, not unexpectedly, were incensed by this proposed policy. "If the US military develops Air-Sea Battle to deal with the People's Liberation Army," a Chinese colonel was quoted as telling the *Washington Post*, "then the PLA will be forced to develop *anti* Air-Sea Battle." In other words, an arms race could well develop, with all the risk and cost recalled from the last time, half a century ago, when the Soviet Union was the putative enemy.

Nonetheless the pivot, the rebalance, the Air-Sea Battle or its rebranded, kinder and gentler-sounding Joint Concept for Access and Maneuver in the Global Commons, has now become a firm component of American foreign policy, a doctrine for the new Pacific fully accepted by President Barack Obama. The language mandating its accomplishment was inserted into the National Defense Authorization Bills for the first time in 2012. Money is a perpetual problem: bitter ideological fights within the U.S. Congress delay or frustrate some aspects of the plan. Worsening troubles in and around the Middle East also delay matters. But in recent months, the first manifestations of the pivot are becoming visible, if only vaguely, in the outer reaches of the ocean.

In northern Australia, for example, there is now a semipermanent garrison of U.S. Marines operating out of an Australian army barracks near Darwin. The first group of 250 arrived in 2013. The eventual plan is for a quarter-century arrangement, with fully 2,500 battle-ready marines and a number of heavy-lift helicopters on station for six months each year. It is a fair-weather deployment, however: all but a skeleton group will go back to Hawaii during the midsummer wet season, when the tropical Australian weather makes training nearly impossible.

In Western Australia, there are now talks about allowing the U.S. Navy to use its Stirling navy base outside Perth and, if agreements are reached, of extending the base runways on Garden Island to allow the deployment of American bombers, and of enlarging the docks to allow the supply and repair of American aircraft carriers.

Singapore has agreed to allow the temporary stationing at its port of four of the U.S. Navy's newest close-in fighting craft, the so-called littoral combat ships. These vessels (small, stealthy, fast, lightly armed), with helicopters and room for small detachments of marines, are specifically designed to deal with the kind of area-denial abilities that the Chinese are developing nearby. They are believed to be the best answer to the new realities of the South China Sea, and the agreement to station them in Singapore is ideal, now that Subic Bay has been denied to America since 1991.

(Except that Subic may now not be entirely denied. Since the beginning of 2013, the Philippine government has agreed in principle to let some limited numbers of American ships use the facilities at Subic, facilities that have been somewhat refurbished in the years since the Pinatubo eruption. What remains there today is decidedly a Philippine base; but under intense diplomatic pressure, and on the payment of substantial sums of money, Manila has recently agreed to play its own part in the American Pacific pivot. Some ships will be allowed to operate, if only on probation, from this elderly but once again crucial naval headquarters.)

So the list of enhancements goes on. More American attack submarines are to be based in the far west of the ocean. More SSBNs (super-secret ballistic missile subs) are being sent on deep-water patrols in the Pacific.* A marine amphibious-ready group will be brought from the Atlantic into the Pacific. New

---

* Seldom does one see an American ballistic missile submarine—unlike in Russia, where Delta-class strategic nuclear missile subs are plainly on show at the Russians' North Pacific base in Petropavlovsk-Kamchatsky. I was at Pearl Harbor when an enormous Ohio-class boat arrived, for unexpected repair. A massive security blanket was suddenly imposed; heavily armed sentries were posted all around the dock where the great black craft was moored. Cameras were forbidden, roads were closed, passersby were told to *keep moving, nothing to see here* . . .

protective missiles, such as the Patriot and THAAD systems, will provide umbrellas for forces in Okinawa and Guam. Talks have begun—and the irony is inescapable—with the government of Vietnam, to see if America might once again use the wartime facilities it built long ago at Cam Ranh Bay.

And within the area, there is an additional proposal for an even newer balance, for a pivot *within* the pivot. The Center for Strategic and International Studies, which is often (but not invariably) in lockstep with seer employees such as Andrew Marshall, and tries to provide coherent and measured advice for the Pentagon on such matters, declared in August 2012: "Current U.S. force posture is heavily tilted toward Northeast Asia, to Korea and Japan, where it focuses properly on deterring the threats of major conflicts on the Korean peninsula, off Japan, and in the Taiwan Strait. *However, as evidenced by recent Chinese activities in the South China Sea and throughout the Pacific islands, the stakes are growing fastest in South and Southeast Asia. To be successful, U.S. strategic rebalancing needs to do more in those areas*" (my italics). Hence Singapore. Hence Vietnam. Hence Australia. And hence its most critical base, now being reclaimed from the volcanic ashes, the Republic of the Philippines.

Hence the American relegation of the Northeast Asian problems (specifically those relating to North Korea and to the Chinese-Japanese arguments over the Senkaku Islands) almost to the status of sideshows. Many academics still consider the North Korean problem to be the greatest threat posed in the western Pacific. These same specialists also consider the possibility of a war between Japan and China, and over disputed island territory, to be similarly exacting. Both these possibilities are seen as alarming in Dubuque and Oshkosh, and Baton Rouge and New York, simply because America is treaty-bound to become involved. Because of this, both stories have lately occupied American newsmagazines; television news producers who are obsessed with the Orwellian imagery of goose-stepping soldiers and stone-faced dictators devote much airtime to the nightmare possibilities. And to be sure, both problems are real—they may well become acute, they may well flare up, they may well result in an abundance of casualties.

But more cautious planners believe that, in the greater scheme of things, such possibilities are no more than distractions from the core of the region's present-day challenges. And that core is undeniably now represented by China. The more than 1.3 billion people of China—together with their nation of such formidable antiquity, equipped as it is with a deep understanding of the history and the context and the cycles of human time—are now central to understanding the Pacific's future. What makes China so vital now are of course the familiar and visible things: the country's new-made wealth, its relentless expansion, its very real power, its fast-growing and subtly directed influence, its sense of pride and self-esteem, its disdain for those outside and beyond its pale, and its stated need for a just measure of Pacific lebensraum.

Most dangerous of all, as Washington sees things, is China's rising philosophical challenge to the current notion that America, and America only, is the nation that best deserves, most needs, and most firmly demands to run the business of the Pacific Ocean, as it has done for the century past.

And here China's long-term wish runs precisely counter to America's long-held strategic objective in the Pacific, which is, to quote officially, "to maintain a balance of power that prevents the rise of any hegemonic state from within the region that could threaten U.S. interests by seeking to obstruct American access or dominate the maritime domain."

The unyielding American belief is that its dominance of the maritime domain has been central to the success of the various economies of the region. Naval officers at Pacific Command headquarters in Hawaii are keen to show off slides of the glittering skylines of Singapore, Hong Kong, Tokyo, Seoul—and yes, of

Shanghai and Guangzhou, too—and then to assert that only by dint of the U.S. Navy's keeping the trade routes free and the sea-lanes open has such prosperity become achievable.

American protection of trade, American suppression of menace, American export of its values—these are key, they say, to the recent wild success of Asia's tiger economies. Abandon America's sovereign care of the Pacific, the message continues—dare to surrender the care of the ocean to another hegemonic state, to a state capable of such excesses as the events of Tiananmen Square, and you risk allowing Asia to sink, deflate, decline, decay, and die.

This kind of assumption, though, is these days being questioned, as the Pacific, now a test bed of such new political and geostrategic ideas, breaks free of what an Australian writer, Coral Bell, recently called "the Vasco da Gama era." For five hundred years, since this one Portuguese adventurer left India and the Malay Peninsula, the East as a whole has existed under the influence of an omnium-gatherum of Western powers—of the Spanish and the Portuguese at first; then the Dutch, the French, the British; and finally, today, the apparently unassailable power of the United States.

These days the planet is witnessing a sudden and wholesale redistribution of world power, one that is unprecedented in its speed. It is experiencing a shift in emphasis that suggests that this Western dominance, especially in the regions where such was both unquestioned and unquestionable, may now, and quite rapidly, be coming to an end.

Old assumptions that were once calmly made are being cast aside, wholesale. As recently as the middle of the last century, for example, it was thought entirely appropriate for American ships to operate for a thousand miles up the Yangtze River inside China. To repeat an age-old question: Can anyone imagine the Chinese navy one day thinking it appropriate to run its new corvettes along the Mississippi, up to Hannibal, past Des Moines, to dock, without asking, in St. Louis? And then to insist that any of its sailors, after a drunken brawl, be subject not to American laws, but only to those back home in China?

The answer is nowadays, and in America, an unequivocal no. But can one imagine the Chinese navy exercising its ships, firing its missiles, testing its submarines, patrolling the shipping lanes, not just off the Philippines, or off Guam, Palau, or Vanuatu, but off Hawaii, say? Or off the coast of California, within sight of the skyscrapers of San Diego or San Jose? Quite possibly, one has to suppose. Even, given the rate of expansion of China's navy, quite probably.

And would that necessarily be a bad thing? Would the Chinese navy have a lesser interest in protecting the sea-lanes than the Americans do today? Further, might not a policy of Asia for the Asians—with all local differences eventually set aside, enabling all in Asia to focus more directly on the competing hegemonies of the non-Asian world—offer greater stability for the region and beyond? We in the West surely do not doubt the competence of those who inhabit the East, so why not allow them to have their turn, to direct the future of the ocean where East and West so obviously meet? Do we somehow not trust them to run it as well? Do we feel offended that our power is not as keenly wanted now as we once believed it to be? Is it envy, or a simple stubborn clinging to the accumulated assumptions of empire, of power, and, dare one say it, of racial supremacy?

To suggest such a thing may be a little short of heresy.

And yet it would have been an equal heresy in 1991 for anyone to suggest that Mount Pinatubo would ever explode, and unimaginable that the consequences of its doing so would have so profound an effect on the region's politics. But in history, just as in geology, the unthinkable and the unimaginable tend to have a way of their own. The same can be said in China, the Philippines, or America. Or, for that matter, anywhere in the immensity of the oceanic Pacific lying in between.

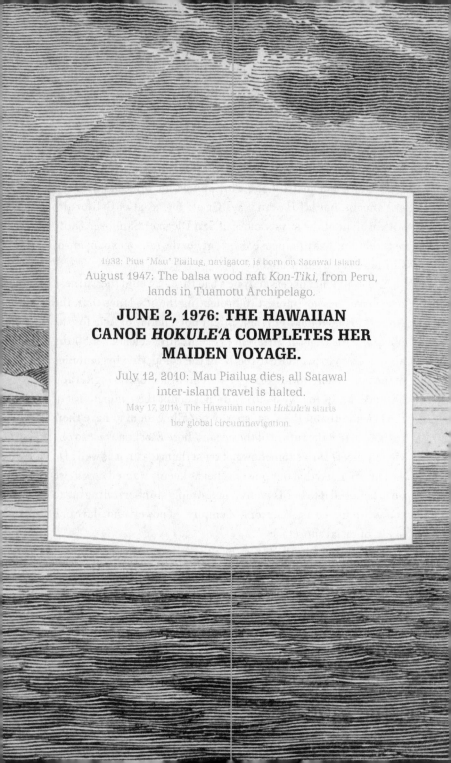

1932: Pius "Mau" Piailug, navigator, is born on Satawal Island.

August 1947: The balsa wood raft *Kon-Tiki*, from Peru, lands in Tuamotu Archipelago.

JUNE 2, 1976: THE HAWAIIAN CANOE *HOKULE'A* COMPLETES HER MAIDEN VOYAGE.

July 12, 2010: Mau Piailug dies; all Satawal inter-island travel is halted.

May 17, 2014: The Hawaiian canoe *Hokule'a* starts her global circumnavigation.

EPILOGUE:
THE CALL OF THE
RUNNING TIDE

———◦○◦———

The ancient Hawaiian phrase *malama honua* speaks of humanity's duty "to care for our island earth." On a warm Saturday evening in mid-May 2014, a venerable sailing craft charged with spreading this simple message pulled away from her dock on the island of Oahu, her twin sails filling on a steadily gathering breeze. She was bent on a mission that would take her the next three years, making passage out from Polynesia and traveling clear around the world.

She was the sixty-ton, forty-year-old, twin-hulled deep-sea sailing canoe *Hokule'a*, named for the Hawaiian word for Arcturus, the star of gladness, the brightest star in the northern night sky. The boat had been launched in 1975, hand-built to the traditional design of a Polynesian long-distance *wa'a*. Her planned journey was powerfully traditional also, but in a particularly technical sense.

Hokule'a, a traditional Hawaiian *wa'a*, or sailing canoe, built on Oahu in 1975, was able to show, with a Polynesian crew, that it was possible to cross long reaches of the Pacific without any kind of navigational instruments. In 2014 the craft embarked on an attempt to circumnavigate the world—with neither clock, compass, sextant, nor GPS.

That May evening, when the craft was untied from her dock and, with great fanfare and conch blasts and prayers and other sunset celebrations, set to sailing, she embarked on a venture made under unusual conditions: she and her crew of thirty were sailing for forty-seven thousand miles across all the world's oceans *without the use of any navigational instruments whatsoever.*

They were taking no compass. No sextant. No radar. No radio. And certainly no GPS. They would sail alone, unaided, as their predecessors sailed across this very ocean, many centuries before.

The crew, most of them clear-eyed youngsters from Hawaii, but with weather-beaten old salts aboard to offer counsel, had trained exhaustively for the challenge. They are still sailing as I am writing, their doughty little craft dipping her way steadily westward to the north of Australia, now nine thousand miles from home, and with nearly forty thousand miles left to go before sighting Diamond Head again. They have left the Pacific behind. The crew have now to divine their way across seas—the Indian Ocean, the Mediterranean, the Atlantic Ocean, the Caribbean Sea—that are very different from their home waters. They will pass beneath skies and patterns of stars quite strange to them.

Whether or not they succeed, those aboard all keenly believe that their simple attempt will serve as a powerful reminder of the sea's singular importance. That is what all on the boat and back in Hawaii believe lies at the heart of their venture. *Malama honua*: that all should be urged to care for a body of water that nourishes every living thing on earth, that gave it life in the first place, and yet that is now wearily compelled to absorb all the excesses of the humans who live beside and around it.

Back in 1975, when the *Hokule'a* was being built in a beachfront boatyard at Kualoa, on Oahu's windward side, much uncertainty still lingered about how this area of the central Pacific had first been populated. Thor Heyerdahl's voyage in his Peruvian balsa raft *Kon-Tiki* was still fresh in many minds. His theory that the mid-Pacific islands had been populated accidentally, by drifting boats from South America, was beguiling—for how else could the American sweet potato, for instance, have wound its way so firmly into the Polynesian diet?

But by the 1970s, his assertions were losing ground; his ideas were being ridiculed, in the face of much other evidence, both genetic and linguistic, together with a growing belief that Pacific islanders had once had supreme navigational skills. These skills, it was thought, could well have brought the Melanesians, the Micronesians, and most especially the Polynesians to their mid-ocean homes not by accident of drifting, but by deliberate feats of good seamanship. Moreover, their navigation had brought them not from the far shores of the Americas, but from Asia.

If this was true, the feat of these seaman was considerable, the stuff of legend. Five thousand years BP (using the new dating convention with which I began this book; 3000 BC as it used to be), islanders from the archipelagos of the Philippines and in the South China Sea had started moving both westward and eastward in canoes, populating as they did islands in the Indian Ocean as far away as Madagascar, and in the Pacific Ocean as distant as Easter Island. The evidence for their having done so is overwhelming: pottery shards, imported animal types, and words of common origin, together with the kinds of boats they used, of which images appear in ancient pictographs. They sailed, they paddled, they established communities, they planned return voyages, they traded, they planted gardens, and they fished.

The southwestern Pacific Ocean was one of these migrants' major destinations and was where, thanks to the accident of tectonic movement explained on earlier pages, the islands are arranged in rough diagonal lines, somewhat more tightly bunched together than elsewhere. To get from one group to another seldom requires a voyage of more than three hundred miles. To the early Melanesians and Papuans, as these first migrants turned out to be, such may well have seemed formidable expeditions, but the distances were not insuperable. We do not think of those who made these island-to-island voyages as mid-ocean rovers. Rather, we see them as they see themselves: as a landed people separated by expanses of sea.

But the achievements of the early people of Polynesia are quite different, and are such as to set them very much aside. The distances for them were truly prodigious. Their present-day homeland, roughly a triangle bounded by Hawaii in the north, Easter Island to the east, and the hulking mass of New Zealand to the west (or, to use the Polynesians' names for these last two, Rapa Nui and Aotearoa), is almost entirely marine, nearly all aqueous. Perhaps a new word should be coined for where such people live: they inhabit not a homeland but a *home sea*, in this case an expanse of fourteen million square miles of ocean, dusted with islands that are few and almost all far between. With the single exception of New Zealand, one statistic underlines the way in which Polynesia is so different: for every two square miles of land, there are fully a thousand square miles of sea, making it remarkable that so much of the land is inhabited. For how on earth (though perhaps not the best turn of phrase) did all these people get there?

Captain James Cook had an early inkling. In the Tahitian Islands, on his first voyage to the South Seas in 1769, he took aboard a Raiatean priest named Tupaia and discovered that the holy man had a quite extraordinary knowledge of all the surrounding islands, and an uncanny ability to tell directions from one to the other. Joseph Banks, the *Endeavour*'s resident scientist, was amazed at Tupaia's talents: "[W]hat makes him more than anything else desirable is his experience in the navigation of these people and Knowledge of the islands in these Seas; he has told us the names of above 70, the most of which he has been at."

Tupaia, who traveled with Cook all the way to Indonesia, then astonished the captain still further by drawing him a map of the ocean that extended for more than two thousand miles, side to side, and showed all the major island groups (Fiji, Tonga, the Marquesas) that Cook hadn't thus far found himself. This remarkable man died of fever in Jakarta (Batavia, as it then was). But he left Cook and Banks with a firm impression that there were navigators on these islands who had special skills, who could get themselves around the ocean without the use of clocks, sextants, compasses, or any of the newfangled devices that were enabling Westerners to explore and conquer the world. Men such as Tupaia did it all by looking at the sea, at the stars, and at the passing fish and birds. From the evidence of nature and nature alone, men like him could wander their world at ease and settle its islands with deliberate purpose.

But for the more than two centuries that followed, the West was openly scornful of this notion. No one could be so clever, they thought, as to sail with purpose across seas so vast as these—certainly not people so backward and savage as the near-naked islanders they encountered in mid-ocean. No doubt they were excellent boat builders, their canoe constructions of great elegance and strength, their sails and paddles and arrangements of outriggers enabling them to slide through the waves with elegance and speed. But as for navigating across thousands of miles of open water to settle themselves? That was just impossible to imagine.

Even Captain Cook, on his subsequent voyages, said he thought that most of Polynesia had been settled not on purpose, but by castaways. The island populations he and his brother explorers happened upon in all their later expeditions could have reached such places as Hawaii and Easter Island only by drifting there, by allowing the currents and the winds to take them, and they had settled down only by luck and happenstance. Their journeys were invariably one-way only; were haphazard in their routes and random in their destinations. This was the fixed belief into modern times; *Kon-Tiki* was an attempt—a failed one, as it happens—to prove that the islands had been populated by drifters from the Americas. Accidental driftings from somewhere—that was the belief.

This thinking was careless and patently absurd: the currents and winds would actually make it impossible for a Polynesian craft coming from the south to happen upon Hawaii. But it seldom occurred to those who discounted the islanders' acumen. The stubborn minds of most Western intruders simply could not grasp—largely because of their belief in their own racial superiority—that any skills other than their own were worthy of consideration or respect.

But as it happens, navigating was a well-honed skill in Polynesia; had been so for hundreds of years; was an integral part of the sagas, songs, and poetry of the islanders; was taught and learned in schools; had a name (the skill known as *ppwo*, its initiated practitioners as *ppalu*); and was key to all travel and settling within the Polynesian triangle. Yet ironically it was a skill that had swiftly diminished as Western empire builders ever more firmly impressed themselves on the Pacific, and mainly for reasons of colonizers' administrative convenience.

The locals' navigational skills started to vanish—a self-fulfilling prophecy—for the simple reason that incoming Westerners banned inter-island canoe travel by their subjects. The Germans, the Japanese, and in some places also the British and the French decided that it was far too dangerous to have island boats set off at will on voyages that were not sanctioned in advance. As well as being risky, incurring maybe a charge on the parish or on insurance firms, it might well require the carrying by crewmen of passports or laissez-passers or other permissions. One Polynesian island might be run by the German authorities, another by the British, a third by the French—it would create mayhem if islanders were to paddle and sail between them willy-nilly.

All of a sudden the informal and unstructured Polynesian nation, which inhabited areas of ocean as large as Russia, or the size of the North and South American continents combined, was quite swept away. Acting, as they would say, in the best interests of order and good governance, the new European masters kept their islanders on the islands to which they, the Europeans, felt they belonged—and as a result, for the islanders, something of their precious essence died.

The main casualty was the skill of the navigator. The elders—much revered in Pacific cultures, since they were the possessors of what some today call "the original instructions" on the conduct of traditional life—allowed their knowledge of the sea to wither away. By the 1970s it was said to be essentially moribund. Pacific Islanders had become involuntary components of twentieth-century culture, and their ancient sailing skills—wayfinding, as it was known—had all but vanished.

Except that, as a group of eager young Hawaiians discovered in the early 1970s, there were one or two *ppalu* still living who knew the secrets. One was a forty-year-old Yapese named Pius Piailug, who had been initiated in the mysteries of wayfinding when he was six years old. He was revered as a master navigator on the island of Satawal, in the Carolines. Though he was still young enough and spry enough to think of imparting his knowledge to others, few on his tiny island—only four hundred people then lived there, eking out a living fishing and gathering coconuts—seemed to want to learn, and he fully expected that his knowledge would die with him. He would be among the last of the navigators in the Polynesian world.

Then came a curious concatenation of circumstances. He took a job as a seaman on an inter-island steamship—by now he had assumed the nickname Mau, signifying his strength as a master outrigger handler. While at sea, journeying among Yap and Chuuk and Kosrae and a host of smaller Micronesian islands, he met and became friends with an American Peace Corps worker named Mike McCoy. The two talked often of the secret mysteries of the noninstrumental navigation that Mau had learned. McCoy, fascinated, then wrote an essay about the technique. The essay attracted the attention of an American anthropologist in Honolulu, Ben Finney, who was just then playing his part in the so-called Second Hawaiian Renaissance, a somewhat faltering movement trying to remind native Hawaiians of their heritage.

The connection led to an unexpected invitation that landed one day in 1974 at Mau Piailug's front door, and that had to be dictated to him, since he read no English. Would he care to be flown up and across to Honolulu, to help plan Hawaii's celebration of the American Bicentennial? For it happened that a new-formed group of Hawaiians, with Ben Finney among them, and all eager to display themselves as supporters of the Polynesian cultural diaspora, had decided on an expedition. They would try to make a twenty-six-hundred-mile journey from Hawaii to Tahiti in a traditional sailing canoe, using no navigational instruments at all. No such thing had been accomplished for six hundred years. Yet having a skilled navigator such as Mau Piailug aboard should give the project a good chance of success.

Mau jumped at the chance. He agreed, and flew right away to Hawaii, first to meet the men and women who had established the Polynesian Voyaging Society, which had organized the building of the *Hokule'a* and had then convinced the state government that, with an inaugural celebratory navigation to Tahiti, a renewed sense of pride could be instilled in the native Hawaiian people. If this voyage succeeded, the Voyaging Society argued, thousands of potentially disaffected native Hawaiians (the young, most especially—members of what back in the 1970s was a forlorn underclass of Hawaiian Islands society) might well rally to the cause. The venture had romance and beauty to it, after all. Some might indeed become navigators themselves, and all might reassert their newfound personae not so much as Americans, but now as reborn Polynesians.

To that extent, the *Hokule'a* was to be a symbol. Mau Piailug would be the man who would first help the craft and the crew sail the twenty-six hundred miles south to Tahiti. He might also teach his skills to the canoe's crew and to others, so that the

Hokule'a could evolve from being simply a symbol and become a catalyst for a profound and ocean-wide cultural reaction.

The basis for noninstrumental navigation, hard to learn but impossible to forget, is a deep knowledge of the sea. Navigators had to know the *feel* of the ocean, how its waves and swells pressed against the hull of a boat. Those who knew Mau when he was back down in the Caroline waters said he was often to be found lying, apparently asleep, deep down in his boat's bilges. When approached, he would raise a hushing finger to his lips: he was carefully listening, hearing, *feeling* the waters rumble beneath him, learning from their sounds how they were going, which way, at what strength. There was much more to it of course. A navigator had to have an intimate knowledge of the stars, had to be familiar with all winds and the caprices of weather, had to know much about the wildlife (birds, most especially) that was to be seen in and above the ocean.

All were equally important—except the stars were the most important. And in that one regard, Mau Piailug was for a while found wanting. The stars he knew, and the ones his brother navigators in Hawaii and the Carolines knew, were those of the Northern Hemisphere. Tahiti, however, was south of the equator, and the skies there were different, and wholly unknown to the man from Satawal.

Yet there was an easy solution. The Bishop Museum in Honolulu had a planetarium, and for several weeks in 1975, Mau and his would-be navigator colleagues went there to study the stars of the Southern Hemisphere, as the projector whirled them across the horizons and through the zeniths and meridians and along their azimuthal arcs. The misshapen cruciform pattern of the Southern Cross became the new lodestar, replacing Polaris of the northern horizon. By the early summer of 1976, Mau pronounced himself sufficiently familiar with the skies to begin the journey.

The *Hokule'a* was ready, too. In appearance she was entirely traditional: she had been built to resemble two large voyaging canoes pictured in Captain Cook's journals, which he had seen in Tonga and Tahiti two centuries before. But the construction was not entirely traditional: performance accuracy had been the byword during her building, so while there was plenty of bamboo and oak, sailcloth and coir, there was also nylon and fiberglass, Dacron and plywood. Her hulls, however, had been given their enormous strength by immense timbers sent down to the Hawaiians by a group of Alaskan Indians, in a gesture of pan-Pacific kinship.

On May 1, 1976, her twin claw sails were unbrailed, and after ceremonial quaffing of 'ava—their last alcohol, because from now on, both strong drink and the company of women were forbidden—the craft set sail from Maui. There were two brawny Hawaiians at the steering sweeps, and the diminutive Mau Piailug at the helm. He was to guide them over the huge distance into the Southern Hemisphere and to the gathering of French possessions that were their intended target. They were seventeen men aboard, together with a dog and a pig (which were both spectacularly seasick for the first week of the trip), and two chickens. They also took, as gifts, plants they hoped would flourish on Tahiti: breadfruit seedlings, yams, tiny coconut palms, and mulberry tree seeds. Their diaries and the ship's log they kept have a curious absence about them: they were never able to record the times of any of the happenings aboard, since they had no watches or clocks. As with all the instruments that could possibly help them—and from an analog watch, of course, one can always deduce direction—these had been left behind, on land.

Mau Piailug assembled a hammock of knotted cords and slept outside, at the boat's stern, for the entire journey. He spent his days contemplating the swells and the clouds, feeling the breeze, watching for birds. He would occasionally nudge the steersmen to turn their sweeps a little this way or that. At night he was focused on the stars, noting where they rose and set, measuring with his hands the angles at which they curved their way across the sky. Once the canoe had passed across the equator and the four arms of the Southern Cross began to appear on the horizon, he murmured with pleasure: it was just as the Bishop planetarium had predicted, the sky not as unfamiliar as he had feared.

He judged the speed by eye, the distance traveled by no more than dead reckoning. Each day at noon—the sun would tell him it was noon, in its brief overhead pause at its zenith in the sky—he would announce to the rest of the crew just where they were, usually in reference to a cape or a headland left far behind. As the voyage progressed, he forecast the number of days left before the *Hokule'a* left the trade winds, or entered the doldrums; and then he finally offered up a bearing for the Tuamotus, and then for Tahiti. The crew was perpetually astonished: the ancient technique still worked, flawlessly. Mau could guide them over all these thousands of miles of trackless water, knowing all the while exactly where he was, and almost precisely when he would reach his destination.

It took exactly a month for the craft to make it first landfall. The swells were suddenly interrupted. A pair of terns flew by. A line of black was etched onto the horizon shortly before dawn on June 1, 1976. The razor-sharp reefs, adroitly avoided, encircled the tiny Tuamotuan island of Mataiva, which was reached on time and exactly as predicted. Tahiti was just 170 miles away from here, and the *Hokule'a* did the journey in one more day, reaching the port of Papeete and a wild celebration of sheer ecstasy, as every pure Polynesian within a hundred miles came down to the dock to offer congratulations and profound thanks for a journey well accomplished. Seventeen thousand people, half the island's population, turned out.

Perhaps no other state in the Union could have offered a more satisfying celebration that Bicentennial year—yet it was more a celebration of Polynesian than American identity. This point was not lost on the Hawaiians when the boat returned later that summer: the spirit of the Hawaiian Renaissance, inaugurated so recently, took immediate hold, and still flourishes today.

The *Hokule'a* has made many journeys in the years since. A young native Hawaiian named Nainoa Thompson, who helped bring the boat back to Hawaii after that first voyage (and who used navigational instruments on the return trip, to test the mixing of the two techniques), would become Mau Piailug's heir and successor. Under his captaincy, the *Hokule'a* has made ten more journeys, most as successful as the first,* though to much more distant destinations. Nainoa Thompson made his own first solo voyages to and from Tahiti in 1980; after that there were journeys across to Tonga; to the west coasts of America and Canada; a very trying trip over to Easter Island in 2000; then, in 2007, to Yap and to Satawal, where the now ailing Mau Piailug—he would die in 2010—was presented with an award for having initiated all this; and then on to Japan.

Here came an epiphany for all. For as the boat slid into and out of port after port along the southern Japanese coast, she transmitted a singular message, if unwittingly, to the Japanese people. The message transcended the simple novelty of the event, remarkable though it might be, of a ship being sailed all the way from Micronesia to Yokohama without anyone aboard having a compass, a sextant, or a watch. What so impressed the Japanese was the realization, the reminder, that they were a Pacific people, too, just as were the Polynesians and the Microne-

---

* The second attempt to reach Tahiti, in 1978, ended in tragic failure when the canoe capsized. A young Oahu surfer named Eddie Aikau tried to find help by surfing the ten miles to shore. While he was gone, a passing Hawaiian Airlines passenger jet saw the *Hokule'a*'s distress flares and summoned the Coast Guard, who rescued everyone—except for Eddie Aikau, who was never seen again. To avoid such recurrences, the *Hokule'a* has since always made her long-distance voyages with a support ship in distant attendance— in the case of her current circumnavigation, she was first followed some miles away by the *Hikianalia*, a similar-looking canoe, but one equipped with modern instruments and radios. For the Indian Ocean, an even bigger craft was chartered as a safety boat, and the *Hikianalia* was brought home.

sians, the Papuans and the Melanesians. They were all somehow united—practically, yes, but mystically also—by an ocean that should rightly be regarded as more of a bridge than a barrier.

Many who came to see the little boat seemed mesmerized by that view: a Japanese people who had configured themselves for centuries as Asian, since Asia lay at their back door, were now transfixed by a newer notion: that they were also a people connected with the ocean that lapped so temptingly at their nation's front door.

Now, after all the forays into the huge home ocean, the *Hokule'a* is embarked on her *malama honua* voyage, one-quarter completed at the time of writing, and with the hope of transfixing the entire world. Hoping, on one level, to astonish outsiders with the thought that such navigation is possible; but also to remind them of familiar things—that there is but one world ocean; that there is just one world; and that all of us are one, passengers on the same planet, challenged by matters that are common to us all. Familiar messages—and I mean no disrespect by setting them to one side.

It seems to me there is even more potent symbolism to the *Hokule'a*'s journey, symbolism that relates quite specifically to the ocean where the boat was born, where her crew members revived and then learned their skills, and from where she came to venture out to the rest of the planet. The Pacific occupies a unique position among the world's seas; the *Hokule'a*'s journey has served as a reminder of why.

I suspect that anyone who learns of this little boat's past doings, and of her present journey, comes away amazed. The thought that hidden still within our present world's technological triumphs (triumphs that most Eastern cultures are naturally eager to embrace), an ancient skill survives that can help humans perform a task of such infinite complexity must surely come as a revelation. The achievements of the navigators seem little short of miraculous. They cause us to pause and draw breath. We revere what they have done, revere and respect the existence of a kind of natural wisdom that we mostly imagined had been exhausted by time and progress.

Yet the simple fact that it does still exist, and here in this one ocean, prompts a greater question.

Mankind—or, more specifically, Western mankind, legatees of the belief that the planet is theirs by unchallengeable right—has spent his past three thousand years on a civilizing tear, ever bound toward the sunset. He has gone from the Fertile Crescent to the Nile, from the Levant to the Pillars of Hercules, from the ports of the Old World to the shores of New. More recently, empowered by the mandates of Manifest Destiny, and careless of those who inhabited the newly found lands of the Americas, he lurched still farther on, to the shores of the Pacific.

Balboa saw the ocean, Magellan embarked on its first crossing—and for the five succeeding centuries of relentless expansion, Western man sought to dominate, exploit, and own some or all of the vast new fetch of peoples and cultures and attitudes and gods who lived on and around this sea. A gathering of peoples who had been living there, and generally quite peaceably and contentedly, for the same thousands of years during which the Westerners planned and executed their own expansion.

It is worth reminding ourselves that those in the East, be they Chinese or Japanese, Koreans or Filipinos, Australian aboriginals or long-distance-sailing Polynesians, seldom sought in ancient times, at least in any significant way, to make similarly expansive imperial journeys of their own. Extraordinary journeys, yes—the ancient Polynesian wayfindings most memorably of all. But these journeys were not as bent on conquest and dominion as were most of ours. We tend to deride them for this, for their demonstrable parochialism, for what we might see as complacent and incurious insularity. The rhetorical inquiries that follow—"Why didn't the Chinese find us before we encountered them?" being the most familiar—tend to have a condescending presumption about them, freighted with the implication of a Western belief that our simply having journeyed to them is proof positive of our right to claim them and their territories as our own. So our empires flourished, our churchly missions grew, our wealth accumulated, and our assertion of a prescriptive right to rule held sway for centuries.

Today, of course, much has changed. Empire has now become a tawdrier notion, many of the legacies of European rule have come to be seen as ruinous, and apologies have since been sought and occasionally given. Some religious zeal clings on: America's Mormon Church, for example, is ever more active in the Pacific these days, with its elders energetic and everywhere, and with authoritative-looking texts to be quoted justifying their presence. And commercial enterprise is inevitably everywhere, too, peddling everything from iPads to Spam, Coke to Cheerios, and unstoppably. But otherwise we have largely withdrawn from our superintendence of the ocean's western peripheries, and we are examining now, if somewhat reluctantly, the need to retrench and reconsider our hold on the waters of the sea itself. And as we do so, in the ebb tide we seem to be assuming a more contemplative take on matters.

We begin to encounter and appreciate such unexpected things as the mysteries of long-haul journeying. We begin to apprehend anew something of the miracles of the kind of knowledge and wisdom—Eastern wisdom and knowledge, we perforce must call it—that we have for so long quite overlooked, in our imperial rush, in our hitherto unshakable belief in our own kingly virtues.

The Pacific is the ocean where, quite literally, East meets West—though the ironic reality of its geography has the East in the west and the West in the east, playing a kind of compass trick on us. It is a trick that adds to the slightly unsettling paradox that the Pacific is in fact the least pacific of all the oceans: storm and war and seismic mayhem have roiled it to a far greater degree than poor Magellan, its christener, ever imagined. Much in these parts, in short, is not at all what it seems.

As we begin our slow backing out from this territory, much of what Westerners (Americans, most particularly) say and seem to believe presently displays a ring of defiance. We speak— indeed, this very book's subtitle speaks, and it does so because this is a reality of today's Pacific—of the fear of a coming collision between East and West, that there is challenge in the air, the sound of clashing swords and angry words. Some of us continue to cleave to the view that East and West are natural foes, peoples about whom Kipling so infamously said two centuries ago, "never the twain shall meet."

The existence of the entity we call Polynesia, with her generally peaceable past, her wealth of undiscovered skills, the long survival of her people, hints that such beliefs as ours, in the permanent immiscibility of far-flung peoples, is a racial assumption that need not be so. We should perhaps amend our view, allow it to evolve, or else discard it. The Pacific should perhaps not be a place where, after years of conquest and dominion, we now only fear confrontation and collision. There has to be another model. Our new and more contemplative take on matters suggests what this should be.

With his long reach across this immense ocean, Western man, coming full circle, has now all but ended his journey around the planet. The two most historically divided peoples now stand face-to-face, wondering, waiting, evaluating, considering. Polynesia suggests that what those of us on the ocean's eastern side should now be doing, rather than considering competition, is something radically different. We should be *learning*.

Instead of *aircraft carriers* and *pollution*, *garbage gyres* and *coral bleaching* being the bywords of our presence, there should now be a fresh kind of lexicon. *Respect, reverence, accommodation, ad-*

miration, and *awe* for much that the East stands for—all these should now be the new watchwords. For from these ancient, calming cultures, there is very much more to learn and absorb than there is to fear and resist. The benefits of Western modernity are quite obvious and should be sought after by all. But the wise benefits of antiquity should not be discounted, either, and we should be readier to embrace them, as counterpoise, as leavening. This is what the Pacific should teach us.

Meanwhile, the *Hokule'a* lumbers her way slowly across the ocean, with half a world to cover before she heads for home. All the while, on her way back to the Pacific, she will be spreading the message central to us all, and which was born on this great sea, of *malama honua*: Take care of where we live. It is all we have or ever will have. It is precious. Learn from it, respect those who know and sense it already, and take good care.

The Pacific has the words for it: *aloha* and *mahalo*.

ACKNOWLEDGMENTS

Charting the complex fascinations to be found in the sixty-four million square miles of the Pacific Ocean was a task made immeasurably easier thanks to the help generously offered to me by Charles Morrison, president of the East-West Center in Honolulu. He gave me an office and administrative assistance during the six weeks I spent in Hawaii in the winter of 2014; and the use of these facilities—most notably the excellent library, which together with the Pacific Collection at the University of Hawaii, just up the road, makes for a truly incomparable resource—rooted me to my subject in a manner that would have been well-nigh impossible to fashion elsewhere. So my thanks must go first to Dr. Morrison, and to his staff and colleagues—June Kuramoto, Anna Tanaka, Phyllis Tabusa, Karen Knudsen, Elisa Johnston, Scott Kroeker, and Carol Fox in particular—who helped me lay the foundations for the making of this book.

I am also most grateful to the U.S. Navy for assistance offered in Hawaii and beyond. Commander Jason Garrett was the point man at Pacific Fleet headquarters at Pearl Harbor; and Lieutenant Colonel Eric Bloom at the all-forces Pacific Command headquarters at Camp Smith, nearby. Both officers bent over backward to offer help and access to various nooks and crannies of the byzantine world that is the U.S. armed forces in the region, and so far as I recall, not a single thing I asked for was denied.

The commander of the Pacific Fleet during my time in Hawaii was Admiral Harry Harris, who was later promoted to take over the entire Pacific Command: his courtesy and help to me were warm and personal, and I am most thankful. One of his senior advisers, Commander Jon Duffy, now at the White House, was helpful both at the time in Hawaii and subsequently when posted to Washington; he took time to read the passages concerning the rise of modern China, and made many useful and constructive comments. Naturally any errors of fact or interpretation are mine alone, and neither his words of advice, nor those of any others mentioned here, should be thought of as suggesting an official endorsement.

My visit to Kwajalein Atoll and to the U.S. Army's missile range operations there was arranged by Michael Sakaio and Shannon Paulsen, both of whom were hospitable to a fault during my stay with them. If I write critically of the local treatment of the Marshall Islanders, Mr. Sakaio and Ms. Paulsen will both appreciate, I am sure, that it is the policies that I fault, and not the personnel, who in their cases were kindness personified.

Scientists at the Woods Hole Oceanographic Institution on Cape Cod, Massachusetts, were especially obliging in sharing with me their research and knowledge of the ocean. I am grateful to Carl Peterson, a longtime trustee of the WHOI, for arranging visits and contacts; and to Susan Avery, the director. With Jayne Iafrate's assistance I was able to spend valuable time with, among others, Daniel Fornari and Adam Soule, discussing hydrothermal vents; with Maurice Tivey, an expert on deep-sea mining;

and with Ken Buesseler, a specialist on the sea-borne radiation effects of atomic testing and from nuclear-related accidents.

JAMSTEC, the Japanese government's principal meteorological research agency, could not have been more helpful; and for arranging visits to see some of its teams of remarkable weather scientists I must especially thank Ms. Mizue Ijima—who nearly missed the plane that would take her and her husband on a well-deserved holiday, to make sure I received timely information that I needed before my visit. The JAMSTEC scientists whose work I found especially relevant to this book were Kentaro Ando, Satomi Tomishima, and Takeshi Doi—the last an expert on the workings of Earth Simulator 2, the homegrown NEC supercomputer that endeavors to solve some of the more complex of the Pacific's weather conundrums.

Officials past and present at the Sony Corporation were unfailingly helpful when I was attempting to piece together the story of Masaru Ibuka and the company's first transistor radios. Hiroko Onoyama worked for many years as chief assistant to Akio Morita in New York; and Hiroko Maeda is with Sony USA today: both made introductions for me and arranged visits to Sony offices, archives, and showrooms in Tokyo that were enormously useful. I also grateful to John Nathan, professor of Japanese cultural studies at the University of California, Santa Barbara, who has written what remains probably the best popular history of the corporation.

The men and women of the Polynesian Voyaging Society, eager, enthusiastic, and brave, were tremendously welcoming and helpful whenever I chanced by their boatyard offices, as they prepared for their epic *malama honua* voyage of 2014. Marisa Hayase, who then acted as communications coordinator for the planned circumnavigation, was subsequently unfailing with her advice and timely assistance. May fair winds continue to attend all who are involved in the expedition.

Among the many individuals to whom I owe much for their

encouragement or assistance or both, I must thank the following: Kate Andrews, an Australian environmentalist friend of many years, who looked after me in Darwin and then read and helped tweak the relevant Australian chapters; Sasha and Marina Belousov, geologists in Kamchatka, who kindly took me to see the Zhupanovsky volcano while it was in full eruption; Simon Bowden and Dana Yee, for allowing me affordable use of their apartment in Honolulu; Mark Bradford, senior meteorologist on Kwajalein, and fund of information on tropical cyclones; Mark Brazil, who from his base on Hokkaido travels the world pursuing his environmental interests; David Christian, director of the Big History Institute at Macquarie University in Sydney; Gavan Daws, the writer, long based in Hawaii, who is a walking treasure-house of Pacific Island matters; John Dvorak, who runs a large university telescope on the summit of Mauna Kea, and is the author of a fine account of the San Andreas fault; the wise writer Gretel Ehrlich, who now lives in Hawaii with her husband and my old NPR friend Neal Conan; John Elias of Nautilus Minerals; Mary Hagedorn, an expert on hot-water corals; Kevin Hamilton, an atmospheric scientist and former director of the International Pacific Research Center at the University of Hawaii; Louise Hancock at the Pitt Rivers Museum in Oxford; Hiroshi Hasegawa, single-handed savior of the short-tailed albatross and a true hero of the avian world; Laurie Irvine of Soil Machine Dynamics in Newcastle upon Tyne; Elizabeth Kapu'uwailani Lindsey, an old friend, a true Hawaiian, and student of the *original instructions* of the world's elders; Kurt Matsumoto of Pulama Lanai; the talented writer Jon Mooallem, who made many noble attempts to unravel today's Lana'i story; my old friend and former Hong Kong government official Peter Moss; Jack Niedenthal, who from his base in Majuro acts as liaison for the displaced people of Bikini atoll; the geochronologist Professor Paula Reimer at Queen's University, Belfast; Kylie Robertson, a great Australian friend and publicist currently based in New

York; Tom Roelans, general manager of the Four Seasons resorts on Lana'i; Philip Smiley, one of the last British colonial officers in the Solomon Islands; Lori Teranishi, who is Larry Ellison's spokesperson in Hawaii; Kazuyoshi Umemoto, formerly Japan's ambassador to the UN, now in Rome; Charlie Veron, the world-revered champion of corals; Julianne Walsh, an expert on the Marshallese people at the Pacific Islands Studies Center at the University of Hawaii; and my son Rupert Winchester, of London and Phnom Penh, who kindly proofread the near-finished book and offered a wealth of corrections and invaluable suggestions.

This was a challenging book, both to research and to write; but the task was made much less daunting by the clear-eyed and wise counsel of my friend and HarperCollins editor Henry Ferris, for whose lexico-surgical skills I have the greatest admiration. I continue to believe that a readable book is the result of intimate team-work between editor and writer; and if this book comes to be regarded as readable, then it will stand as testament to the hard work that Henry Ferris put into it to help make it so. Both he and I were greatly aided by Nick Amphlett, his stellar editorial assistant at Harper, who attended to the myriad nuts and bolts of this project with great good humor and forbearance. So to all in the HarperCollins team in New York, as well as to my splendid London editor, Martin Redfern, and his colleagues there, I raise a glass, or several, in salute.

As I do also to my agents at William Morris: in New York to the redoubtable Suzanne Gluck, to her incomparable assistant Clio Seraphim; and across in London, to my great friend Simon Trewin. My sincerest thanks, and blessings to you all!

Simon Winchester
Sandisfield, Massachusetts
July 2015

NOTE ON SOURCES

Fuller details of the books I mention here are to be found in the bibliography. However, on those few occasions where I mention that a particular researcher or institution (such as Woods Hole Oceanographic) has produced "many publications" that are relevant to a particular topic, I have decided to leave it to the interested reader to undertake the necessary Internet search. To include references to everything written would consume a great deal of valuable space.

PROLOGUE

Discussions relating to the creation of time zones and the positioning of the International Date Line in the Pacific Ocean are to be found both in Clark Blaise's biography of Sir Sandford Fleming, *Time Lord*, and in the account *Greenwich Time and the Discovery of the Longitude*, by Derek Howse. For readers fascinated by the more technical aspects of the field, and by the robust arguments between the affected countries, these topics are also well covered in the published *Proceedings* of the 1884 Meridian Conference, held in Washington, DC.

The complicated and sometimes distressing condition of many native residents of some of the Marshall Islands (Kwajalein most notably) are bravely (and controversially) told by Julianne Walsh in her *Etto Nan Raan Kein: A Marshall Islands History*. The history of the early twentieth-century Japanese influence on islands in the western Pacific is to be found in *Nanyo: The Rise and Fall of the Japanese in Micronesia*, by Mark Peattie.

AUTHOR'S NOTE

The definitive document calling for the establishment of January 1950 as the beginning of the standard reference year for dating purposes is the appeal by Richard Flint and Edward Deevey in the foreword to the 1962 issue of the journal *Radiochemistry*. Eric Wolff at the British Antarctic Survey in Cambridge and Paula J. Reimer, the director of the Centre for Climate, Environment, and Chronology at Queens University, Belfast's School of Geography, have also written on the topic, encouraging the acceptance of 1950 as the "present" in the new dating system that has replaced AD and BC with BP.

CHAPTER 1: THE GREAT THERMONUCLEAR SEA

Details of the crucial conversations between President Truman and his CIA director, Admiral Souers, which led to the decision to develop fusion weapons, later to be tested on

Bikini and Enewetak atolls, can be found in Richard Rhodes's classic study of the development of hydrogen bombs, *Dark Sun*. The matter of then selling the test program to the Bikinians is more than amply covered in Holly Barker's *Bravo for the Marshallese*, Connie Goldsmith's *Bombs over Bikini*, and Jack Niedenthal's *For the Good of Mankind*—the last being the cynically persuasive argument put forward by the generals and admirals who came a-courting the eagerly patriotic islanders.

Jonathan Weisgall's *Operation Crossroads* gives a full account of the principal Pacific tests of the postwar fission weapons; for the subsequent tests of the much more powerful fusion bombs, the best accounts are to be found in the official reports of what was then named the U.S. Defense Nuclear Agency, particularly those relating to the dangerously mismanaged Castle series.

Film clips showing the spectacular upending of the entire twenty-six-thousand-ton battleship *Arkansas*, by the force of the Crossroads *Baker* shot, can be found at Sonicboom.com.

Work on the notoriously accident-prone "Demon Core" of plutonium hemispheres is meticulously covered by Louis Hempelmann in his paper on radiation effects, published by the Los Alamos National Laboratory in 1962.

Chapter 2: Mr. Ibuka's Radio Revolution

I found that of all the many works on the early days of the Sony Corporation, the best and most enjoyably disinterested is John Nathan's *Sony: The Private Life*. Publications put out by the company itself, especially the two book-length obituaries of Akio Morita and Masaru Ibuka, certainly offer a useful and accurate catalogue of the firm's milestones, but are understandably somewhat hagiographic. Miss Hiroko Onoyama, formerly Akio Morita's assistant in New York, offered much private information, which proved most helpful. The company's original prospectus, written by Ibuka, is on display at the Sony Archives in Tokyo.

The Bell Labs work that resulted in the invention of the transistor is lucidly explained in a paper published in the *Journal of the American Physical Society*, vol. 9, part 10, in 2000. The *New York Times* account of the warehouse robbery in Queens that put Sony's early transistor radios on the map appears on page 17 of the issue of January 17, 1958.

Chapter 3: The Ecstasies of Wave Riding

Few history books can be as beguiling or as seductive as Matt Warshaw's *The History of Surfing*, which manages to be both beautiful in appearance and comprehensive in scope: I turned to it ceaselessly, until its pages were ragged with overuse. Scott Laderman's *Empire in Waves: A Political History of Surfing*, is somewhat more sober, but I found it useful nonetheless. And Jack London's writings, both in his *Voyage of the Snark* and in his famous essay in the *Woman's Home Companion* of October 1907, are powerfully suggestive of the passion with which devotees took to the new sport.

The story of George Freeth, claimed in parts of his native Ireland to have been the original "king of surfing," is very well told in a motion picture, *Waveriders*, made in 2008 by the Irish director Joel Conroy.

That California in the early years of the twentieth century so swiftly became America's first mainland surfing paradise is a phenomenon chronicled in loving detail by William Friedricks in his 1992 book *Henry E. Huntington and the Creation of Southern California*. The strange story of Grubby Clark and his abrupt closure of Clark Foam, with all of its myriad unintended consequences, is told in many issues of *Surfing* magazine. Yvon Chouinard, founder of the outdoor gear and clothing company Patagonia, describes his attitude to wave-dictated flextime in his amusing book *Let My People Go Surfing*.

CHAPTER 4: A DIRE AND DANGEROUS IRRITATION

I first came across the improbable story that North Korea had been almost accidentally created by no less than an American soldier, Colonel Charles H. Bonesteel III, in Dean Rusk's otherwise rather dull autobiography, *As I Saw It*. Max Hastings fleshes out the yarn somewhat in his definitive *The Korean War*, still unarguably the best book on this miserable and pointless conflict, and which left so poisonous a legacy.

Jack Cheevers and Ed Brandt both wrote well-received books on the capture of the USS *Pueblo*; the ship's captain, Lloyd Bucher, wrote his own account of his wretched months in captivity, *Bucher: My Story*. Lest anyone have doubts about the savagery of the successive governments that have ruled North Korea since the 1953 armistice, the *Report of the [UN] Commission of Inquiry into Human Rights in the Democratic People's Republic of Korea*, published in 2013, should be essential reading.

CHAPTER 5: FAREWELL, ALL MY FRIENDS AND FOES

I relied on the considerable storytelling abilities of the former journalist Brian Izzard, and his book *Sabotage*, which tells, in almost hour-by-hour detail, the tragic and fiery end of the adored Cunarder RMS *Queen Elizabeth*. Issue 189 of the *Socialist Review*, published in September 1995, offers a sweeping and coherent analysis of General Gracey's near-unimaginable rearming of defeated Japanese soldiers to help him fight against Vietnamese nationalists in postwar Saigon. Those wishing further detail should know that the general's extensive collection of papers is lodged at the Liddell Hart Centre for Military Archives at King's College, London.

Two books in particular proved essential to my understanding something of the Battle of Dien Bien Phu: Ted Morgan's *Valley of Death* and Bernard Fall's *Hell in a Very Small Place*. The consequent Communist rout and conquest of the totality of Vietnam twenty years later is covered in exquisite and painful detail in *Last Days in Vietnam*, a film made by Rory Kennedy and released in 2014.

Few colonial territories now exist in the Pacific. One of the last to be returned to its rightful owners was the former British enclave of Hong Kong, written about in its new postcolonial identity by Jan Morris, in the classic *Hong Kong: Xianggang*.

The four Pitcairn Islands are all that now remains of Britain's once immense Pacific empire. Dea Birkett managed to make herself most unpopular by writing the vastly informative *Serpent in Paradise*; and Kathy Marks confirmed the underlying rottenness of the place in her coverage of the sexual abuse trials, which I consulted to write this chapter.

CHAPTER 6: ECHOES OF DISTANT THUNDER

I made great use of *Warning*, by Sophie Cunningham, to build my account of the devastating Cyclone Tracy, following my own visit in 2014 to the now wholly restored and largely rebuilt Australian tropical city of Darwin.

Information on the formation and explosive growth of Typhoon Haiyan, which inflicted so much Darwin-like destruction in the Philippines four decades later, came largely from publications written by the team at the U.S. Joint Typhoon Warning Center at Pearl Harbor.

Kerry Emanuel's large-format book *Divine Wind*, an analysis of the atmosphere's endless capacity for high-velocity ferocity, proved invaluable for my writing of this rather complicated chapter.

Kevin Hamilton, director of the International Pacific Research Center at the University of Hawaii, has written numerous technical papers on the El Niño Southern Oscillation (ENSO); and Mark Bradford, chief meteorologist on Kwajalein Atoll, is similarly to be regarded as a voice of authority on El Niño and its related complexities. JAMSTEC, the Japan

Agency for Marine-Earth Science and Technology in the Tokyo Bay port city of Yokosuka, also issues streams of data on its El Niño researches.

Much science is being performed throughout the Pacific on this ever more crucial topic. Yet quietly, in the background of all this hubbub, stands the magisterial figure of Sir Gilbert Walker, the ultimate discoverer of the phenomenon: the lengthy entry in the *Oxford Dictionary of National Biography* provides a justly sympathetic portrait of this eccentric and unforgettable, yet near-forgotten, figure.

CHAPTER 7: HOW GOES THE LUCKY COUNTRY?

Paul Kelly's *The Dismissal* remains the finest account of the unprecedented sacking of the Australian prime minister, Gough Whitlam. Both Whitlam (*The Truth of the Matter*) and his nemesis John Kerr (*Matters for Judgment*) wrote their own, understandably partisan, accounts of the saga, adding to an immense literature on an event that is precious little known beyond Australia's shores.

The equally complicated and nuanced story of the building of the Sydney Opera House is perhaps best related in a long-forgotten BBC documentary film *Autopsy on a Dream*, made by the Australian director John Weiley. The film, highly critical of Sydney's treatment of the building's Danish architect, was shown in Britain, but before it could be screened in Australia, it was destroyed, chopped to pieces with a meat cleaver. Thirty years later a misfiled early cut of the film was discovered in London and sent to be shown Down Under, to a mixture of acclaim and cringe. Likewise, an Australian-made TV documentary provided me with some insight into the story of the brief rise to prominence of the politician Pauline Hanson: the ABC *60 Minutes* profile of Miss Hanson, who was interviewed with rapier-like skill by journalist Tracey Curro, remains a legendary moment in television.

CHAPTER 8: THE FIRES IN THE DEEP

The tireless work of *Alvin* and her sister submersibles has been the subject of countless reports and publications issued by the Woods Hole Oceanographic Institution. A fine summary of the main findings, of hydrothermal vents and of smokers, black and white, is to be found in *Discovering the Deep*, by Daniel Fornari, Jeffrey Karson, Deborah Kelley, Michael R. Perfit, and Timothy M. Shank.

I commend to readers Stephen Hall's incisive essay on the deep-sea mapmaker Marie Tharp, published in the *New York Times Magazine* in December 2006.

Anyone with an interest in the finer points of tectonic theory, which underpins a science central to the formation of the Pacific, could do no better than to read *Plate Tectonics*, by Naomi Oreskes, published in 2002 and now a classic of the field.

Robert Ballard's extended essay on deep-sea exploration, *The Eternal Darkness*, published in 2000, tells much about *Alvin* and her work. And Colleen Cavanaugh's interview with the *Harvard Gazette* offers considerable information about the role of sulfur in the origination and sustenance of deep-sea life-forms.

CHAPTER 9: A FRAGILE AND UNCERTAIN SEA

Personal communications with coral expert Charlie Veron enabled me to fill out the picture of the disastrous beginning of coral bleaching on the Great Barrier Reef. Iain McCalman's *The Reef: A Passionate History* takes the story further, placing this enormous living creature, now very much a threatened creature, in a wider context, both biological and cultural.

Mary Hagedorn of MarineGEO in Hawaii, a Smithsonian-supported oceanographic research center, is behind a project to try to help coral populations survive the relentless rise in ocean temperature that seems to be doing them so much harm. Her publications, in aggregate, make for fascinating and informative reading. Similarly, papers put out by the Great Barrier Reef Marine Park Authority round out the story.

It was Mark Brazil's excellent small book *The Nature of Japan* that first led me to the work of Hiroshi Hasegawa and his heroic rescue of the albatross population of Torishima. Oxford's Pitt Rivers Museum publishes an excellent online monograph describing the Hawaiian ceremonial cloaks in its possession.

Jon Mooallem's article on Larry Ellison's purchase of and plans for the Hawaiian island of Lana'i in the *New York Times Magazine* of September 28, 2014, raised many hackles; much of what he observed confirms the impression I gained when I visited the island six months earlier.

CHAPTER 10: OF MASTERS AND COMMANDERS

Three books in particular set the scene for the current Chinese expansion of influence in the far western Pacific and the fretfulness it is causing in Washington: Bill Hayton's *The South China Sea*; Robert Haddick's *Fire on the Water*; and, by the always reliable Robert Kaplan, *Asia's Cauldron*. However, the situation in the region is changing so rapidly that all these studies are in danger of becoming dated—so that interested readers would do well to monitor the near-constant streams of publications from such bodies as the Center for Strategic and International Studies, the Asia Society, and the International Crisis Group, among others, to keep properly abreast.

The International Crisis Group Paper 229 (2012), "Stirring Up the South China Sea," presents useful background; and for truly ultra-deep background, there are always new editions available of Alfred Mahan's classic *The Influence of Sea Power upon History, 1660–1783*—essential reading for those who might be curious about how these new naval challenges are likely to play themselves out.

And as a coda: the complexities of the Pacific as battle space are nicely explained in an essay in the *Washington Post* of August 1, 2012, by the paper's defense writer Greg Jaffe.

EPILOGUE

The journey of the Hawaiian *wa'a Hokule'a* is being reported until 2017, and in great daily detail, by the Polynesian Voyaging Society in Honolulu. Much background to the story can be found in Sam Low's book on the Hawaiian Renaissance, *Hawaiki Rising*; in Ben Finney's explanations of Polynesian navigation, *Sailing in the Wake of the Ancestors*; and most fascinating of all, in David Lewis's account, *We, the Navigators*.

BIBLIOGRAPHY

Armitage, David, and Alison Bashford, eds. *Pacific Histories: Ocean, Land, People*. New York: Palgrave, 2014.

Bain, Kenneth. *The Friendly Islanders*. London: Hodder, 1967.

Ballard, Robert. *The Eternal Darkness: A Personal History of Deep-Sea Exploration*. Princeton, NJ: Princeton University Press, 2000.

Barker, Holly. *Bravo for the Marshallese: Regaining Control in a Post-Nuclear, Post-Colonial World*. Belmont, CA: Wadsworth, 2013.

Barlow, Thomas. *The Australian Miracle: An Innovative Nation Revisited*. Sydney: Picador, 2006.

Barrie, David. *Sextant: A Voyage Guided by the Stars and the Men Who Mapped the World's Oceans*. London: William Collins, 2014.

Beaglehole, J. C. *The Exploration of the Pacific*. Stanford, CA: Stanford University Press, 1934.

Bentley, Jerry H., et al., eds. *Seascapes: Maritime Histories, Littoral Cultures, and Transoceanic Exchanges*. Honolulu: University of Hawaii Press, 2007.

Bergreen, Laurence. *Over the Edge of the World: Magellan's Terrifying Circumnavigation of the Globe*. New York: Harper, 2003.

Birkett, Dea. *Serpent in Paradise: Among the People of the* Bounty. New York: Doubleday, 1997.

Blaise, Clark. *Time Lord: Sir Sandford Fleming and the Creation of Standard Time*. New York: Vintage, 2002.

Borneman, Walter R. *The Admirals: Nimitz, Halsey, Leahy, and King—the Five-Star Admirals Who Won the War at Sea*. New York: Little, Brown, 2012.

Brandt, Ed. *The Last Voyage of USS* Pueblo. New York: W. W. Norton, 1969.

Brazil, Mark. *The Nature of Japan*. Sapporo: Japan Nature Guides, 2014.

Bryant, Nick. *The Rise and Fall of Australia: How a Great Nation Lost Its Way*. New York: Bantam, 2014.

Bucher, Lloyd. *Bucher: My Story*. New York: Doubleday, 1970.

Butler, Robert. *The Jade Coast: The Ecology of the North Pacific Ocean*. Toronto: Key Porter Books, 2003.

Cameron, Ian. *Magellan and the First Circumnavigation of the World*. London: Weidenfeld, 1974.

Carson, Rachel, ed. *The Sea Around Us*. New York: Oxford University Press, 2003.

Chandler, Alfred D. *Inventing the Electronic Century: The Epic Story of Consumer Electronics and Computer Industries*. Cambridge, MA: Harvard University Press, 2001.

Cheevers, Jack. *Act of War: Lyndon Johnson, North Korea, and the Capture of the Spy Ship* Pueblo. New York: Penguin, 2013.

Chouinard, Yvon. *Let My People Go Surfing: The Education of a Reluctant Businessman*. New York: Penguin, 2005.

Collins, Donald E. *Native American Aliens: Disloyalty and the Renunciation of Citizenship by Japanese Americans in World War 2*. Westport, CT: Greenwood Press, 1985.

Cooper, George, and Gavan Daws. *Land and Power in Hawaii*. Honolulu: University of Hawaii Press, 1990.

Cox, Jeffrey. *Rising Sun, Falling Skies: The Disastrous Java Sea Campaign of World War 2*. Oxford: Osprey, 2014.

Cralle, Trevor, ed. *Surfin'ary: A Dictionary of Surfing Terms and Surfspeak*. Berkeley: Ten Speed Press, 1991.

Cramer, Deborah. *Ocean: Our Water, Our World*. New York: HarperCollins/Smithsonian, 2008.

Cullen, Vicky. *Down to the Sea for Science: 75 Years of Ocean Research, Education, and Exploration at the Woods Hole Oceanographic Institution*. Woods Hole: WHOI, 2005.

Culliney, John L. *Islands in a Far Sea: The Fate of Nature in Hawaii*. Honolulu: University of Hawaii Press, 2006.

Cunningham, Sophie. *Warning: The Story of Cyclone Tracy*. Melbourne: Text Publishing, 2014.

Cushman, Gregory T. *Guano and the Opening of the Pacific World*. New York: Cambridge University Press, 2013.

Daniel, Hawthorne. *Islands of the Pacific*. New York: Putnam, 1943.

Danielsson, Bengt. *The Forgotten Islands of the South Seas*. London: Allen and Unwin, 1957.

——. *The Happy Island*. London: Allen and Unwin, 1952.

Daws, Gavan. *A Dream of Islands: Voyages of Self-Discovery in the South Seas*. New York: W. W. Norton, 1980.

Denoon, Donald, et al. *A History of Australia, New Zealand, and the Pacific*. Oxford: Blackwell, 2000.

Dobbs-Higginson, Michael. *Asia Pacific: A View on Its Role in the New World Order*. Hong Kong: Longman, 1993.

Dodd, Edward. *The Rape of Tahiti*. New York: Dodd, Mead, 1983.

Dolin, Eric Jay. *When America First Met China: An Exotic History of Tea, Drugs, and Money in the Age of Sail*. New York: Norton, 2012.

Dower, John W. *Embracing Defeat: Japan in the Wake of World War Two*. New York: W. W. Norton, 1999.

Durschmied, Erik. *The Weather Factor: How Nature Has Changed History*. London: Hodder, 2000.

Dvorak, John. *Earthquake Storms: The Fascinating History and Volatile Future of the San Andreas Fault*. New York: Pegasus, 2014.

Ellis, Richard. *The Encyclopedia of the Sea*. New York: Knopf, 2006.

Emanuel, Kerry. *Divine Wind: The History and Science of Hurricanes*. New York: Oxford University Press, 2005.

Etulain, Richard W., and Michael P. Malone. *The American West: A Modern History, 1900 to the Present*. Lincoln: University of Nebraska Press, 1989.

Evans, Julian. *Transit of Venus: Travels in the Pacific*. London: Secker, 1992.

Fagan, Brian. *Beyond the Blue Horizon: How the Earliest Mariners Unlocked the Secrets of the Oceans*. New York: Bloomsbury, 2012.

Fall, Bernard B. *Hell in a Very Small Place: The Siege of Dien Bien Phu*. New York: Lippincott, 1966.

Finney, Ben. *Hokule'a: The Way to Tahiti*. New York: Dodd, Mead, 1979.

———. *Sailing in the Wake of the Ancestors*. Bishop Museum, 2003.

Fischer, Steven Roger. *A History of the Pacific Islands*. Basingstoke: Palgrave, 2002.

Fisher, Stephen, ed. *Man and the Maritime Environment*. Exeter, UK: University of Exeter Press, 1994.

Fornari, Daniel J., et al. *Discovering the Deep: A Photographic Atlas of the Seafloor and Ocean Crust*. Cambridge, UK: Cambridge University Press, 2015.

Freeman, Donald B. *The Pacific*. London: Routledge, 2010.

Friedricks, William. *Henry E. Huntington and the Creation of Southern California*. Columbus: Ohio State University Press, 1991.

Garfield, Brian. *The Thousand-Mile War: World War II in Alaska and the Aleutians*. Fairbanks: Alaska University Press, 1995.

Garnaut, Ross. *Dog Days: Australia After the Boom*. Melbourne: Redback, 2013.

George, Rose. *Ninety Percent of Everything: Inside Shipping*. New York: Henry Holt, 2013.

Gibney, Frank. *The Pacific Century: America and Asia in a Changing World*. New York: Scribner, 1992.

Gillis, John R. *The Human Shore: Seacoasts in History*. Chicago: University of Chicago Press, 2012.

Glacken, Clarence J. *Traces on the Rhodian Shore*. Berkeley: University of California Press, 1967.

Glavin, Terry. *The Last Great Sea: A Voyage Through the Human and Natural History of the North Pacific Ocean*. Vancouver: Greystone Books, 2000.

Goldsmith, Connie. *Bombs over Bikini: The World's First Nuclear Disaster*. Minneapolis. Twenty-First Century Books, 2014.

Greely, Adolphus Washington. *Handbook of Alaska*. New York: Scribner, 1925.

Grimble, Arthur. *A Pattern of Islands*. London: John Murray, 1952.

———. *Return to the Islands*. London: John Murray, 1957.

Gurnis, Michael, et al. *Oceans: A Scientific American Reader*. Chicago: University of Chicago Press, 2007.

Gwyther, John. *Captain Cook and the South Pacific: The Voyage of the Endeavour, 1768–1771*. Cambridge, MA: Houghton Mifflin, 1954.

Haddick, Robert. *Fire on the Water: China, America, and the Future of the Pacific*. Annapolis, MD: Naval Institute Press, 2014.

Haley, James L. *Captive Paradise: A History of Hawaii*. New York: St. Martin's, 2014.

Hamilton-Paterson, James. *The Great Deep: The Sea and Its Thresholds*. New York: Random House, 1992.

Harwit, Martin. *An Exhibition Denied: Lobbying the History of Enola Gay*. New York: Copernicus, 1996.

Hastings, Max. *The Korean War*. London: Michael Joseph, 1987.

Hattendorf, John B., ed. *The Oxford Encyclopedia of Maritime History*. 4 vols. New York: Oxford University Press, 2007.

Hayton, Bill. *The South China Sea: The Struggle for Power in Asia*. New Haven, CT: Yale University Press, 2014.

Henderson, Bonnie. *Strand: An Odyssey of Pacific Ocean Debris*. Corvallis: Oregon State University Press, 2008.

Hersh, Seymour. *The Target Is Destroyed*. London: Faber, 1986.

Heyerdahl, Thor. *Fatu-Hiva: Back to Nature*. New York: Doubleday, 1974.

———. *Kon-Tiki: Across the Pacific by Raft*. New York: Rand McNally, 1950.

Hill, Ernestine. *The Territory: The Classic Saga of Australia's Far North*. Sydney: Angus and Robertson, 1951.

Hinrichsen, Don. *The Atlas of Coasts and Oceans*. London: Earthscan, 2011.

Holt, John Dominis. *On Being Hawaiian*. Honolulu: Kupa'a Publishing, 1964.

Horwitz, Tony. *Blue Latitudes: Boldly Going Where Captain Cook Has Gone Before*. New York: Henry Holt, 2002.

Howarth, David Armine. *Tahiti: A Paradise Lost*. London: Harvill, 1983.

Howse, Derek. *Greenwich Time and the Discovery of the Longitude*. Oxford: Oxford University Press, 1980.

Ienaga, Saburo. *The Pacific War, 1931–1945*. New York: Pantheon, 1978.

Igler, David. *The Great Ocean: Pacific Worlds from Captain Cook to the Gold Rush*. New York: Oxford University Press, 2013.

Ilyichev, V. I., and V. V. Anikiev. *Oceanic and Anthropogenic Controls of Life in the Pacific Ocean*. Dordrecht: Kluwer Publishers, 1992.

Inada, Lawson Fusao. *Only What We Could Carry: The Japanese American Internment Experience*. Berkeley: Heyday, 2000.

Izzard, Brian. *Sabotage: The Mafia, Mao, and the Death of the Queen Elizabeth*. Gloucester, UK: Amberley, 2012.

Johnson, R. W. *Shootdown: The Verdict on KAL 007*. London: Chatto and Windus, 1986.

Kaplan, Robert D. *Asia's Cauldron: The South China Sea and the End of a Stable Pacific*. New York: Random House, 2014.

Kashima, Tatsuden. Foreword. *Personal Justice Denied: Report of the Commission on Wartime Relocation and Internment of Civilians*. Seattle: University of Washington Press, 1982.

Kelly, Paul. *The Dismissal: Australia's Most Sensational Power Struggle—The Dramatic Fall of Gough Whitlam*. Sydney: Angus and Robertson, 1983.

Kennedy, David M. *Freedom from Fear*. New York: Oxford University Press, 1999.

Kerr, John. *Matters for Judgment*. Sydney: Macmillan, 1978.

King, Ernest J. (Fleet Admiral). *Official Reports: U.S. Navy at War 1941–45*. Washington, DC: U.S. Navy, 1946.

King, Samuel, and Randall Roth. *Broken Trust: Greed, Mismanagement, and Political Manipulation at America's Largest Charitable Trust*. Honolulu: University of Hawaii Press, 2006.

Klein, Bernhard, and Gesa Mackenthun, eds. *Sea Changes: Historicizing the Ocean*. New York: Routledge, 2004.

Kyselka, Will. *An Ocean in Mind*. Honolulu: University of Hawaii Press, 1987.

Laderman, Scott. *Empire in Waves: A Political History of Surfing*. Berkeley: University of California Press, 2014.

Lal, Brij V., and Kate Fortune, eds. *The Pacific Islands: An Encyclopedia*. Honolulu: University of Hawaii Press, 2000.

Lavery, Brian. *The Conquest of the Ocean: An Illustrated History of Seafaring*. New York: Dorling Kindersley, 2013.

Levy, Steven. *Insanely Great: The Life and Times of Macintosh*. New York: Penguin, 1994.

Lewis, David. *We, the Navigators: The Ancient Art of Landfinding in the Pacific*. Honolulu: University of Hawaii Press, 1972.

Linklater, Eric. *The Voyage of the Challenger*. New York: Doubleday, 1972.

Linzmayer, Owen W. *Apple Confidential 2.0: The Definitive History of the World's Most Colorful Company*. San Francisco, CA: No Starch Press, 2004.

London, Jack. *The Cruise of the Snark*. New York: Macmillan, 1911.

Low, Sam. *Hawaiki Rising: Hokulea, Nainoa Thompson, and the Hawaiian Renaissance*. Waipahu, HI: Island Heritage Publishing, 2013.

Lyons, Nick. *The Sony Vision*. New York: Crown, 1976.

Macintyre, Michael. *The New Pacific*. London: Collins, 1985.

Macintyre, Stuart. *A Concise History of Australia*. 3rd ed. New York: Cambridge University Press, 2009.

Mack, John. *The Sea: A Cultural History*. London: Reaktion Books, 2011.

Mahan, Alfred Thayer. *The Influence of Sea Power upon History, 1660–1783*. 1890; Reprint. New Orleans: Pelican Publishing, 2003.

Malcolmson, Scott L. *Tuturani: A Political Journey in the Pacific Islands*. London: Hamish Hamilton, 1990.

Manjiro, John, ed. *Drifting Toward the Southeast: The Story of Five Japanese Castaways*. New Bedford, MA: Spinner Publications, 2003.

Marks, Kathy. *Pitcairn: Paradise Lost*. Sydney: Harper Australia, 1988.

Mason, R. H. P., and J. P. Caiger. *A History of Japan*. Rutland, VT: Tuttle, 1997.

Matsuda, Matt K. *Pacific Worlds: A History of Seas, Peoples and Cultures*. New York: Cambridge University Press, 2012.

McCalman, Iain. *The Reef: A Passionate History*. Melbourne: Penguin, 2013.

McCune, Shannon. *The Ryukyu Islands*. Newton Abbot: David and Charles, 1975.

McEvedy, Colin. *The Penguin Historical Atlas of the Pacific*. New York: Penguin, 1998.

McLean, Ian W. *Why Australia Prospered: The Shifting Sources of Economic Growth*. Princeton, NJ: Princeton University Press, 2013.

Megalogenis, George. *The Australian Moment: How We Were Made for These Times*. Melbourne: Penguin, 2012.

Melville, Herman. *Typee, Omoo, and Mardi*. New York: Library of America, 1846–49.

Michener, James. *Hawaii*. New York: Random House, 1959.

Mitchell, General William L. *The Opening of Alaska*. Anchorage: Cook Inlet Historical Society, 1982.

Moore, Michael Scott. *Sweetness and Blood. How Surfing Spread from Hawaii and California to the Rest of the World, with Some Unexpected Results*. New York: Rodale, 2010.

Moorehead, Alan. *The Fatal Impact: An Account of the Invasion of the South Pacific, 1767–1840*. London: Hamish Hamilton, 1966.

Morgan, Ted. *Valley of Death: The Tragedy at Dien Bien Phu That Led America into the Vietnam War*. New York: Presidio Press, 2010.

Morison, Samuel Eliot. *The Two-Ocean War*. Boston: Little, Brown, 1963.

Morita, Akio. *Made in Japan: Akio Morita and Sony*. New York: Dutton, 1986.

Morris, Jan. *Hong Kong: Xianggang*. New York: Penguin, 1988.

Motteler, Lee S. *Pacific Island Names: A Map and Name Guide to the New Pacific*. Honolulu: Bishop Museum Press, 2006.

Muir, John. *Travels in Alaska*. Boston: Houghton Mifflin, 1915.

Nathan, John. *Sony: The Private Life*. Boston: Houghton Mifflin, 1999.

Nicholls, Henry. *The Galápagos: A Natural History*. New York: Basic Books, 2014.

Niedenthal, Jack. *For the Good of Mankind: A History of the People of Bikini and Their Islands*. Majuro: Micronitor/Bravo Publishers, 2001.

Niiya, Brian, ed. *Japanese-American History: An A–Z Reference, 1868 to the Present*. New York. Facts on File, XXXX.

Olson, Steve. *Mapping Human History: Genes, Race and Our Common Origins*. New York: Houghton Mifflin, 2003.

Oreskes, Naomi. *Plate Tectonics: An Insider's History of the Modern Theory of the Earth*. Westview Press, 2003.

Oosterzee, Penny van. *A Natural History of Australia's Top End*. Marleston: Gecko Books, 20014.

Paik, Koohan, and Jerry Mander. *The Superferry Chronicles: Hawaii's Uprising Against Militarism, Commercialism and the Desecration of the Earth*. Kihei, HI: Koa Books, 2009.

Paine, Lincoln. *The Sea and Civilization: A Maritime History of the World*. New York: Knopf, 2013.

Parker, Bruce. *The Power of the Sea: Tsunamis, Storm Surges, Rogue Waves, and Our Quest to Predict Disasters*. New York: Palgrave Macmillan, 2010.

Peattie, Mark. *Nanyo: The Rise and Fall of the Japanese in Micronesia, 1885–1945*. Honolulu: University of Hawaii Press, 1988.

Pembroke, Michael. *Arthur Phillip: Sailor, Mercenary, Governor, Spy*. Melbourne: Hardie Grant Books, 2013.

Philbrick, Nathaniel. *Sea of Glory: America's Voyage of Discovery, the U.S. Exploring Expedition*. New York: Penguin, 2003.

Prados, John. *Islands of Destiny: The Solomons Campaign and the Eclipse of the Rising Sun*. New York: Penguin, 2012.

Pratt, H. Douglas, et al. *The Birds of Hawaii and the Tropical Pacific*. Princeton, NJ: Princeton University Press, 1987.

Pyle, Kenneth B. *Japan Rising: The Resurgence of Japanese Power and Purpose*. New York: Public Affairs, 2007.

Rankin, Nicholas. *Dead Man's Chest: Travels After Robert Louis Stevenson*. London: Faber and Faber, 1987.

Regan, Anthony J. *Light Intervention: Lessons from Bougainville*. Washington, DC: U.S. Institute of Peace Press, 2010.

Reid, Anna. *The Shaman's Coat: A Native History of Siberia*. New York: Walker, 2002.

Rhodes, Richard. *Dark Sun: The Making of the Hydrogen Bomb*. New York: Simon and Schuster, 1995.

———. *The Making of the Atomic Bomb*. New York: Simon and Schuster, 1986.

Ricketts, Edward F., et al. *Between Pacific Tides*. Stanford, CA: Stanford University Press, 1939.

Riesenberg, Felix. *The Pacific Ocean*. London: Museum Press, 1947.

Roberts, Callum. *The Unnatural History of the Sea*. Washington, DC: Island Press, 2007.

Robertson, Geoffrey. *Dreaming Too Loud: Reflections on a Race Apart*. Sydney: Random House, 2013.

Romoli, Kathleen. *Balboa of Darién: Discoverer of the Pacific*. New York: Doubleday, 1953.

Rusk, Dean. *As I Saw It*. New York: W. W. Norton, 1990.

Safina, Carl. *Song for a Blue Ocean*. New York: Henry Holt, 1997.

Segal, Gerald. *Rethinking the Pacific*. Oxford: Oxford University Press, 1990.

Sherry, Frank. *Pacific Passions: The European Struggle for Power in the Great Ocean in the Age of Exploration*. New York: Morrow, 1994.

Sims, Eugene C. *Kwajalein Remembered: Stories from the Realm of the Killer Clam*. Eugene, OR: Eugene C. Sims, 1993.

Sloan, Bill. *Given Up for Dead: America's Heroic Stand at Wake Island*. New York: Bantam, 2003.

Spate, O. H. K. *The Pacific Since Magellan*. 3 vols. Canberra: Australian National University Press, 1979.

Stanton, Doug. *In Harm's Way: The Sinking of the USS* Indianapolis. New York: Henry Holt, 2001.

Starck, Walter. *The Blue Reef: A Report from Beneath the Sea*. New York: Knopf, 1978.

Stark, Peter. *Astoria: John Jacob Astor and Thomas Jefferson's Lost Pacific Empire*. New York: Ecco, 2014.

Starr, Kevin. *Golden Gate: The Life and Times of America's Greatest Bridge*. New York: Bloomsbury, 2010.

Stevenson, Robert Louis, and Fanny Stevenson. *Our Samoan Adventure*. New York: Harper and Brothers, 1955.

Stuart, Douglas T. *Security Within the Pacific Rim*. Aldershot: Gower, 1987.

Talley, Lynne, et al. *Descriptive Physical Oceanography: An Introduction*. London: Elsevier, 2011.

Tennesen, Michael. *The Next Species: The Future of Evolution in the Aftermath of Man*. New York: Simon and Schuster, 2015.

Tess, Leah. *Darwin*. Sydney: University of New South Wales Press, 2014.

Theroux, Paul. *The Happy Isles of Oceania: Paddling the Pacific*. Boston: Houghton Mifflin, 1992.

Thomas, Nicholas. *Cook: The Extraordinary Voyages of Captain James Cook*. Toronto: Viking, 2003.

———. *Islanders: The Pacific in the Age of Empire*. New Haven, CT: Yale University Press, 2010.

Tink, Andrew. *Australia, 1901–2001: A Narrative History*. Sydney: NewSouth, 2014.

Toops, Connie, and Phyllis Greenberg. *Midway: A Guide to the Atoll and Its Inhabitants*. Naples, FL: LasAves, 2012.

Turvey, Nigel. *Cane Toads: A Tale of Sugar, Politics, and Flawed Science*. Sydney: University of New South Wales Press, 2013.

United Nations. *Report of the Commission of Inquiry on Human Rights in the Democratic People's Republic of Korea*. Geneva: United Nations, 2014. Available at http://www.ohchr.org/EN/HRBodies/HRC/CoIDPRK/Pages/CommissionInquiryonHRinDPRK.aspx.

U.S. Department of Defense. Defense Nuclear Agency. *Castle Series*. Washington, DC: U.S. Government Printing Office, 1982.

Veron, J. E. N. *A Reef in Time: The Great Barrier Reef from Beginning to End*. Cambridge, MA: Harvard University Press, 2008.

Visher, Stephen Sargent. *Tropical Cyclones of the Pacific*. Honolulu, HI: Bishop Museum, 1925.

Viviano, Frank. *Dispatches from the Pacific Century*. New York: Addison-Wesley, 1993.

Walsh, Julianne. *Etto Nan Raan Kein: A Marshall Islands History*. Honolulu, HI: Bess Press, 2012.

Warshaw, Matt. *The History of Surfing*. San Francisco: Chronicle Books, 2010.

Weisgall, Jonathan. *Operation Crossroads: The Atomic Tests at Bikini Atoll*. Annapolis, MD: Naval Institute Press, 1994.

Wertheim, Eric. *The Naval Institute Guide to Combat Fleets of the World*. Annapolis, MD: Naval Institute Press, 2013.

Whelan, Christal. *Kansai Cool: A Journey into the Cultural Heartland of Japan*. Rutland, VT: Tuttle, 2014.

Whistler, W. Arthur. *Plants of the Canoe People: An Ethnobotanical Journey Through Polynesia*. Lawai, Kauai: National Tropical Botanical Garden, 2009.

Whitlam, Gough. *The Truth of the Matter: An Autobiography*. Melbourne: Penguin, 1979.

Wilson, Derek. *The Circumnavigators*. New York: M. Evans and Co., 1989.

Wilson, Dick. *When Tigers Fight: The Story of the Sino-Japanese War, 1937–1945*. London: Viking, 1892.

Wilson, Rob. *Reimagining the American Pacific*. Durham, NC: Duke University Press, 2000.

Winchester, Simon. *Pacific Rising: The Emergence of a New World Culture*. New York: Simon and Schuster, 1991.

Withey, Lynne. *Voyages of Discovery: Captain Cook and the Exploration of the Pacific*. Melbourne: Hutchinson, 1987.

Wood, Gillen D'Arcy. *Tambora: The Eruption That Changed the World*. Princeton, NJ: Princeton University Press, 2014.

Wozniak, Steve. *iWoz: Computer Geek to Cult Icon—How I Invented the Personal Computer and Co-Founded Apple*. London: W. W. Norton, 2006.

Wright, Ronald. *On Fiji Islands*. London: Viking, 1986.

Yoshihara, Toshi, and James R. Holmes. *Red Star over the Pacific: China's Rise and Challenge to U.S. Maritime Strategy*. Annapolis, MD: Naval Institute Press, 2010.

Young, Louise B. *Islands: Portraits of Miniature Worlds*. New York: W. H. Freeman, 1999.

Zweig, Stefan, ed. *Decisive Moments in History*. Riverside, CA: Ariadne Press, 1999

———. *Magellan*. London: Pushkin Press, 2011.

INDEX

Page numbers in *italics* refer to maps and illustrations.

ABOUT THE AUTHOR

SIMON WINCHESTER is the acclaimed author of many books, including *The Professor and the Madman*, *The Men Who United the States*, *The Map That Changed the World*, *The Man Who Loved China*, *A Crack in the Edge of the World*, and *Krakatoa*, all of which were *New York Times* bestsellers and appeared on numerous best and notable lists. In 2006 Winchester was made an officer of the Order of the British Empire (OBE) by Her Majesty the Queen. He resides in western Massachusetts.

ALSO BY SIMON WINCHESTER

THE MEN WHO UNITED THE STATES
America's Explorers, Inventors, Eccentrics and Mavericks,
and the Creation of One Nation, Indivisible
Available in Paperback, E-book, Audio, and Large Print

ATLANTIC
Great Sea Battles, Heroic Discoveries, Titanic Storms,
and a Vast Ocean of a Million Stories
Available in Paperback, E-book, Audio, and Large Print

THE MAN WHO LOVED CHINA
The Fantastic Story of the Eccentric Scientist Who
Unlocked the Mysteries of the Middle Kingdom
Available in Paperback and E-book

A CRACK IN THE EDGE OF THE WORLD
America and the Great California Earthquake of 1906
Available in Paperback and E-book

KOREA
A Walk Through the Land of Miracles
Available in Paperback and E-book

OUTPOSTS
Journeys to the Surviving Relics of the British Empire
Available in Paperback and E-book

KRAKATOA
The Day the World Exploded: August 27, 1883
Available in Paperback and E-book

THE MAP THAT CHANGED THE WORLD
William Smith and the Birth of Modern Geology
Available in Paperback and E-book

THE FRACTURE ZONE
My Return to the Balkans
Available in Paperback and E-book

THE PROFESSOR AND THE MADMAN
A Tale of Murder, Insanity, and the Making of The Oxford English Dictionary
Available in Paperback and E-book